Infrared Absorption Spectroscopy

second edition

KOJI NAKANISHI
PHILIPPA H. SOLOMON

INFRARED ABSORPTION SPECTROSCOPY

—Second Edition—

INFRARED ABSORPTION SPECTROSCOPY

—Second Edition—

Koji Nakanishi
Professor of Chemistry
Columbia University

Philippa H. Solomon
Lecturer
Douglass College, Rutgers University

HOLDEN-DAY, INC., San Francisco,
Düsseldorf, Johannesburg, London,
Panama, Singapore, Sydney

Library of Congress Catalog Card Number: 76-27393
ISBN: 0-8162-6251-9

Printed in the United States of America

1234567890 0987

FOREWORD

It is not often that one has the opportunity of writing a foreword for a book fifteen years after having first encountered it. It is with relish, pride and pleasure that I do so now.

When I first saw Professor Nakanishi's original Japanese version of his "Infrared Absorption Spectroscopy" monograph in the late 1950's, IR was still the most widely used physical method of the organic chemist, NMR spectroscopy having entered the organic chemist's armamentarium only a few years before. I encouraged him to prepare an English version because I felt that he had accomplished a rare feat—the production of a small, easily readable monograph that could be used by a student as well as a practicing chemist to learn and appreciate by himself the scope and subtleties of IR spectroscopy. The large collection of problems with associated solutions in the appendix greatly added to the "self teaching" utility of the book. It was not surprising, therefore, that when the English version appeared in 1962 it proved to be an immediate success.

Fifteen years have intervened and much has happened in organic chemical methodology, notably in the use of physical methods. Proton NMR spectroscopy and mass spectrometry have become powerful routine tools and CMR spectroscopy is about to become equally widely used. Present day organic chemists, especially those who entered graduate school during the past ten years, have a tendency to use NMR and mass spectrometry almost indiscriminately without realizing that much useful and sometimes unique information can be gathered from older well-established methods, such as infrared spectroscopy. The appropriate instrumentation is considerably less expensive and its availability ubiquitous; furthermore, the instrument is very easy to operate and with relatively little practice interpretation of the data can be accomplished easily by the user himself.

It is high time that the tendency of always resorting only to the most recent and frequently also most expensive technique is counterbalanced by a critical presentation of older and time-tested methods. Infrared spectroscopy is an example *par excellence*, and having Professor Nakanishi—an internationally recognized authority in IR as well as NMR and chiroptical methods—make this case is particularly persuasive. The new edition of his book represents a fairly complete revision of the original text together with a new chapter on laser Raman spectroscopy as well as many new problems. I expect it to receive as enthusiastic a reception as the original version, and I only hope that fifteen years from now I shall again have the opportunity of writing a foreword for the third edition.

Stanford University
March, 1977

Carl Djerassi

PREFACE

For the effective usage of infrared spectroscopy in qualitative problems, it is important to become acquainted with the appearance as well as the range of absorption of the various characteristic absorption bands. The simplest way to achieve this is naturally to become familiar with actual curves. This book is set up as follows to accomplish this purpose. After a short introductory chapter, Chapter 2 presents tables of qualitative data. These are followed by Chapters 3 and 4 describing the factors that influence band positions and intensities. To supplement and enlarge on the material presented in the tables, the problem section gives a wide variety of curves to be analyzed. The answer section gives assignments of individual bands as well as interpretations of curves.

Circled figures in the "Figs." column of tables denote the problem number in which the particular band can be found.

The figures in parentheses following explanations of respective bands in the answer section refer to the standard range of absorption of that group.

The infrared and nuclear magnetic resonance data are complementary in the detection of groups. For example, the various methyl groups such as *gem*-dimethyl, methoxyl, and N-methyl are more easily detected by NMR, and accordingly a table of chemical shifts has been added (Appendix).

The book was first published in Japanese in 1960 (Nankodo, Tokyo). The qualitative section of the present book is a translation with minor revisions, while the problem and answer sections have been largely rewritten with free inclusion of curves from the IRDC Cards (cf. p. 9).

I wish to acknowledge the great help provided by the comprehensive standard texts by Bellamy, and Jones and Sandorfy. A great deal of information was also derived from the articles contributed by many authors to the "IR Spectra" (cf. p. 8).

I am very much indebted to Professors D. H. R. Barton and C. Djerassi, under whose suggestion I undertook publication of this English version. I wish particularly to extend my warmest thanks to Professor Djerassi for contributing a Foreword.

I am deeply indebted to all of my colleagues here for generous donation of time and effort, especially to A. Terahara who spent the hot summer vacation proofreading, to Mrs. J. Inoue and Miss K. Takahashi for their help in typing the manuscript, and to Phillip A. Hart, Stanford University, for reading the proofs and checking the English. Although Japanese wives are usually not acknowledged,

I think the time is ripe for this custom to be changed; I thus thank my wife Yasuko for handling expertly the noise in central Japan.

Thanks are due to the Infrared Data Committee of Japan for granting permission to reproduce a number of curves, and to the editors of *Annalen der Chemie, Annals of the New York Academy of Science, Bulletin of the Chemical Society of Japan, DMS Cards, Helvetica Chimica Acta, Journal of the American Chemical Society, Nippon Kagaku Zasshi,* and *Spectrochimica Acta* for permission to redraw diagrams. Finally, I thank the staffs of the Nankodo Company and Holden-Day in handling the numerous problems, for example, my laziness.

11th August, 1962
Tokyo

Koji Nakanishi

★ ★ ★ ★ ★

When the first edition of this book was published in 1962, following the suggestions of Professors D. H. R. Barton and C. Djerassi, I had no idea how it would be received by the chemical community. However, I was greatly relieved to see that it was accepted quite favorably, and for this I am very grateful. I had some thoughts of revising the book but various factors have contributed in giving me an excuse for not doing so.

I am greatly indebted to Dr. P. H. Solomon who has kindly undertaken the task of revising the first edition. The main characteristic frequencies had more or less been well identified by the early 1960's and hence although a computer search was undertaken for data updating, the bulk of the material has more or less been retained.

The major changes introduced in this edition consist of updating miscellaneous sections throughout the book, replacement and addition of tables and spectra, and addition of a brief section on Raman spectroscopy.

We are indebted to Professor C.F. Hammer, Georgetown University, for a critical review of the revised manuscript, to Professor R.H. Callender, City College of New York, for numerous discussions and suggestions concerning the Raman Chapter, to Professor B.J. Bulkin, Brooklyn Polytechnic Institute, for measurements of some of the Raman spectra, and to Professor C. Djerassi for kindly consenting to write another overgenerous foreword. Finally I thank my wife again for being a strong Japanese wife over the years.

December 31, 1976
New York City

Koji Nakanishi

It only remains for me to add my thanks to K.N. and to my husband Dan. Their forbearance and encouragement made this task far less fearsome than suggested in K.N.'s description above.

December 31, 1976
New York City

Philippa Heggs Solomon

TABLE OF CONTENTS

Foreword

Preface

QUALITATIVE DATA.

Chapter 1/	The Infrared Spectrum	1
Chapter 2/	Tables of Characteristic Frequencies	10
2.1.	Position of Absorption Bands	10
2.2.	Intensity	10
2.3.	Assignments	11
Table 1.	Alkanes	14
Table 2.	Alkenes	17
Table 3.	Aromatics	19
Table 4.	X≡Y, X=Y=Z groups	23
Table 5.	Alcohols and phenols	26
Table 6.	Ethers and related groups	32
Table 7.	Amines and ammonium salts	34
Table 8.	Carbonyl groups	39
Table 9.	Nitro, nitroso, etc.	46
Table 10a.	Heterocyclic compounds	48
Table 10b.	Pyridine derivatives	50
Table 11.	Sulfur groups	51
Table 12.	Phosphorus groups	53
Table 13.	Silicon groups	55
Table 14.	Halogens and miscellaneous groups	56
Table 15.	Inorganic salts	57
Chapter 3/	Band Positions and Intensity	58
3.1.	Band positions	58
3.2.	Deuteration	65
3.3.	Band Intensities	66
Chapter 4/	Example of Absorption Band Shifts	68

RAMAN SPECTROSCOPY

Chapter 5/	Laser Raman Spectroscopy	75
5.1.	Introduction	75
5.2.	Examples of Raman Spectra	83

PROBLEMS 91
ANSWERS 149
APPENDICES
Appendix I/ NMR 275
Table 1. Approximate chemical shift of methyl, methylene, and methine protons 275
Table 2. Chemical shift of miscellaneous protons 276
Appendix II/ Wave-number Wavelength Conversion Table 277
INDICES
General Index 283
Index of Compounds 285

QUALITATIVE DATA

CHAPTER 1 / THE INFRARED SPECTRUM

The infrared spectrum is generally regarded as one of the more characteristic properties of a compound.

The range from 0.75 micron (one micron is 10^{-4} cm and is expressed by μ) to 200 microns, namely, from just outside the visible region and extending up to the microwave region, is called the infrared (Fig. 1.1). However, the so-called infrared region usually covers only the range between 2.5 and 40 to 50 μ; the shorter and longer wavelength regions are called, respectively, the near- and far-infrared regions. The far-infrared region extends out to 10 cm^{-1} and affords information mainly on vibrations involving heavy atoms or bond torsions.* However, it is still a rather specialized field.

The wavelength of infrared light is most frequently expressed in terms of wave-numbers, which are the reciprocal of wavelengths expressed in centimeter units. The unit of the wave-number is thus cm^{-1}. For example, the range 2.5~25 μ corresponds to 4000~400 cm^{-1}. Two types of spectrophotometers are available, those linear in wavelengths and those linear in wave-numbers. The wave-number unit is more widely used today.

All molecules are made up of atoms linked by chemical bonds. The movement of atoms and chemical bonds can be likened to that of a system comprised of springs and balls in constant motion. Their motion can be regarded as being composed of two components, the stretching and bending vibrations. The frequencies of these vibrations are not only dependent on the nature of the particular bonds themselves, such as the C–H or C–O bonds, but are also affected by the entire molecule and its environment. This situation is similar to that encountered in the spring-ball system in which the vibration of a single spring is under the influence of the rest of the system. "The internal motion of this system will become greater if energy is transferred to it." Similarly, the vibrations of bonds, which accompany electric vibrations, will increase their amplitude if an electromagnetic wave (infrared beam) strikes them. The difference between a molecule and the spring-ball system is that the vibrational energy levels of the

* J. W. Brosch, Y. Mikawa, and R. J. Jakobsen, "Chemical Far Infrared Spectroscopy," *Applied Spectroscopy Reviews*, 1, 187 (1968).
K. B. Whetsel, "Near-Infrared Spectrophotometry," *Applied Spectroscopy Reviews*, 2, 1 (1968).

former are quantized; therefore, only the infrared beam with a frequency exactly corresponding to that required to raise the energy level of a bond will be absorbed, *viz.*, the amplitude of the particular vibration is increased suddenly by a certain amount and not gradually. When the sample is irradiated by an infrared beam whose frequency is changed continuously, the molecule will absorb certain frequencies as the energy is consumed in stretching or bending different bonds. The transmitted beam corresponding to the region of absorption will naturally be weakened, and thus a recording of the intensity of the transmitted infrared beam versus wave-numbers or wavelength will give a curve showing absorption bands. This is the infrared spectrum.

It was mentioned above that the frequencies of the respective bonds in a molecule are affected by the whole molecular environment. However, certain bonds have distinguishing characteristics: multiple bonds are stronger than single bonds, and X–H type bonds (N–H, O–H, C–H, etc.) have the especially light terminal hydrogen atoms. These correspond to springs that are especially strong or those connecting especially light terminal balls. Like the spring-ball system, the vibrations of these bonds are affected by the rest of the molecule to a relatively small extent. Thus the stretching frequencies of these specific bonds appear within a range characteristic for the respective type of bonds; collectively, they appear in the range of 3600~1500 cm^{-1} (Fig. 1.1). The C=O stretching frequency appears between 2000 and 1500 cm^{-1} and is very sensitive to differences in structure and environment, a fact that makes the carbonyl absorptions extremely useful in organic chemistry.

In the region below 1600 cm^{-1} there appear bands due to the stretching of single bonds, C–C, C–N, C–O, C-halogen, etc., and also those due to the bending of various bonds. Single bonds have bond strengths of roughly the same order, and furthermore, they are usually linked cumulatively, e.g., C–C–C–O.

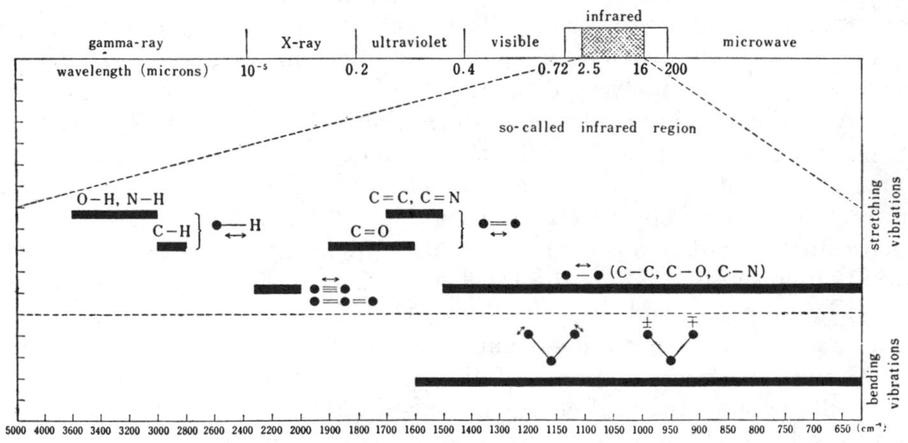

Fig. 1.1 Stretching and bending vibrations

This results in stronger mutual interaction, wider range of appearance of the bands, and great sensitivity of band positions to even minor structural changes.

The frequency range of the infrared spectrum below 1300 cm^{-1} is known as the **fingerprint region** and each compound has its own specific pattern.

However, caution must be exercised when using the fingerprint region to establish the identity of compounds, since different compounds may have very similar spectra (e.g., solution spectra of homologous fatty acids), whereas a single compound may give different spectra due to differences in sampling conditions, **polymorphism**, etc. (see Problem (25)).

We shall next consider the factors that determine the position of absorption bands. The stretching frequency ν of a diatomic molecule composed of atoms with masses m and m′ can be expressed by equation (1.1)

$$\nu = \frac{1}{2\pi c}\sqrt{\frac{f(m + m')}{mm'}} \qquad (1.1)$$

where c is the velocity of light, and f is the force constant (bond strength or bond order, corresponding to Hooke's constant of springs). The stretching frequencies of X–H bonds and multiple bonds in a larger molecule can also be approximated by the same expression. It can be seen that the position is determined by the bond strength and mass of the atoms linked by the bond. The stronger the bonds and the smaller the masses, the higher the absorption frequency of that particular bond, i.e., the more energy is required to vibrate the bond. For example, the bond strengths increase from single to double to triple bonds and the stretching frequencies also increase from 700~1500 cm^{-1} to 1600~1800 cm^{-1} to 2000~2500 cm^{-1} (Fig. 1.1). Thus a bond of higher order absorbs at higher frequency, provided the masses of the bonded atoms are identical. Also, to a first approximation, the heavier the atoms concerned, the lower the frequency. Thus the stretching vibration of the O–H bond is at 3600 cm^{-1} but is lowered to 2630 cm^{-1} in the O–D bond, where the bond strength is the same and only the mass has been increased. Qualitatively, the bending frequencies can also be treated in the same fashion.

The characteristic bands most frequently used in the interpretation of molecular structure appear within a certain range which is independent of the rest of the molecule, since they arise from bands which contain the especially light hydrogen atom or the especially strong multiple bonds. However, since individual bonds never vibrate completely independently with respect to the remaining part of the molecule, the band positions vary in a complex fashion from case to case. This variation in positions is a factor of major importance in qualitative analysis.

A vibration is not necessarily accompanied by an infrared absorption band. An absorption occurs only when the vibration causes a change in the charge distribution within the molecule. The larger the change the stronger the absorption. Accordingly, bands of hydrocarbons, which are composed only of carbon and hydrogen atoms, are weak, but bands associated with bonds connecting atoms that differ considerably in their electronegativities, e.g., C–N, C–O, C=O, C≡N, are usually quite strong. Although it was mentioned above that bending frequencies and stretching frequencies of single bonds appear in the same region, the C–O and C–N stretching bands can be detected rather easily because they are stronger than the C–C stretching bands.

Absorption bands that appear with a relatively high intensity in a range characteristic for a certain group and that are useful for the identification of that group are called **characteristic frequencies** or **characteristic absorption bands.** A tremendous amount of data regarding these bands have been accumulated since the late 1940's and many bands in the fingerprint region are now also effective for characterizing the various groups.

Fig. 1.2 shows the characteristic absorptions of the alcoholic hydroxyl group. These bands appear no matter in what molecules the hydroxyl group is contained. The O–H stretching band (1) because of its typically high frequency, and the C–O stretching band (3) because of its intensity are especially useful for identification. The bending frequencies (2) and (4) are also employed, although seldom, for identification. However, since the vibration of a specific bond is under the subtle influence of other nearby bonds, both the position and the intensity of these characteristic frequencies are different according to the structure of the molecule and the state of measurement. For instance, the appearance of the four hydroxyl absorptions shown in Fig. 1.2 will depend on: whether the hydroxyl group is attached to a primary, secondary, or tertiary carbon atom; whether the group is free, intermolecularly hydrogen-bonded, or intramolecularly hydrogen-bonded; the strength of the hydrogen-bond, etc. Thus a comparison

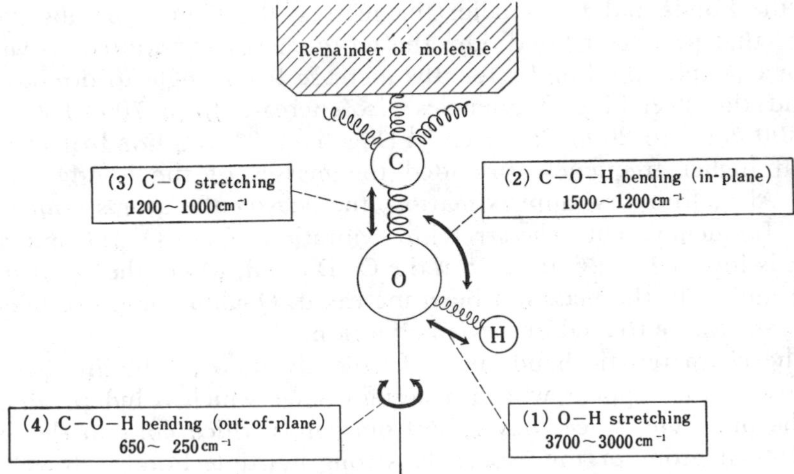

Fig. 1.2 Vibrations and absorptions of the alcoholic hydroxyl group.

of the bands with those of similar and known compounds would enable one to guess what sort of hydroxyl exists and in what state. This guess will be strengthened if changes in measuring conditions (solvents, etc.) or chemical structure (oxidation to ketone, acetylation, etc.) are accompanied by the appropriate changes in absorption bands. For example, if a compound having the isopropenyl group is reduced, its characteristic frequencies should disappear and instead, frequencies characteristic of the isopropyl group should appear; this would provide strong support for the preliminary assumption.

$$-C{\begin{array}{l}\diagup\!\!\!\!\diagup CH_2\\ \diagdown CH_3\end{array}} \xrightarrow{[H]} -CH{\begin{array}{l}\diagup CH_3\\ \diagdown CH_3\end{array}}$$

Characteristic frequencies of most groups are now known thanks to the self-recording instruments that became available in the late 1940's. These data have been compiled in book, table, and card form and are indispensable for the identification of various groups. This **qualitative** information is one of the most important benefits of infrared spectroscopy. The absorption bands are a direct reflection of the state of molecular bonds and this aspect affords considerable information of **theoretical** interest. The band intensity is proportional to the amount of sample present and this leads to another important application of infrared spectroscopy, **quantitative** analysis.

MEASUREMENT

A polystyrene film is commonly used for calibration of wave-numbers (Fig. 1.4). An alternative is calibration using the spectrum of atmospheric H_2O (Fig. 1.3). For highly critical measurements, the International Union of Pure and Applied Chemistry (IUPAC) has formulated wave-number calibration tables based on specified mixtures of indene, cyclohexanone and camphor.‡ The cell thickness can be measured conveniently by the interference fringes (cf. Problem 1). Solvents for solids should not damage the sodium chloride cell windows and also should not react with the sample; the curves will be distorted in regions where the transmission is less than 35%. Nonpolar solvents such as carbon tetrachloride (and carbon disulfide) are most suitable (Figs. 1.6 and 1.7), but the more polar chloroform is also widely used for its greater solvent power (Fig. 1.8). Polar solvents such as dioxane and tetrahydrofuran are employed for special purposes (Figs. 1.9 and 1.10).

Solids may also be measured as a mull with Nujol (mineral oil) or as KBr pellets. The two types of measurement are complementary since the regions obscured by Nujol absorptions (Fig. 1.5) are clear in the KBr spectra. The latter, however, usually contain bands due to adsorbed water in the regions 3300 cm^{-1} (medium intensity, O–H stretch) and 1640 cm^{-1} (weak, H–OH bend). For removal of water see p. 58).

Some tables and curves are given below.

TABLE 1.1 PRISM MATERIALS *

Material	Optimum range (μ)	Suitable range (μ)	Solubility (20°C) (g/100g H_2O)
Glass	0.3~ 2	0.3~ 2	insol.
Quartz	0.3~ 3.5	0.2~ 4	insol.
Lithium fluoride	0.6~ 6	0.2~ 6.5	0.27
Calcium fluoride	0.2~ 9	0.2~ 9.5	0.016
Sodium chloride	2~15.5	0.2~17	36.0
Potassium bromide	10~25	0.2~25	54.0
Silver chloride	2~20	0.2~22	0.00015
KRS-5 (TlBr–TlI)	2~40	1~40	0.05
Barium fluoride	0.2~13	0.2~13.5	0.17
Cesium bromide	10~37.5	1~38	124.0

* Taken from Perkin-Elmer, *Instruction Manual* **3B** 9 (1954).
The limit for usage as cell windows is at 1~2 μ longer wavelength than that shown in the Table.

‡ R. N. Jones and A. Nadeau, *Canadian Journal of Spectroscopy 20*, 33 (1975).

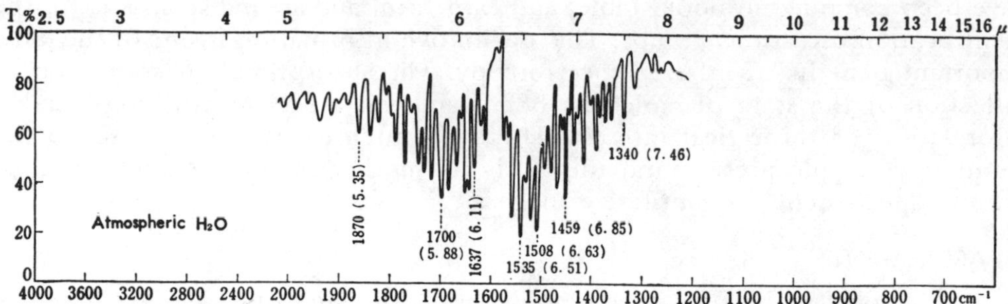

Fig. 1.3 Measured by single beam apparatus. Conspicuous bands are employed for wave-number calibration.

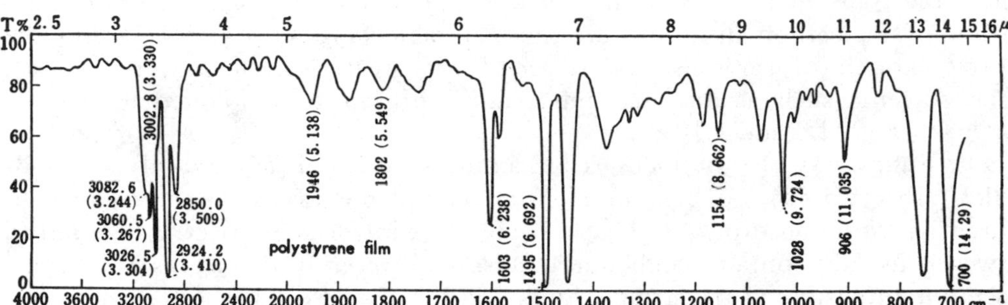

Fig. 1.4 The wave-number is calibrated with the 3026, 3003, 2924, 1603, 1495, 906 cm^{-1}, and other bands. The film is usually mounted on a handy frame and is used also for adjusting the running conditions of the spectrophotometer.

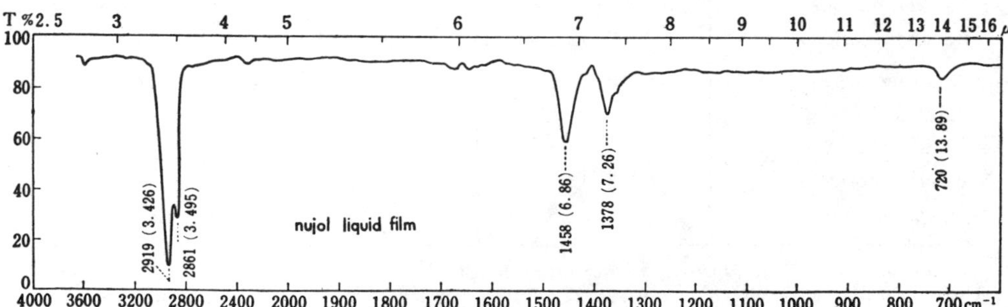

Fig. 1.5 For solid samples. Bands are due to methylene and methyl groups.

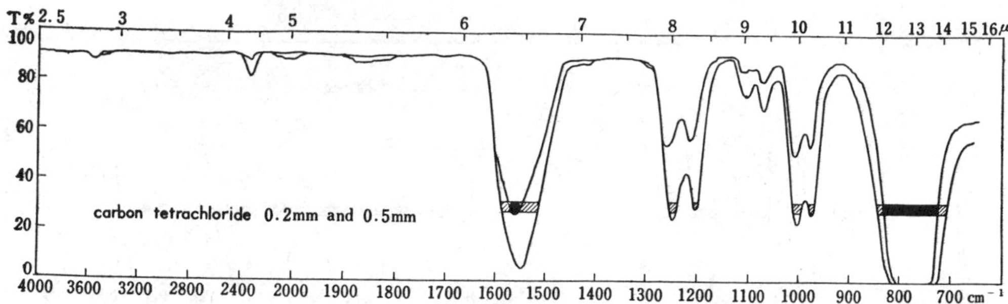

Fig. 1.6* General solvent. Specially suited for high frequency region.

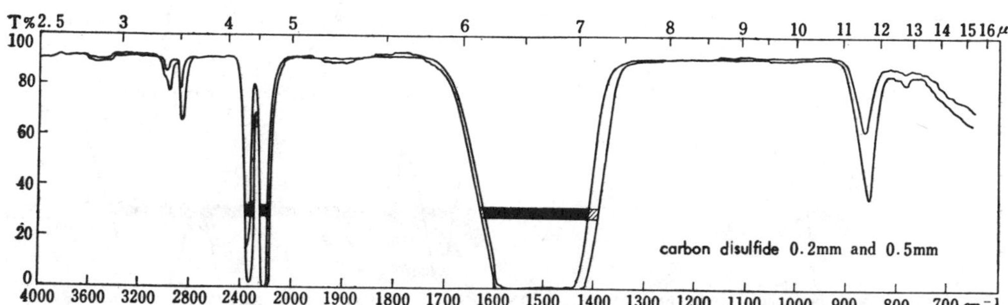

Fig. 1.7* General solvent. Specially suited for low frequency region. Best used in a chamber because of its toxicity and inflammability. Primary and secondary amines occasionally react with carbon disulfide to give alkyl dithiocarbamates. [R. N. Jones and C. Sandorfy, "Chemical Applications of Spectroscopy," p. 511 (1956).]

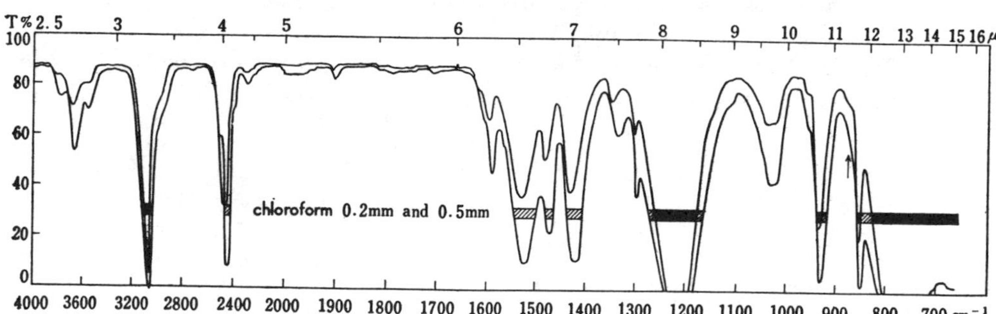

Fig. 1.8* General solvent. Commercial chloroform contains 1~2% ethanol as stabilizer and this can be easily detected by appearance of small band at ca. 880 cm^{-1} (shown by arrow in Fig.). When measuring samples that react or form hydrogen-bonds with the ethanol, the latter should be removed beforehand by passing through alumina or silica gel columns or other methods.‡ The ethanol-free chloroform can be stored for about a week; formation of phosgene is detected by appearance of band at 1810 cm^{-1}.

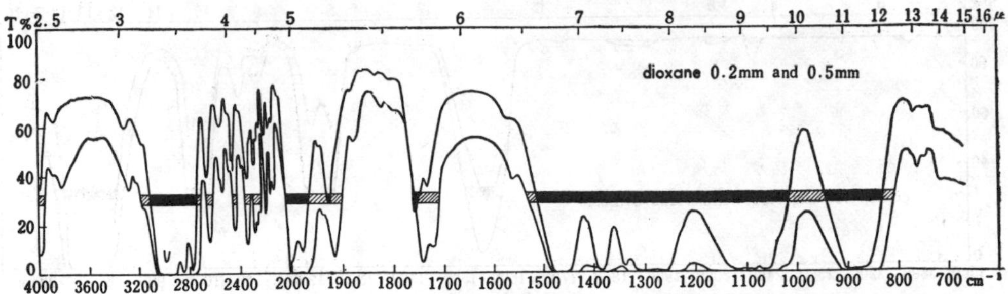

Fig. 1.9* Special solvent for hydrogen-bond studies, etc.

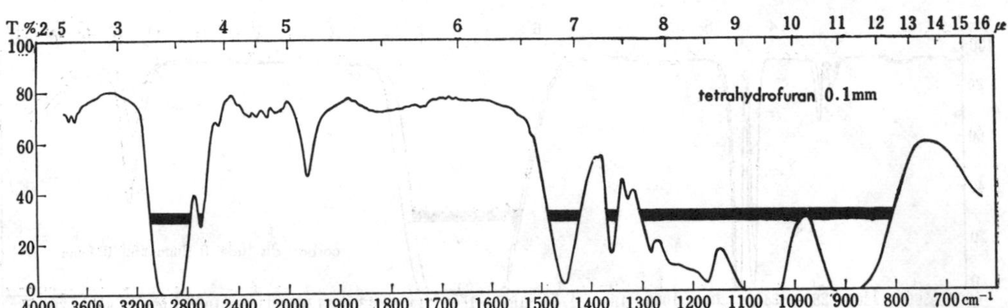

Fig. 1.10* Special solvent for hydrogen-bond studies, etc.

CALIBRATION

For accurate calibration of prism and grating spectrometers, one should consult the following monograph.

IUPAC Commission on Molecular Structure and Spectroscopy. "Table of Wavenumbers for the Calibration of Infra-red Spectrometers," Butterworths, London, 1961.

---

* Ranges in which the percent transmission is less than 35% are shadowed in Figs. 1.6~1.10 and the respective solvents cannot be used in these particular regions.

‡ D. D. Perrin, W. L. F. Armarego and R. Perrin, "Purification of Laboratory Chemicals," p. 110, Pergammon Press, Oxford, 1966.

References

I. General

[1] L. J. Bellamy, "The Infra-red Spectra of Complex Molecules," 3rd, Methuen, London, 1975.

[2] L. J. Bellamy, "Advances in Infrared Group Frequencies," Methuen, London, 1958.

[3] F. F. Bentley, L. D. Smithson and A. L. Rozek, "Infrared Spectra and Characteristic Frequencies ~700 to 300 cm^{-1}, A Collection of Spectra, Interpretation and Bibliography," Interscience, Div. of Wiley, New York, 1968.

[4] N. B. Colthup, *J. Opt. Soc. Am.*, **40**, 397 (1950).

[5] N. B. Colthup, L. H. Daly and S. E. Wiberley, "Introduction to Infrared and Raman Spectroscopy," Academic Press, New York, 1964.

[6] N. B. Colthup, "Interpretation of Infrared Spectra," ACS Audio Course, American Chemical Society, Washington, 1971.

[7] R. T. Conley, "Infrared Spectroscopy," Allyn and Bacon, Boston, 2nd ed., 1972.

[8] A. D. Cross, "Introduction to Practical Infra-red Spectroscopy," 3rd ed., Butterworth, London, 1969.

[9] K. Dobriner, E. R. Katzenellenbogen and R. N. Jones, "Infrared Absorption Spectra of Steroids," Interscience, New York, 1953.

[10] A. Finch, P. N. Gates, K. Radcliffe, F. N. Dickson and F. F. Bentley, "Chemical Applications of Far Infrared Spectroscopy," Academic Press, New York, 1970.

[11] M. St. C. Flett, "Characteristic Frequencies of Chemical Groups in the Infra-Red," Elsevier, Amsterdam, 1963.

[12] A. R. Katritzky, ed., "Physical Methods in Heterocyclic Chemistry," Vol. II, Academic Press, New York, 1963.

[13] C. E. Meloan, "Elementary Infrared Spectroscopy," Macmillan, New York, 1963.

[14] R. G. J. Miller, ed., "Laboratory Methods in Infrared Spectroscopy," Heyden, London, 1965.

[15] K. Nakamoto, "Infrared Spectra of Inorganic Coordination Compounds," Wiley, New York, 1970.

[16] J. P. Phillips, "Spectra-Structure Correlation," Academic Press, New York, 1964.

[17] W. J. Potts, Jr., "Chemical Infrared Spectroscopy, Vol. I Techniques," Wiley, New York, 1963.

[18] C. N. R. Rao, "Chemical Applications of Infrared Spectroscopy," Academic Press, New York, 1963.

[19] F. Schienmann, ed., "An Introduction to Spectroscopy Methods for the Identification of Organic Compounds, Vol. I," Pergammon Press, 1970.

[20] J. E. Stewart, "Far Infrared Spectroscopy," in *Interpretive Spectroscopy*, S. K. Freeman, ed., New York, Reinhold, 1965.

[21] H. A. Szymanski, "I.R. Theory and Practice of Infrared Spectroscopy," Plenum Press, New York, 1964.

[22] H. A. Szymanski, "Correlation of Infrared and Raman Spectra of Organic Compounds," Hertillon Press, Cambridge Springs, Pa. 1969.

II. Reference Spectra

[1] Richard A. Nyquist and Ronald O. Kagel, "Infrared Spectra of Inorganic Compounds (3800–45 cm^{-1})," Academic Press, New York, 1971.

[2] Charles J. Pouchert, "The Aldrich Library of Infrared Spectra," 2nd ed., Aldrich Chemical Co., Milwaukee, 1975.

[3] ASTM – Am. Soc. for Testing and Materials, Alphabetical List of Compound Names, Formulae and References to Published Infrared Spectra, an Index to 92,000 Published Infrared Spectra, Amer. Soc. for Testing and Materials, Philadelphia, Pa., 1969; there is also a Molecular Formula List and a Serial No. List.

[4] IRDC – Infrared Data Committee of Japan; Nankodo, Haruki-cho, Hongo, and Tokyo.

[5] SADTLER – Catalog of Infrared Spectrograms, Sadtler Research Laboratories, 3314-20 Spring Garden Street, Philadelphia, Pa.

[6] *Documentation of Molecular Spectroscopy* (DMS), Butterworths Scientific Publications, London; and Verlag Chemie GMBH, Weinheim/Bergstrasse, West Germany.

[7] API – Selected Infrared Spectral Data, American Petroleum Institute, Research Project 44, Department of Chemistry, Texas A & M University, College Station, TX 77843.

CHAPTER 2 / TABLES OF CHARACTERISTIC FREQUENCIES

2.1 POSITION OF ABSORPTION BANDS

Unless otherwise stated the bands usually appear within ±10 cm^{-1} of the position shown. The data are mostly for compounds measured as dilute solutions in nonpolar solvents, and the external and internal factors such as solvation, inter- or intra-molecular hydrogen-bonding, electric effects of nearby groups, steric strain, and other steric effects will shift the bands from their normal positions.

Bands that are not practically useful are written in italics. Some have been completely omitted, e.g., skeletal vibration in the fingerprint region.

2.2 INTENSITY

In qualitative consideration of infrared spectra, intensity is usually expressed as vs (very strong), s (strong), m (medium), w (weak), etc. These notations, which are often made subjectively by visual inspection of the spectrum, correspond to the ϵ^a values given in the table below.*

TABLE 2.1 INTENSITIES

Intensity	ϵ^a
very strong (vs)	200
strong (s)	75–200
medium (m)	25–75
weak (w)	5–25
very weak (vw)	0–5

In the following tables the apparent molecular coefficient has been given whenever possible, besides the conventional notations. It is defined as:

$$\epsilon^a = \frac{1}{cl} \log_{10}\left(\frac{T_0}{T}\right)_v$$

* Private communication, Professor C. F. Hammer, Georgetown University, Washington, D.C.

ϵ^a: apparent molecular absorption coefficient
c: concentration, moles per liter
l: cell thickness, cm
T_0 and T: intensity of entering and transmitted light. Measurements are not made under strictly monochromatic conditions because of the slit width. Accordingly, $\log_{10} (T_0/T)_\upsilon$ is the apparent optical density obtained when the instrument is set to a wave-number of υ, and it is not the true absorptivity involved in the BEER-LAMBERT rule.

The apparent integrated area is defined as:

$$B = 2.303 \int \epsilon^a \cdot d\upsilon$$

Methods for converting apparent intensities into true molar absorption coefficients and integrated areas (A) are described in reference (1).

The ϵ^a values given in the tables give only a rough idea of the intensities since the values vary according to instrument, slit width, temperature, solvent and other experimental conditions. The difference between the true **molar absorption coefficient** ϵ, and ϵ^a is also variable (ϵ is larger than ϵ^a by 0~20%). Recent advances in instrumentation, especially the widespread use of grating instruments, have made it possible to obtain more reliable ϵ^a values; however, remeasurement of ϵ^a values has received little priority. There thus remains a dearth of reliable ϵ^a measurements, despite their potential usefulness in functional group analysis.[2]

While not extensively used for qualitative analysis, intensity measurements have been widely used in quantitative analysis.[3] A more accurate method for determining the integrated absorption intensity was recently proposed.[7]

2.3 ASSIGNMENTS

A molecule of n atoms has $3n$ degrees of freedom, three of which are assigned to translational, another three to rotational (two in linear molecules such as acetylene), and the remaining $3n-6$ ($3n-5$ if linear) to vibrational motions. These various vibrations (fundamental vibrations) absorb, respectively, the infrared beam at characteristic frequencies and give rise to absorption bands. However the number of absorption bands does not coincide exactly with the number of fundamental vibrations. Thus, if the vibration does not induce any change in the dipole moment of the entire molecule, the vibration will not absorb the infrared energy and the number of infrared peaks will be decreased accordingly.

Raman spectra complement infrared data, since absorption of energy in the **Raman spectrum** is dependent only on a change in polarizability when energy is absorbed, rather than a change in dipole moment. For example, the vibration of the central C=C bond of the ethylene molecule will not absorb infrared energy (infrared inactive) whereas the ethylene C=C and other symmetrical C=C bonds invariably show a Raman band.

Fundamental vibration bands can be divided into those caused by stretching vibrations and those caused by deformation vibrations. Vibrations of the methylene group are shown as an example:

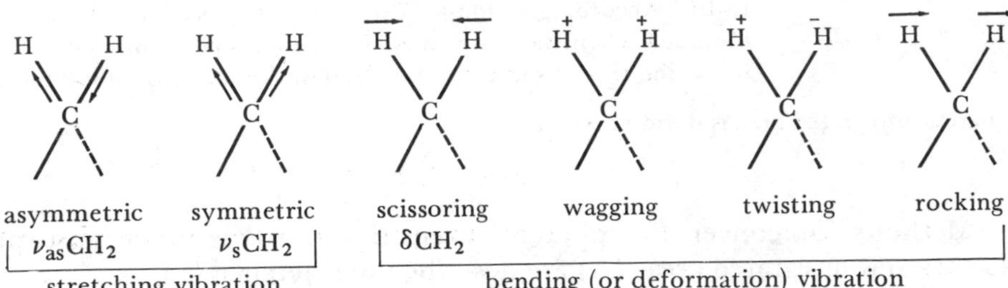

Vibrations of the methylene group (the + and – signs denote vibrations in the direction perpendicular to the paper).

Note: Particular asymmetric stretching vibrations in which two bonds move in the opposite direction, such as the asymmetric stretching of the methylene group, are also called antisymmetric stretching vibrations.

Other groups having similar arrangements of atoms, e.g.:

amino nitro carboxylate

also possess fundamental vibrations corresponding to those of the methylene group. In certain cases, vibrations other than the normal ones will appear.

Overtone: Overtones having twice the frequency of the normal vibration will occasionally be observed as weak bands. For instance, the overtone of the carbonyl 1715 cm^{-1} absorption may appear around 3430 cm^{-1}, and may be confused with the hydroxyl absorption.

Combination tone: These are weak bands occasionally appearing at frequencies that are the sum or difference of two or more fundamental bands. Thus, fundamental bands at x and y cm^{-1} may give rise to weak bands at $(x + y)$ or $(x - y)$ cm^{-1}.

The following phenomena also occur in the spectra.

Coupling: Very often a 1:1 correspondence between a bond vibration and an absorption band cannot be assumed. The coupling phenomenon occurs when two bonds of the same symmetry are closely located within the molecule, the respective bands are strong, and the absorptions occur in the same region; it is highly dependent on molecular geometry. Coupling of vibrations results in a shift of the absorptions from the characteristic frequencies, one to higher frequency and one to lower. The coupled bands constitute useful characteristic absorption bands for such groups as acid anhydride ((61)), carboxylate ((55)), nitro ((15) , (60)) and amides (amide I, II, III bands (75) , (76) , (77)).

Fermi resonance: When an overtone or combination band is located near a fundamental frequency of the same symmetry, the band intensity of the former may be anomalously enhanced, or bands may be split (ethylene carbonate, benzoyl chloride). Unexpected doubling of a band is not always due to Fermi resonance; for example, alkyl vinyl ethers (n-butyl vinyl ether, (30)) show two C=C bands due to rotational isomerism about the –C–O– bond.[4] Fermi resonance may be detected by the methods of isotopic substitution or solvent variation.[5,6]

References

[1] R. N. Jones and C. Sandorfy, "Chemical Applications of Spectroscopy," p. 271, Interscience (1956). K. S. Seshadri and R. N. Jones, *Spectrochim. Acta,* **19**, 1013 (1963).

[2] Arthur S. Wexler, "Integrated Intensities of Absorption Bands in Infrared Spectroscopy," *Applied Spectroscopy Reviews,* **1**, 29 (1967).

[3] John A. Perry, "Quantitative Analysis by Infrared Spectroscopy," *Applied Spectroscopy Reviews,* **3**, 229 (1970).

[4] A. D. H. Claque and A. Danti, *Spectrochim. Acta,* **24A**, 439 (1968).

[5] L. J. Bellamy and R. L. Williams, *Trans. Faraday Soc.,* **55**, 14 (1959).

[6] J. F. Bertran, L. Bollester, L. Dobrihalova, N. Sanchez and R. Arrieta, *Spectrochim. Acta,* **24**, 1765 (1968).

[7] L. van Haverbeke and M. A. Herman, *Appl. Spectrosc.,* **28**, 580 (1974).

TABLE 1 ALKANES

(Non-diagnostic bands in italics)

Group	Band (cm^{-1})	Intensity ($mole^{-1} \cdot l \cdot cm^{-1}$)	Assignment	Remarks	Figs.
1) $-CH_3$	2960	s 70	ν_{as} CH_3	C=C–CH_3 moved to higher frequency	
	2870	m 30	ν_s CH_3	2830~2815 (ϵ^a 35~75) in R–OMe, Ar–OMe[1]	(31)
				2820~2730 (ϵ^a 15~21) in –NMe,[2] lacking in –NEt.	(66)
				2820~2710 in methylenedioxy[3]	(47)
				C=C–CH_3 moved to higher frequency and split into doublet	
	1460	m 15	δ_{as} CH_3		(2)
	1380	m 15	δ_s CH_3	Doublet in *gem*-dimethyl groups (see 5~7). See also (11)~(13). Shift on attachment to other atoms depends on electronegativity and C–X bond distance, e.g., $-SCH_3$ (1330), CH_3I (1250), $-SiCH_3$ (1255), $-BCH_3$ (1322±7).[3a]	(2)
2) $-CH_2-$	2925	s 75	ν_{as} CH_2		
	2850	s 45	ν_s CH_2		
	1470	m 8	CH_2 scissor	1445 in cyclopentanes, 1450 in cyclohexanes. See also (14)~(15).	
	725~720[4]	s 3	$(CH_2)_n$ rock	Sometimes doublet in solid state. Present when $n \geqslant 4$. Higher with smaller n. Pr 743~734; Et 790~770. $-CH_2-$ also has wag and twist band at ca. 1300 (ϵ^a 1). Solid state spectra of long chain with terminal polar group (acids, esters, amides) show regular series of bands at 1350~1180 (CH_2 wag).	(2), (54)
3) $-\overset{\mid}{\underset{\mid}{C}}-H$	*2890*	w	ν CH	Of no practical use.	
	1340	w	δ CH	Of no practical use.	
4) $-(CH_2)_4-O-$	742~734		CH_2 rock	Corresponds to 720 band of (2).	
5) $CH_3 \diagdown$ $\quad \rangle CH-$ $CH_3 \diagup$	1170	s Weaker than the 1380 (ϵ^a 15) doublet	skeletal	Doublet at 1380 suggests *gem*-dimethyl group; identification of this group then made by skeletal vibration (5)~(7).	(10)
	1145			1145 band is shoulder on 1170 band.	
6) $CH_3-\overset{CH_3}{\underset{CH_3}{C}}-$	1255	s ”	”	Position more constant than 1210 band.	
	1210	s ”	”	Absorption also at 930~925.	
7) quat. C $-\overset{CH_3}{\underset{CH_3}{C}}-$	1215	s ”	”	Position of 1195 band more constant; 1215 band forms shoulder on 1195 band.	(20), (40)
	1195	s ”			
8) CH_2 [5] $\rangle C$—$C\langle$ R H C $\rangle C$—$C\langle$	3070±10	m	ν_{as} CH_2	Lacking when ring carries no CH_2.	(42)
	3005±5	m	ν_{as} CH		
	1020~1005	m 20~80	skeletal	Confirmatory only; often absent or obscured by other strong bands (hydroxyl, ether, etc.).	
	860				

Group	Band (cm^{-1})	Intensity ($mole^{-1}\cdot l\cdot cm^{-1}$)	Assignment	Remarks	Figs.
9) epoxide ring (O bridging >C—C<H(H))	3000	20~60	ν CH of methine	Shifted to 3040~3030 if ring strain of epoxy increased, e.g., fused to cyclopentanes[6].	(34)
(cf. Table 6)	3050	30	ν_{as} CH_2 of methylene		
10) aziridine ring (NH bridging >C—CH_2)	3050		ν_{as} CH_2	Ref. 7.	

Some characteristic CH bending frequencies around $1400 cm^{-1}$

Group	Band	Remarks	Figs.
11) O-CO-CH_3	1380~1365	Characterized by high intensity. Upon acetylation the 1380 band usually becomes stronger than 1460 band. Same with enol and phenol acetates. Occasionally doublet.	(58),(66)
12) -CO-CH_3	1360~1355	Lower than ordinary CH_3 bending (1380), strong and sharp. Asymmetric bending at usual position (1460), but lower for terpenes (1420~1425).	(43),(44)
13) -$COOCH_3$ [8]	1440~1435	δ_{as} CH_3. Bands also at 1135, 1155, 790~760.	(39)
	1365~1356	$\delta_s CH_3$	
14) -CH_2-CO-[10]	1440~1400	Bending of all active methylenes appear as strong sharp bands between ordinary 1465 and 1380 bending bands. Same with -CH_2-SO_2-, etc. Integrated intensity may give number of active methylenes.	(40),(54) (90),(93)
15) -CH_2-C=C -CH_2-C≡C-	1445~1430	Not so strong. Double bond can be aromatic. Lowered further if -CH_2- is between two multiple bonds.	(9),(22) (17)
16) -CH_2-N^+[9]	1440~1400	-CH_2- adjacent to the electron-attracting N^+ is also shifted lower as in (14) and (15). If new band appears in this region upon conversion of amine to its salt, -CH_2-N- group is present.	(39),(72)

References

[1] H. B. Henbest, G. D. Meakins, B. Nicholls and A. A. Wagland, *J. Chem. Soc.* 1462 (1957).

[2] R. D. Hills and G. D. Meakins, *J. Chem. Soc.* 760 (1958); J. T. Braunholtz, E. A. V. Ebsworth, F. G. Mann and N. Sheppard, *ibid.* 2780 (1958).

[3] L. H. Briggs and L. D. Colebrook, *Anal. Chem.* 29, 904 (1957).

[3a] L. J. Bellamy, *Advances in Group Frequencies*, pp. 10–13, Methuen, London (1968).

[4] J. M. Martin, R. W. B. Johnston, and M. J. O'Neal, *Spectrochim. Acta* 12, 12 (1958).

[5] cf. S. A. Liebman and B. J. Gudzinowicz, *Anal. Chem.* 33, 931 (1961).

[6] H. B. Henbest, G. D. Meakins, B. Nicholls, and K. J. Taylor, *J. Chem. Soc.* 1459 (1957).

[7] Ju. N. Scheinker, E. M. Peresleni, and G. I. Bras, *J. Phys. Chem. Moscow* 29, 518 (1955) (see DMS card No. 649 and 650).

[8] A. R. H. Cole and D. W. Thornton, *J. Chem. Soc.* 1007 (1956).

[9] K. Nakanishi, T. Goto and M. Ohashi, *Bull. Chem. Soc. Japan* 30, 403 (1957).

[10] A. R. Katritsky and S. Øksne, *Spectrochim. Acta,* 17, 1286 (1961).

ABSORPTIONS OF ALKANES

Since organic compounds often contain methylene and methyl groups most IR spectra have bands at 2950~2850 cm^{-1} (νCH, usually two with the resolution obtained by NaCl prism), ca. 1465 cm^{-1} (δCH of methylene and methyl), and 1380 cm^{-1} (δCH of methyl). The spectrum of paraffin oil (nujol) shows these bands quite typically (Fig. 1.5).

In general, the groups shown in Table 1 are more readily characterized by proton and carbon magnetic resonance.

Absorption in 3000 cm^{-1} Region

Alkane groups have their CH stretching at frequencies lower than 3000 cm^{-1}, whereas alkenes and aromatics have them higher than 3000 cm^{-1}; the cyclopropane methylene and $-CH_2-$halogen also absorb at 3050 cm^{-1}. Thus the shape of absorption around 3000 cm^{-1} gives a rough idea of the groups present. The region is much better resolved with a grating instrument. Quartz, lithium fluoride or calcium fluoride prisms can be used for the same purpose. The spectrum of polystyrene (Fig. 1.4) then shows a deep cut between the bands arising from the saturated and unsaturated portion of the polymer.

gem-Dimethyl Groups (Table 1, 5~7)

The asymmetric CH_3 bending at 1380 cm^{-1} is split into a doublet whose normal separation is 20 cm^{-1}. This separation is decreased by the presence of an electronegative substituent e.g., it is 12 cm^{-1} in an isopropyl ether. The appearance of this doublet is also dependent on the angle between the two methyl groups; when the angle is large, as in 1,1-dimethylolefins, only a single band is seen. If the weighted contribution of the *gem*-dimethyl group is small, as in steroids containing the side-chain isopropyl group, it will not be easy to detect with sodium chloride optics. A grating instrument or calcium fluoride prism gives much better resolution; solid samples should be measured as KBr discs or carbon tetrachloride solutions.

Bands due to groups other than alkanes may appear in the same region to form an apparent doublet. The next step is to charactcrize the *gem*-dimethyl groups from the 1150~1250 cm^{-1} bands. However, since the band intensities are not so great, and since many other vibrations have absorptions in this area, characterization is possible only when the molecule is rather simple.

Groups Having Characteristic Absorption Around 1400 cm^{-1} (Table 1, 11~16)

As exemplified by the nujol spectrum (Fig. 1.5), organic compounds usually have bands at 1460 cm^{-1} and 1380 cm^{-1}, the former being the stronger. Reversed intensity relations would suggest an acetoxyl; a rather sharp band on the lower frequency side of the 1380 cm^{-1} band would suggest a methyl ketone or a methoxycarbonyl; and a band near 1410 cm^{-1} between the two usual absorptions would suggest $-CH_2-CO-$, $-CH_2-N^+-$, and the like, to be present.

Other Absorptions

The 720 cm^{-1} band (CH_2 rock) almost always appears in compounds with at least four contiguous CH_2 groups and is diagnostically useful. Besides the 720 cm^{-1} band, and bands in the regions of 3000 cm^{-1} (νCH) and 1400 cm^{-1} (δCH), alkane groups show absorption in the 1300~1100 cm^{-1} region. These are caused by CH_2 wagging, CH_2 twisting, and CH_2 rocking, but they are usually weak; moreover, C–C stretching and other skeletal vibrations appear in this region. These 1300~1100 cm^{-1} bands are therefore not used in qualitative organic chemistry excepting those associated with the *gem*-dimethyl groups (Table 1, 5~7). However, when the molecule contains a polar group they are sometimes unusually strong, in certain cases the strongest in the spectrum.

TABLE 2 ALKENES (Non-diagnostic bands in italics)

CH stretching frequencies (ϵ^a in parentheses)

1) $=CH_2$	3080, m (30) 2975, m	ν_{as} CH_2 ν_s CH_2	2975 band overlaps with alkane absorption. Bands higher than 3000 suggest presence of unsaturated =CH– (alkenes, aromatics).
2) =CH–	3020, m	ν CH_2	

Substitution type (ϵ^a in parentheses)

Type	Overtone of δ CH (out-of-plane)	ν C=C	δ CH (in-plane)	δ CH (out-of-plane)	Figs.
3) R/CH=CH_2 (term. vinyl)	*1860~1800* m (30)	1645, m (40)	1420, m (10~20) Same region as in Table 1, 14~16. *1300*, m~w	990 s (50), and 910 s (110); 990 ⊕H–C=C–H⊕, H⊖ 910 H–C=C–H⊕, H⊕	5, 19, 24
4) R(R)C=CH_2 (term. methylene)	*1800~1750* m (30)	1655, m (35)	1415, m (10~20) Same region as in Table 1, 14~16.	890 s (100~150) C=C–H⊕, H⊕	3, 20
5) R(H)C=C(R′)(H) (*cis*)		1660, m (10)	1415, m (10~20)	730–675 m (40), ambiguous and variable. ⊕H–C=C–H⊕	5, 6
6) R(H)C=C(H)(R′) (*trans*)		*1675*, w (2)		965 s (100) ⊕H–C=C–H⊕	5
7) R(R′)C=C(R″)(H) (tri-subs.)		1670, w~m		840–800 s (40). Vibration of single H; not as useful as others.	3, 19
8) R(R′)C=C(R″)(R‴) (tetra-subs.)		*1670*, w Intensified if bonded directly to O or N.		No out-of-plane bending (H absent), but >C=C–CH_2– may be detected by δ CH_2 bending (Table 1, 15).	

Conjugated double bonds

Group	ν C=C	δ CH (out-of-plane)	Figs.
9) Diene	1650 and 1600	Position not affected much by conjugation. The *trans* 965 band is sometimes shifted to 990. The *cis* band is usually found at 720, occasionally as a group of bands.	4
10) Triene	*1650 and 1600*; sometimes only one band; sometimes additional shoulder on 1650 band.	*trans-cis-trans* and *cis-trans-trans* systems have bands at 990 s, 960 m, and 720.	5
11) Polyene	*Broad band at 1650~1580.*	990~970 if *trans*-double band present.	6
12) Enones	– see Table 8 (p. 16).		

ABSORPTION OF ALKENES

As described already, unsaturated compounds can be detected by virtue of the bands higher than 3000 cm^{-1} if they are not hidden by the stronger main band below 3000 cm^{-1}. It is advantageous to use grating instruments or calcium fluoride or lithium fluoride prisms. Presence of the double bond is further checked by looking for the C=C stretching frequency in the 1650 cm^{-1} region, and the mode of substitution is identified by the C–H out-of-plane bending bands. Overtones and C–H in-plane bending bands also may be of assistance. The tetrasubstituted double bond usually cannot be detected unless an oxygen or nitrogen atom is directly attached to it.

Substitution of electronegative groups for R groups attached to the double bond dramatically increases the intensity of the C=C stretching band; the frequency of the absorption band decreases, and for the heavier elements (e.g., Br, I, S) falls to 1600~1580 cm^{-1}. Out-of-plane bending (δCH) frequencies are also affected. For the C=C twist of trans substituted ethylenes (965 cm^{-1}) and vinyl groups (990 cm^{-1}), substitution of an **electronegative** group causes a **decrease** in the frequency of the observed band (see 1 and 2). For the $=CH_2$ wag of vinyl (910 cm^{-1}) and methylene (890 cm^{-1}) groups the frequency is **decreased** when the substituent is an **ortho-para** director (see 2) and **increased** when it is a **meta** director (see 3).

1	CH_3, H / C=C / H, CH_3 964 cm^{-1}	CH_3, H / C=C / H, Cl 926 cm^{-1}	Cl, H / C=C / H, Cl 892 cm^{-1}
2	CH_3, H / C=C / H, H 986, 908 cm^{-1}	CH_3–O, H / C=C / H, H 960, 813 cm^{-1}	
3	R_1, H / C=C / R_2, H 890 cm^{-1}	R_1, H / C=C / N≡C, H 950 cm^{-1}	N≡C, H / C=C / N≡C, H 1000 cm^{-1}

Conjugated dienes are undoubtedly better characterized by ultra-violet spectroscopy. The 990~970 cm^{-1} band arising from the **trans** double bond in conjugated systems, however, is quite strong and is useful for its detection.

TABLE 3 AROMATICS

(Non-diagnostic bands in italics)

Phenyls (naphthalenes, phenanthrenes, etc., are similar)

Position	Intensity, ϵ^a	Assignment	Remarks	Figs.
1) ~3030, several	less than 60	ν CH and combination	With NaCl prisms, the bands frequently appear only as a weak shoulder on the main νCH aliph. band. In certain arom. compounds, the main absorption appears below 3000 cm^{-1}.	(7),(9)
2) 2000~1660 several bands (Fig. 2.1)	w, ~5	Overtone of δ CH (out-of-plane) and combination	Group of 2 to 6 bands characteristic of substitution pattern; identified by comparison with standard curves, e.g. alkylbenzenes. Hidden by strong bands due to C=O, etc. Sample concentration should be more than ten times that of normal.	(7),(9)
3) 1600, (1580), 1500, (*1450*)	ϵ^a variable but usually less than 100. The 1600 cm^{-1} band sometimes stronger than C=C band.	Phenyl nucleus	Variable intensity, 1500 usually stronger than 1600. In principle, 1580 band only appears when phenyl is conjugated with unsatd. groups or groups having lone pair electrons. Conj. intensifies all 3 peaks but positions not affected. 1450 cm^{-1} peak overlaps with CH_2 band. Condensed systems absorb at 1650~1600, (1600), 1525~1450 (two) cm^{-1}; pyridines similar to phenyl group.	(7),(9) (11),(31)

In-plane bending (1225~950 cm^{-1} region)

One to several absorptions appear according to the number of hydrogen atoms on phenyl group; weak but sharp. However, polar substituents may intensify them considerably. In-plane bands have only supplementary significance because C–C, C–O and other single bonds absorb in the same region. They are not affected much by conjugation.

Substitution	Position	Figs.
4) 1:2–, 1:4–, 1:2:4–	*1225~ 1175, 1175~1125* (only with 1:2:4– substitution). *1070~1000* (two)	(31),(46)
5) 1–, 1:3–, 1:2:3–, 1:3:5–	*1175~1125, 1110~1070* (absent with 1:3:5– substitution), *1070~1000*	
6) 1:2–, 1:2:3–, 1:2:4–	*1000~960*	

Out-of-plane bending and ring puckering (below 900 cm^{-1})

These bands are all very intense (ϵ^a 100~500). The out-of-plane bending bands can be conveniently divided according to the number of adjacent hydrogen atoms on the phenyl ring. Correlations shown in the following table also hold for condensed systems and pyridines. In pyridines, quinolines, etc., the hetero-atom is treated as a ring substituent. Bands due to isolated H (900~860 cm^{-1}) are weaker than the others in this region; also the position may be shifted down to 800 cm^{-1} (cf 1,3,5-substituted phenyl, Fig. 2.1).

The position of the out-of-plane bending bands is shifted to lower frequencies with an increase in the number of α-hydrogens on alkyl substituents, e.g., toluene, ethylbenzene, isopropylbenzene – 728, 745, 757 cm^{-1} respectively.

Conjugation of a double bond with the phenyl ring raises the frequency, sometimes by as much as 30 cm^{-1}. When the overall substitution of the phenyl ring results in deactivation, the out-of-plane bending bands are no longer useful as group frequencies, since their intensity drops and they are shifted to higher frequencies.

7) Ring puckering	710~690	Present in mono, **1:3–, 1:3:5–** and **1:2:3– substituted phenyls.**	
8) Five adjacent H	770~730		(17), (76)
9) Four " "	770~735		(21), (31)
10) Three " "	810~750		(31)
11) Two " "	860~800	Frequently appears at 820~800 cm^{-1}.	(37), (68)
12) Isolated H	900~860	May be shifted to ca. 800 cm^{-1}.	(7), (10)

ABSORPTION OF AROMATICS

The presence of an aromatic group is shown by the 3030 and 1600~1500 cm^{-1} bands, and substitution pattern is identified by the strong absorption bands below 900 cm^{-1}. The overtone and combination tone bands at 2000~1600 cm^{-1} ("5–6μ band," Fig. 2.1) are sometimes useful. 1225~950 cm^{-1} bands are of secondary importance. The strong CH out-of-plane bending bands and the 5–6μ pattern associated with various types of benzene substitution are shown in Fig. 2.1.

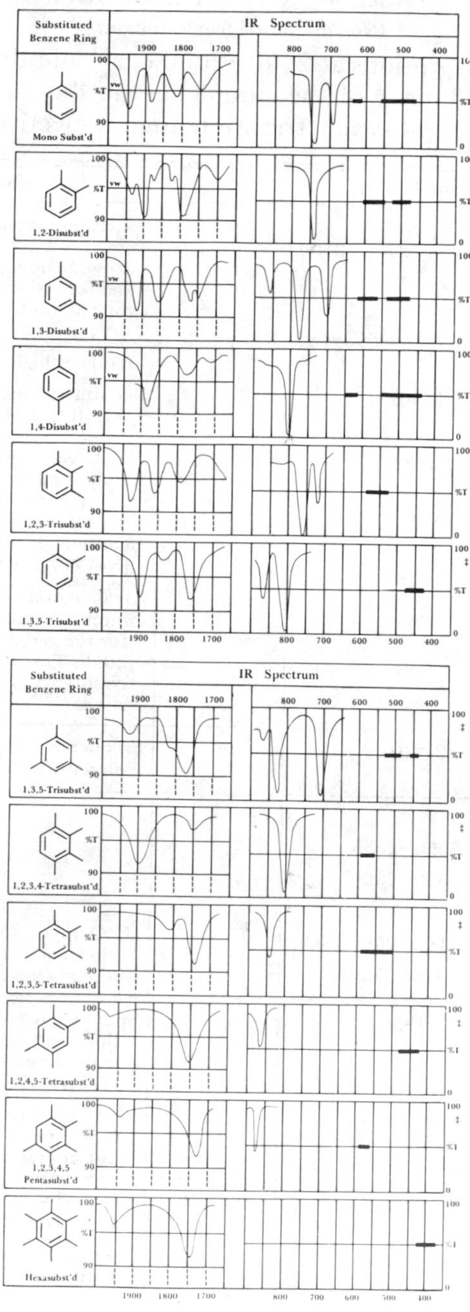

Fig. 2.1 Shape of the aromatic "5–6μ band," out-of-plane binding and torsional frequencies.* The spectra shown are those seen with a prism instrument; the pattern shows additional fine structure with grating instruments (see (7),(33)).

---

* Redrawn with permission of Professor C. F. Hammer, Georgetown University, Washington, D.C.

‡ Less reliable for identification purposes than corresponding bands of disubstituted benzenes.

TABLE 4 X≡Y, X=Y=Z GROUPS

(Non-diagnostic bands in italics)

The strong absorptions at 2300~2000 cm^{-1} and the weak absorptions at ca. 1350 cm^{-1} of the cumulenes are due to XYZ antisymmetric stretching and symmetric stretching, respectively.

Group	Band (cm^{-1})	ϵ^a	Remarks	Ref.	Figs.
1) Acetylene, –C≡C–	terminal position: 2140~2100	5	ν≡CH appears sharply at 3300 cm^{-1}, ϵ^a 100. ≡C–H also shows bending at 700~600, and occasionally its overtone at 1300~1200 cm^{-1}.	1	(16),(76)
	central: 2260~2190	1	Completely absent when symmetry is high.		(17)
	1330±5		Medium to weak, sharp. CH_2 wag of C≡C–CH_2–.		
2) Nitrile, –C≡N	2260~2210	Variable, 10~150	The lower wave-number region indicated is observed when conjugated to other unsatd. groups or when contained in **cyanamide (N–CN)** group.	2 3	(13)
			Alkali metal salts show very intense bands (ϵ_a 10^4~10^5) for the carbanion at 2100~2050. Position varies with counter ion and physical state.	18	
3) Diazonium salt, $-N_2^{\oplus}$	2260±20	m	Frequency depends largely on nature of counter ion.	4	
4) Diazo					
$RCH=N^{\oplus}=N^{\ominus}$	2050~2035				
$R_2C=N^{\oplus}=N^{\ominus}$	2030~2000				
Diazoketone				5,6	
$-CO-CHN_2$	2100~2090		νC=O appears at 1645, 1630~1605 cm^{-1}, respectively, in aliphatic and aromatic.		
$-CO-CRN_2$	2070~2060				
5) Allene, C=C=C	1950		Split when at terminal position and bonded to electron-attracting groups, e.g., $-CO_2H$, –COR.	7,17	(16)
	850 (terminal allene only, δ CH)		Stronger than 1950 cm^{-1} Overtone also at *1700* cm^{-1} but unpractical.		
6) Isocyanate, –N=C=O	2275~2250	1300~2000	Extremely strong, position not affected by conj.	8,9	(11)
	1350		No practical value because of weak intensity and overlapping with CH_2 peak.		
7) Carbodiimide, –N=C=N–	aliphatic: 2140~2130	1340~1580	Extremely high intensity, stronger than carbonyl (ϵ^a 600).	10	(12),(88)
	aromatic: 2145	1380~1680	Values of 3 measurements.		
	2115	400~1560	Cause of splitting unknown.		

Group	Band (cm^{-1})	ϵ_a	Remarks	Ref.	Figs.
8) Azide, $-N_3$	2160~2095		ν_{as}, strong; aromatics usually split due to Fermi resonance with C–N, $\Delta\nu$~60 cm^{-1}.	11,19	(15)
	1340~1180		ν_s, weak.		
9) Thiocyanate, –S–C≡N	aliphatic: 2140 aromatic: 2175~2160	Stronger than isocyanates	Integrated intensity has been measured.	12, 13, 14	(14)
Isothiocyanate, –N=C=S	aliphatic: 2140~1990 aromatic: 2130~2040		Somewhat broad, frequently split or accompanied by shoulders. Other bands: aliph. 1090 (s), arom. 1250 (w) and 930 cm^{-1} (s). [12]		
10) Ketene, C=C=O	~2150 *~1120*			15	
11) Ketenimine, C=C=N	~2000			16	
12) Carbon dioxide O=C=O	2349.3		Strong, antisym. stretching, fine structure.		
	720.5 667.3		Fine structure, all three used for calibration of spectrophotometer.		

References

[1] R. A. Nyquist and W. J. Potts, *Spectrochim. Acta* **16**, 419 (1960).
[2] R. E. Kitson and N. E. Griffith, *Anal. Chem.* **24**, 334 (1952); M. W. Skinner and H. W. Thompson, *J. Chem. Soc.*, 487 (1955); D. G. I. Felton and S. F. D. Orr, *ibid.* 2170 (1955); P. Sensi and G. G. Gallo, *Gazz, Chim. Ital.* **85**, 224, 235 (1955).
[3] J. P. Jessor and H. W. Thompson, *Spectrochim. Acta* **13**, 217 (1958).
[4] M. Aroney, R. J. W. LeFevere and R. L. Werner, *J. Chem. Soc.*, 276 (1955).
[5] P. Yates, B. L. Shapiro, N. Yoda and J. Fugger, *J. Am. Chem. Soc.* **79**, 9756 (1957).
[6] E. Fahr, *Ann.* **617**, 11 (1958).
[7] J. H. Wotiz and D. E. Mancuso, *J. Org. Chem.* **22**, 207 (1957).
[8] H. Hoyer, *Ber.* **89**, 2677 (1956).
[9] G. L. Calcow and H. W. Thompson, *Spectrochim. Acta* **13**, 212 (1958).
[10] G. D. Meakins and R. J. Moss, *J. Chem. Soc.* 993 (1957).
[11] L. J. Bellamy, "Advances in Infrared Group Frequencies," p. 62, Methuen, London, 1968.
[12] cf. E. Lieber, C. N. R. Rao and J. Ramachandran, *Spectrochim. Acta* **13**, 296 (1959).
[13] N. S. Ham and J. B. Willis, *Spectrochim. Acta* **6**, 279, 393 (1960).
[14] P. E. B. Butler, D. R. Eaton and H. W. Thomson, *Spectrochim. Acta* **13**, 223 (1958).
[15] C. L. Stevens and J. C. French, *J. Am. Chem. Soc.* **75**, 657 (1953).
[16] C. L. Stevens and R. J. Gasser, *ibid.* **79**, 6057 (1957).
[17] A. M. Snider, P. F. Krause, F. A. Miller, *J. Phys. Chem.*, **80**, 1262 (1976).
[18] I. N. Juchnovski and A. G. Binev, *J. Organomet. Chem.*, **99**, 1 (1975).
[19] E. Licker, C. N. R. Rao, A. E. Thomas, E. Oftedahl, R. Minnis, C. V. N. Nambury, *Spectrochim Acta.*, **19**, 1135 (1963).

THE 2200cm^{-1} REGION

Absorptions appearing in this region originate from groups of the type X≡Y, X=Y=Z. Although interpretations are easy since no other groups absorb strongly in this region, due consideration should be given to the fact that carbon dioxide absorbs strongly at 2350 cm^{-1}. Thus, the carbon dioxide peak is seen in all infrared spectra measured with single beam instruments; it also appears as small troughs or peaks when measured with double beam instruments in which the sample and reference light paths are not correctly compensated. Furthermore, peaks of nitrile and other groups may become smeared or blanked out under high carbon dioxide concentration. In order to avoid this, the room should be well-ventilated or a carbon dioxide absorbant should be placed within the instrument.

With terminal acetylenic groups C≡C–H, the νCH band appears sharply at a characteristic position and the νC≡C band is also present. Central C≡C bonds, on the other hand, will not appear if the symmetry is good, and thus, absence of the C≡C stretching band will not necessarily mean absence of the group.

Although "immonium bands" (2200~1800 cm^{-1}, Table 7d) also appear in this region, they are always accompanied by broad "ammonium bands" (2500~2300 cm^{-1}) and therefore can be distinguished from cumulene bonds.

TABLE 5 ALCOHOLS AND PHENOLS

(Non-diagnostic bands in italics)

Table 5a O-H Stretching vibration

State of OH	cm^{-1}	ϵ^a	Shape	Remarks	Figs.
Free (monometric)	3640~3610	30~100	sharp	Data measured in so-called non-polar (CCl_4, $CHCl_3$, etc.*) dilute solutions. Water absorption at 3710 cm^{-1} when solution is damp.	(18) (26)
1) *p*–OH	3640	70			
1) *s*–OH	3630	60~50			
1) *t*–OH	3620	45			
1) phenolic OH	3610				
2) –O–O–H, hydro-peroxide	3560~3530[1]				
Intermol. H-bond					
3) dimeric	3600~3500		rather sharp	Usually hidden in polymeric band (*vide infra*) unless polymer formation is hindered by steric hindrance. Absorptions arising from H-bond with polar solvents such as ethers, ketones, and amines also appear in this region.	(18)
4) polymeric	3400~3200	strong	broad	With solids and liquids, this broad absorption is the only one observed; in dilute solutions, accompanied by monomer band. Free and assoc. ν NH of amines and amides also appear at 3500~3200 cm^{-1}. First overtone of $\nu C{=}O$ (ca. 1720 cm^{-1}) also appears at 3500~3400 cm^{-1}, but can be differentiated because of low intensity, $\epsilon^a < 10$. **Water of crystallization:** also at 3600~3100 cm^{-1}, but not so strong and somewhat narrower; also weak band at 1640~1615 cm^{-1} (H–O–H bending vibration).	(19),(22) (84)
Intramol. H-bond					
5) polyvalent alcohols (e.g., 1,2-diols)	3600~3500	50~100	sharper than dimeric band		(26)
6) π H-bond	3600~3500			Many data since 1957[3].	
7) chelation (intramol. H-bond with C=O, NO_2, etc.)	3200~2500		broad	Lower the freq., stronger the intramolecular H-bond. **May be overlooked** even when not overlapping with ν CH, because occasionally broad and weak, and may appear as ill-separated band groups.	(45)

* Intensity of free –OH underwent temperature variation when methanol was measured in CCl_4 under dilute conditions in which no intermolecular H-bond was present. This was regarded as being due to the presence of a weak H-bond between the OH and the solvent C–Cl, and it was inferred that even CCl_4 is not an "inert" solvent: U. Liddel, *Ann. N. Y. Acad. Sci.* 69, 70 (1957).

Table 5b C-O Stretching vibration * βC–Cα / C–Cα' / C–Cα'' → C–OH

Standard position (ϵ^a, 60~200)	Shift	Remarks	Figs.
p–OH: 1050 cm^{-1} s–OH: 1100 cm^{-1} t–OH: 1150 cm^{-1}	α-branching: −15 cm^{-1} α-unsaturation: −30 cm^{-1} ring formation between α, α': −50 cm^{-1} α-unsaturation and α'-branching: −90 cm^{-1} α- and α'-unsaturations: −90 cm^{-1} α-, α'-, and α''-unsaturations: −140 cm^{-1}	Unsatd. bonds can be arom. Branching or unsaturation further than β-position has no effect. Additional α' and α'' branches also lowers by 15 cm^{-1}.	(22) (25), (27) (20)
phenolic OH: 1200 cm^{-1}			(25)

* Deduced empirically from ref. 2, standard cards, and other data. Relation between C–O peak position and various factors affecting the position is quite variable. The Table has been made only to give a rough guide to approximate positions.

Table 5c O-H Bending vibration (no practical value)

	Free OH	Associated OH	Figs.
in-plane	*1250* cm^{-1}, m	*1500~1300*, broad, frequently two, m. Shifted to higher frequencies with stronger H-bond.	(82)
out-of-plane	~*225* cm^{-1}?	*650*, broad, m. Higher frequencies with stronger H-bonds.	

References

[1] M. A. Kovner, A. V. Karijakin, A. P. Efimov, *Optics and Spectroscopy* (Optika i Spectroskopiya), **8**, 128 (1960); C.A. **54**, 16177d (1960).

[2] H. H. Zeiss and M. Tsutsui, *J. Am. Chem. Soc.*, **75**, 897 (1953).

[3] M. Oki and T. Yoshida, *Bull. Chem. Soc. Jap.*, **44**, 1336 (1971), and previous paper.

ABSORPTIONS OF HYDROXYL GROUPS

The hydroxyl absorptions were the subject of early IR investigations, and have been measured since the end of the 19th century. All of the characteristic frequencies are intimately related with hydrogen-bond formation. Usually the most important is the νOH band in the 3300 cm^{-1} region, the position of which is subtly affected by concentration, state of measurement, and temperature. Furthermore, measurements in the 3000 cm^{-1} region of all prism instruments are quite sensitive to temperature variation, and accordingly, the wavelength should be accurately calibrated for rigorous discussions. Because the ≃3600 cm^{-1} band of the free hydroxyl is sharp and appears at a position well-separated from other absorptions*, it is effective for the identification of hydroxyl groups; moreover, accurate measurements (comparison with models) may identify the type of hydroxyl group and give information on its conformation.

The intermolecular hydrogen-bond peak is intensified with increase in concentration, but the intramolecular hydrogen-bond is unaffected by concentration. The same relation holds for partners of the hydrogen-bonds, e.g., C=O,

* Overtones of carbonyl peaks may also appear in the O–H stretching region. Thus, the overtone of a 1715 cm^{-1} C=O band may be present as a medium intensity broad band at ca. 3430 cm^{-1}.

NO_2. For instance, in 1,2-diols, the intensity ratio of the free OH and the intramolecularly bonded OH is constant under low concentrations in which no intermolecular hydrogen-bond occurs (usually CCl_4 solutions of concentrations less than 0.005 mole per liter). When the concentration is increased, only the free OH band is weakened, and is replaced by a band arising from intermolecular hydrogen-bonds at 3500 cm^{-1} or lower, *i.e.*, at frequencies lower than the intramolecular hydrogen-bond peak (3600~3500 cm^{-1}).

The νOH band simply appears as a monomeric or dimeric band if association of the hydroxyl group is sterically prevented or hindered. This can be observed in the following data of hindered phenols measured in nujol:

2–*t*–butyl–4–methylphenol: 3380 cm^{-1} (API 924)
2–methyl–4, 6–di–*t*–butylphenol: 3570 cm^{-1} (shoulder at 3462 cm^{-1}) (API 925)
2, 6–di–*t*–butyl–4–methylphenol: 3510 cm^{-1} (API 926)
2, 6–di–*t*–butyl–4–ethylphenol: 3570 cm^{-1} (API 927)
2, 6–di–*t*–butyl–4–cyclohexylphenol: 3530 cm^{-1} (API 928)

The νC–O band is stronger than the νOH band, but it should be kept in mind that other absorptions also appear in the 1200~1000 cm^{-1} region. Furthermore, the νOH band also appears with moist samples, or samples containing water of crystallization or alcohol. Presence of the hydroxyl group can be ascertained if upon acetylation, the νOH band disappears and is replaced by acetate bands; disappearance of the alcoholic νC–O band cannot be used as a diagnosis for ester formation, because acetoxyl groups also have two absorptions at 1300~1050 cm^{-1} besides the 1735 cm^{-1} νC=O band. **Phenols**[1] naturally give rise to aromatic absorptions in addition to the above-mentioned hydroxyl absorption. Bending vibrations given in Table 5c are only of complementary significance and have no practical value with complex molecules.

Table 5d Steroids and Triterpenes

Steroids (A/B *trans*, C_a-substituent)

Substituent	Hydroxyl [1]	Methoxyl [2]	Acetoxyl [2]
eq. bond	1040	1100	1030
ax. bond	1000	1090	1020

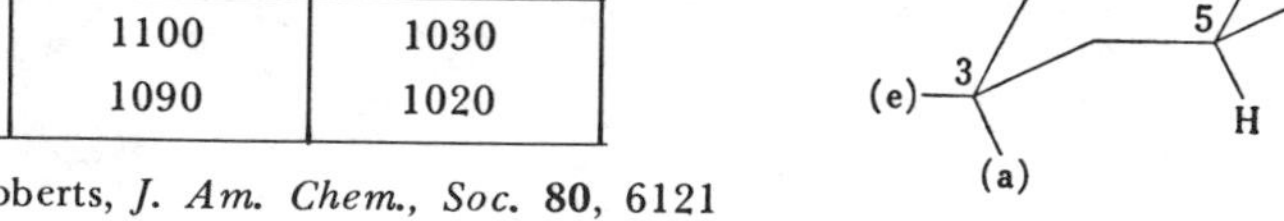

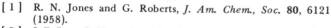

[1] R. N. Jones and G. Roberts, *J. Am. Chem., Soc.* **80**, 6121 (1958).
[2] J. E. Page, *J. Chem. Soc.* 2017 (1955).

[1] IR of alkyl phenols: D. D. Shrewsbury, *Spectrochim. Acta,* **16**, 1294 (1960).

The difference seen between equatorial and axial bonds of 5α–steroids is considered to be due to the fact that the C–O stretching of the former (in the direction of ring A) requires higher energy than that of the latter (perpendicular to ring A). With 5β–A/B *cis* steroids, the difference is small: 3β–OH at 1033~1030; and 3α–OH at 1038~1035 cm^{-1}.

It has also been shown[1] that acetates of the equatorial C_3–OH group have singlet "acetate bands" at 1240 cm^{-1}, whereas acetates of the axial C_3–OH group have multiplet bands in the 1240 cm^{-1} region. The same relations have been noted with cyclic compounds other than steroids such as decalols[2], and have also been used in determining the hydroxyl configuration in the natural products field[3]. However, the position of the equatorial C–O stretch is not invariably at higher frequency than the axial, even in simple systems.[7] The C–halogen bonds of halo-steroids[4] and C–D bonds of deuterated steroids[5] also give rise to differences between equatorial and axial bonds[6].

Triterpenes* (A/B *trans*)

	ν OH	ν C–O**
3,β–OH (e)	3630 (e^a 50~60)	1020 and 1040
3α–OH (a)	3638 (e^a 50~10)	1065

* I. L. Allsop, A. R. H. Cole, D. E. White and R. L. S. Willix, *J. Chem. Soc.* 4868 (1956).

** The fact that axial bonds appear at higher frequencies, a behavior which is contrary to steroids, has been attributed to the presence of the *gem*-dimethyl group at C_4.

In most cases NMR spectroscopy is now used to distinguish axial (a) and equatorial (e) substituents. In six-membered rings with the chair conformation, the α-protons can be distinguished by their coupling constants, since J_{aa} (8~13Hz) differs substantially from J_{ae} (2~6Hz) and J_{ee} (1~5Hz). Where the coupling constants cannot be resolved, the width at half height of the signal (W 1/2) can be used, since it is generally larger than 15 Hz for axial protons and smaller than 12 Hz for equatorial protons.[8]

[1] R. N. Jones, P. Humphries, E. Herling and K. Dobriner, *J. Am. Chem. Soc.* 73, 3215 (1951); A. Fürst, H. H. Kuhn, R. Scotoni, and Hs. H. Günthard, *Helv. Chim. Acta* 35, 951 (1952); A. R. H. Cole, *J. Chem. Soc.* 1952 4969.

[2] W. G. Dauben, E. Hoerger and N. K. Freeman, *J. Am. Chem. Soc.* 74, 5206 (1952).

[3] A. Aebi, D. H. R. Barton, A. W. Burgstahler and A. S. Lindsey, *J. Chem. Soc.* 4659, 1954; A. Stoll, Th. Petrizilka, J. Rutschmann, A. Hofmann and S. H. Günthard, *Helv. Chim. Acta* 37, 2039 (1954); W. Pelletier, D.M. Locke, *J. Am. Chem. Soc.* 79, 4531 (1957).

[4] D. H. R. Barton, A. Da. S. Compos.-Neves and R. C. Cookson, *J. Chem. Soc.* 3500 (1956); E. G. Cummins and J. E. Page, *J. Chem. Soc.* 3847 (1957).

[5] E. J. Corey, R. A. Sneen, M. G. Danaher, R. L. Young and R. L. Rutledge, *Chem. & Ind.* 1294 (1954).

[6] Review: M. Ohashi and K. Nakanishi, "IR Spectra," Vol. 6, p. 1 (1958).

[7] L. K. Dyall and R. G. Moore, *Aust. J. Chem.* 21, 2569 (1968).

[8] L. M. Jackman and S. Sternhell, "Application of Nuclear Magnetic Resonance Spectroscopy in Organic Chemistry," 2nd edition, Pergammon Press, New York (1969), p. 288.

Tables 5e and 5f Sugars

Table 5e D-Glucopyranose

Band \ Assignment \ Pyranose	Type 1	Type 2 (a and b)	Type 3	Figs.
Assignment	Ring vibration similar to that of dioxane.	Bending vibration of C_1–H (equatorial in α-sugars, axial in β-sugars) Fig. 1.	Breathing vibration of pyranose ring, Fig. 2.	
α–	917±13	844±8 (type 2 a)	766±10	(33)
β–	920± 5	891±7 (type 2 b)	774± 9	

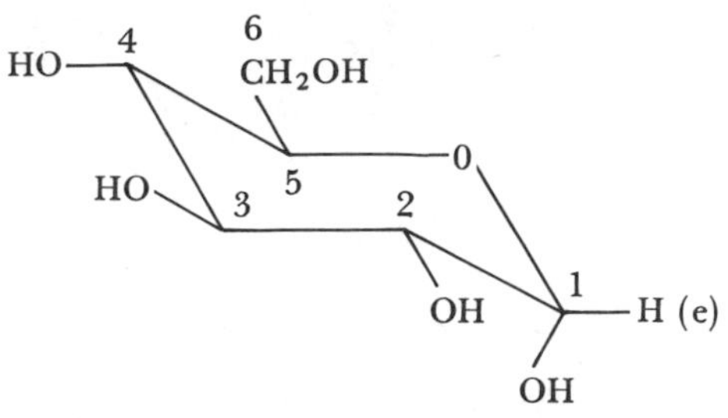

Fig. 1 α-D-Glucopyranose

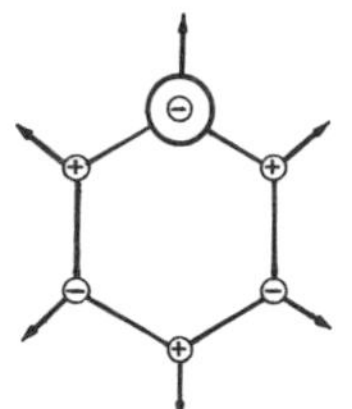

⊖ Oxygen below *xy* plane
⊖ Carbon below *xy* plane
⊕ Carbon above *xy* plane

Fig. 2 Ring-breathing vibration of tetrahydropyran

Table 5f Type 1 and 3 Bands of α-Polysaccharides

Type of band \ Bond	Type 1	Type 3	Type 2a	Figs.
α–1:4 (e.g., starch)	930±4	758±2	844±8	
α–1:6 (e.g., dextrane)	917±2	768±1		
α–1:3		793±3		

The solid state spectra will be discussed, since most infrared spectra of sugars are measured in nujol, KBr or as films. A broad band with shoulders is present in the region of 3800~3200 cm^{-1}, which is due to hydroxyl groups that are hydrogen-bonded in various degrees (the band disappears when all hydroxyls are acetylated). Complex bands attributed to ethereal and hydroxylic νC–O appear between 1200~1030 cm^{-1}; there is of course no absorption in the νC=O region. The infrared curves become simpler in the series, monosaccharides → oligosaccharides → polysaccharides, because of overlapping of many infrared bands; a similar change is observed in the series, amino acids → oligopeptides → proteins. Results shown in Table 5e have been obtained by Barker and co-workers[1] for the fingerprint region of D-glucopyranose.[2] Thus the α- and β-series can be

[1] S. A. Barker, J. E. Bourne, M. Stacey and D. H. Whiffen, *J. Chem. Soc.* 171 (1954); S. A. Barker, E. J. Bourne, R. Stephens and D. H. Whiffen, *ibid.* 3468 (1954); S. A. Barker, *et al., ibid.* 4211, 4550 (1954).
Review of IR of sugars: S. A. Barker, E. J. Bourne and D. H. Whiffen in "Methods of Biochemical Analysis," Vol. 3, p. 213~245, Interscience (1956).

[2] Type 2a band has been observed independently by M. Takahashi: *J. Pharm. Soc. Japan* **74**, 320 (1954); **75**, 237 (1955).

differentiated by means of the type 2a and 2b bands. However, since some type 1 bands of the α-series appear in the region of the type 2b bands, a peak around 890 cm^{-1} does not necessarily imply a β-sugar. The relation shown in Table 5e holds for other hexo- and pento-pyranoses such as galactose, mannose, arabinose, and their respective acetates. The type 2a and 2b bands of di-, oligo-, and polysaccharides also appear in the range shown in Table 5e, and hence, can be used for identifying the α- and β-series. Furthermore, the type 1 and type 3 bands of α-sugars can be used for identifying the glucoside linkage (Table 5f). Since the IR spectra of enantiomeric sugars are identical, the method for differentiating between α- and β-sugars can be applied to the L-glucopyranose series.

TABLE 6 ETHERS AND RELATED GROUPS

(Non-diagnostic bands appear in italics)

ϵ^a of νC–O bands are usually greater than 200.

Group		Band (cm^{-1})	Remarks	Figs.
1) Ethers, general	aliph. and ring C-O-C [1]	1150~1070	C-O-C antisym. stretching. Common with other C-O bands. Effect of α-branching unclear.	(29)
	arom. and vinyl =C-O-C	1275~1200	Antisym. stretching. Common with other =C-O. So-called "1250 cm^{-1} band." νC=C of vinyl ethers intensified.	(30),(41)-f
		1075~1020	Sym. stretching. Weaker than 1250 cm^{-1} band.	
	Methoxyl group has following bands in addition.			
	aliph. $-OCH_3$ [2]	2830~2815 (ϵ^a 50)	CH_3 sym. stretching. Lacks in $-OC_2H_5$. The asym. stretching is at 2990–2970 (present in $-OC_2H_5$) but position is non-characteristic.	
	arom. $-OCH_3$ [3]	2850	CH_3 sym. stretching. Following bands also present, but position non-characteristic: 3000, 2950, 2915 (the 3 appear as triplets), 1460 and 1340 (δCH), 1250 (=C-O-C antisym. stretching), 1180 and 1125 (methyl rocking and arom.?), 1040~1020 (=C-O-C sym. stretching).	(31)
2) Methylenedioxy [3] 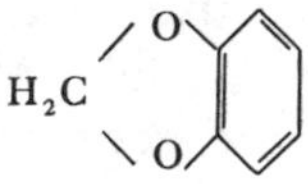		2780	CH_2 sym. stretching.	(47)
		925	Most useful, probably related to C-O stretching.	
		720	Frequently as shoulder on other strong bands. Following peaks also observed: 3010, 2950, 2910 (other νCH), 1480 (δCH_2), *1400* and 1360 (sometimes lacking), 1250 (=C-O-C, sometimes branched), 1130 (sometimes lacking), 1040 (=C-O-C).	
3) Ketals, acetals C-O-C-O-C [4,11]; cyclic C-O-C-O-C [5]		1200~1030	Group of 4 to 5 bands.	
4) Spiroketal of steroidal sapogenins [1,6]		1350~ 650	Many characteristic sharp bands.	
5) Epoxy	[1,7,9,12] 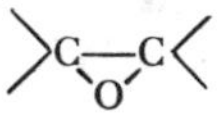	3000 (ϵ^a 40)	Epoxy methine group of chain compounds and strainless rings.	(34)
		3040 (ϵ^a 20)	Epoxy methine group of strained rings.	
		3050 (ϵ^a 30)	Methylene group of terminal epoxy rings.	
		1250	So-called "8 μ band"; sym. stretching of ring.	
		950~ 810	So-called "11 μ band"; asym. stretching of ring.	
		840~ 750	So-called "12 μ band."	
6) Peroxides [10]				
C-O-O-C, aliph.		*890~820*	Both weak and difficult to detect.	(33)
arom.		*1000*	-O-O-H at 3450 cm^{-1}. Aliph. -CO-O-O-CO- at 1820~1810 and 1800~1780 cm^{-1} (2 bands), and arom. -CO-O-O-CO- at 1805~1780 and 1785~1755 cm^{-1} (two bands).	
7) Ozonides [10]		*1060~1040*		

References

[1] Review of natural products containing these bonds: A. Fujino and M. Yamaguchi, "IR Spectra," Vol. 3, p. 101 (1958).
[2] H. B. Henbest, G. D. Meakins, B. Nicholls and A. A. Wagland, *J. Chem. Soc.* 1462 (1957).
[3] L. H. Briggs, L. D. Colebrook, H. M. Fales and W. C. Widman, *Anal. Chem.* 29, 904 (1957).
[4] E. D. Bergmann and S. Pinchas, *Rev. trav. chim.* 71, 161 (1952).
[5] H. Tschamler and R. Leutner, *Monatsh.* 83, 1502 (1952).
[6] R. N. Jones, E. Katzenellenbogen and K. Dobriner, *J. Am. Chem. Soc.* 75, 158 (1953). C. R. Eddy, M. E. Wall and M. K. Scott, *Anal. Chem.* 25, 266 (1953).
[7] J. Bomstein, *Anal. Chem.* 30, 544 (1958).
[8] H. B. Henbest, G. D. Meakins, B. Nicholls and K. J. Taylor, *J. Chem. Soc.* 1459 (1957).
[9] W. A. Patterson, *Anal. Chem.* 26, 823 (1954).
[10] L. J. Bellamy, "The Infrared Spectra of Complex Organic Molecules," 2nd ed., pp. 121–122, Wiley, New York (1958).
[11] K. Nikada, *Spectrochim. Acta,* 18, 745 (1962); J. E. Katon and P. D. Miller, *Appl. Spectrosc.,* 29, 501 (1975).
[12] Donald A. George, *J. Chem. Eng. Data,* 20, 445 (1975).

ABSORPTION OF ETHERS

The mass of the oxygen atom and the strength of C–O bonds do not differ much from those of the carbon atom, and hence the positions of ether absorptions are not characteristic. However, the intensities are large because the change in dipole moment accompanying C–O vibrations is larger than that of C–C. Identification of ether groups from the infrared is not easy because organic compounds frequently contain other C–O bonds, *e.g.*, alcohols, esters, acids. In summary, a peak at $\simeq$1100 cm^{-1} or $\simeq$1250 cm^{-1} only infers the presence of a C–O bond or a =C–O bond, respectively. Band intensities are usually strong with an ϵ^a larger than 200, but quantitative measurements have not been carried out. Of the two ethereal C–O–C symmetrical and antisymmetrical stretching vibrations, only the latter is infrared active.

TABLE 7 AMINES AND AMMONIUM SALTS

(Non-diagnostic bands appear in italics)

Table 7a Stretching vibrations

Amino group	cm^{-1}	Intensity	Remarks	Figs.
Free				
1) *p*-Amine $R-NH_2$ and $Ar-NH_2$	~3500 ~3400	w in $R-NH_2$ ϵ^a~30 in $Ar-NH_2$	The two correspond to ν_{as} and ν_s, resp. Wave-number relation: $\nu_s=345.53+0.876\ \nu_{as}$[1]	(35),(36)
2) *s*-Amine R-NH-R	3350~3310	w	Especially weak in N-heterocyclic compounds, e.g., piperidine.	
Ar-NH-R	3450	ϵ^a 30~ 40		
Pyrroles, indoles, etc.[2]	3490	ϵ^a 150~300	Position and intensity both characteristic. See Table 10-2.	(38),(87)
C=NH	3400~3300		νC=N band at 1690~1640 cm^{-1}.	

Association: The shift from the free position is less than 100 cm^{-1}, although shifts down to 3100 cm^{-1} are sometimes observed. In general bands tend to be weaker, sharper and less likely to change with concentration than the corresponding hydroxyl bands.

Table 7b N-H Bending vibrations (lacking in *t*-amines)

Group	cm^{-1}	Intensity	Remarks	Figs.
3) $-NH_2$	1640~1560	s~m	In-plane bending, corresponds to CH_2 scissoring, same in $R-NH_2$ and $Ar-NH_2$.	(35),(36)
	900~ 650	m (broad)	Out-of-plane bending, corresponds to CH_2 twisting, rather characteristic.	
4) -NH-	*1580~1490*	w	Difficult to detect, especially in Ar-NH- where it is blurred by arom. 1580 cm^{-1} band.	

The N–H bending absorptions are shifted towards slightly higher frequencies under associated conditions, a behavior which is similar to other X–H type bending vibrations.

Table 7c C-N Stretching vibrations

The position does not differ much from C–C stretching vibrations, but the intensity is relatively large because of C–N polarity.

Group	cm^{-1} (intensity)	Remarks	Figs.
5) R-N<H (R), H (R)	1230~1030 (m)	Two in *t*-amines. In cyclic amines, lower range of indicated region.	(35)
6) Ar-N<R, H (R)	1360~1250 (s) 1280~1180 (m)	$C_{arom.}$-N stretching. High position due to increase in double bond character. $C_{aliph.}$-N stretching.	(36)

Table 7d Amine salts

Group	cm^{-1}	Shape (intensity)	Assignment and remarks	Figs.
7) $-NH_4^+$	3300~3030	broad (s)		
	1430~1390	s		
Satd. amine[3]*				
8) $-NH_3^+$	~3000	broad (s)	$\nu_{as}NH_3^+$ and ν_s NH_3^+ (corresponds to the two CH_3 bands), overlaps broadly over νCH band, commonly named "ammonium band."	(35),(72)
	~2500	1~several (m)	Overtones, combination tones. Sometimes lacking.	
	~2000	m	Overtones, combination tones. Sometimes lacking.	(71)
	1600~1575 } 1500 }	s	$\delta_{as}NH_3^+$ and $\delta_sNH_3^+$ (corresponds to the two δCH_3 vibrations).	
9) $-NH_2^+$	2700~2250	broad(s), or group of relatively sharp bands.	$\nu_{as}NH_2^+$ and $\nu_sNH_2^+$. Differentiation from νCH band not so clear-cut as with *t*-amines. "Ammonium band."	
	2000	m	Mostly lacking. Overtones, combination tones.	
	1600~1575	m	δNH_2^+.	(64)
10) $-NH^+$	2700~2250	broad(s), or group of relatively sharp bands.	One is due to νNH^+, others are overtones and combinations. Clearly separated from νCH band. "Ammonium band." δNH^+ band is weak and of no practical value.	(37),(39)
11) $\rangle \overset{+}{N} \langle$			Quaternary ammonium salts lack characteristic bands.	
Unsatd. amine[4] (arom. amine)				
12) $C=\overset{+}{N}-H$	2500~2300	broad(s), or group of relatively sharp bands.	"Ammonium band." Clearly separated from νCH as with *t*-amine salts.	
	2200~1800	1~several (m)	"Immonium band." This is the difference between *t*-amine salts.	
	~1680	m	$\nu C=N^+$. Higher by 20~50 cm^{-1} than free $\nu C=N$ (1690~1640 cm^{-1}, ϵ^a variable from 5~880)[4,5].	(98)

* KBr data of hydrochlorides. It has been reported that the νNH^+ band of the mineral acid salt of a certain *t*-amine is shifted from ca. 2600 to ca. 2800 and ca. 3100 cm^{-1}, respectively, as the anion is changed from Cl^- to Br^- and ClO_4^- [8]. The bands given in the Table are also observed with amino acids and their salts. δCH of the CH_2 adjacent to N^+ is shifted to 1440~1400 cm^{-1}–(Table 1~16).

Group	cm^{-1}	Shape (intensity)	Assignment and remarks	Figs.
13) Guanidiniums[6]** $\left[\begin{array}{l}\rangle N \\ \rangle N\end{array} \rangle C\text{–}N\langle\right]^{\oplus}$	~3300	broad	Guanidines are mostly measured in the form of salts because of strong basicity (pK_a ca. 13). νNH. Appears in free amino region rather than ammonium region. This suggests that plus charge is localized on central C atoms rather than being distributed on N atoms.	(85)
νC=N of free guanidines at ca. 1660.[7] In hydrochlorides:				
mono-substd.	1660 and 1630	s	"Guanidinium I band" and "Guanidinium II band."	
di-substd.	1680 and 1595	s	Similarly I and II bands. Band separation larger than mono-substituted salts.	
tri-substd.	1635	s	Only one guanidinium band.	

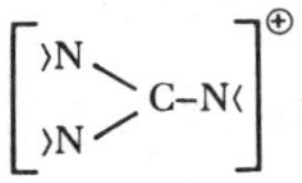

** KBr data of hydrochlorides.

Table 7e The 2800~2700 region

Group of bands that occur on the lower frequency side of main νCH band. Associated with lone-pair electrons of the N atom because they disappear upon salt formation or conversion into N-oxides or amides.

Group	cm^{-1}	ϵ^a	Remarks	Figs.
14) $N\text{–}CH_3$[9]			Lacking in $N\text{–}CH_2CH_3$.	(66)
$Ar\text{–}NH\text{–}CH_3$, 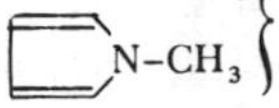 (cyclic) $N\text{–}CH_3$	2820~2815	15~ 40		
$R\text{–}NH\text{–}CH_3$	2805~2780	100~170		
$Ar\text{–}N(CH_3)_2$	2800	60~ 70		
$R\text{–}N(CH_3)_2$	three: 2825~2810, 2775~2765, 2730	100~180, 125~215, 25~ 45		
15) *trans*-Quinolizidine[10] 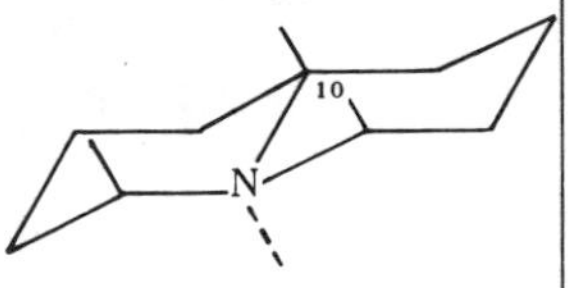	2800~2700 "Bohlman band"		Group of small absorptions; observed when more than two adjacent CH bonds are *trans* to the N lone-pair electrons. Thus, present in 10-methyl-*trans*-quinolizidine, but absent in *cis*-quinolizidines.	(38)
16) Others[3]	2800~2700		Group of small bands are also present in di- and tri-ethyl-amine, piperidine, N-ethylpiperidine, morpholine, etc. Related to *trans*-quinolizidine band?	

References

[1] L. J. Bellamy and R. L. Williams, *Spectrochim. Acta* **9**, 341 (1957).

[2] B. Witkop and J. B. Patrick, *J. Am. Chem. Soc.* **73**, 713, 1558 (1951).

[3] K. Nakanishi, T. Goto and M. Ohashi, *Bull. Chem. Soc. Japan* **30**, 403 (1957).

[4] B. Witkop, *Experientia,* **10**, 420 (1953); *J. Am. Chem. Soc.* **76**, 5597 (1954).

[5] L. J. Bellamy, "The Infra-red Spectra of Complex Organic Molecules," 2nd ed., p. 269, Wiley, New York (1958).

[6] T. Goto, K. Nakanishi and M. Ohashi, *Bull. Chem. Soc., Japan* **30**, 723 (1957).

[7] E. Lieber, D. R. Levering and L. Patterson, *Anal. Chem.* **23**, 1594 (1951).

[8] E. A. V. Ebsworth and N. Sheppard, *Spectrochim. Acta* **13**, 261 (1959).

[9] S. Oseko, *J. Pharm. Soc. Japan,* **77**, 120 (1957); R. D. Hills and G. D. Meakins, *J. Chem. Soc.* 760 (1958); J. T. Braunholtz, E. A. V. Ebsworth, F. G. Mann and N. Sheppard, *ibid.* 2780 (1958). NCH_3 in amino acid hydrochlorides: C. C. Watson, *Spectrochim. Acta* **16**, 1322 (1960).

[10] F. Bohlmann, *Ber.* **91**, 2157 (1958).

AMINES AND AMMONIUM SALTS

The infrared absorptions of these compounds can usually be considered analogous to other groups bearing the same number of H atoms: NH_2 (CH_2), NH (CH), N^+H_2 (CH_2), N^+H (CH). However, the differences in polarity between the nitrogen and carbon atoms will of course be reflected.

The free amines that are easily identified by the infrared method are the primary amines and aromatic amines. Thus:

Primary amines: Two νNH bands are present even in dilute non-polar solvents. The wave-number relation of the two bands can be approximated by the empirical equation:

ν_sNH (lower frequency component)=345.53+0.876 ν_{as} NH (higher frequency component)

The relation was checked with 62 compounds, and the results agreed with a standard deviation of 4.8 cm^{-1}. The relation holds for conditions other than dilute solutions excepting the case when only one of the N—H bonds is involved in hydrogen-bonding. This empirical relation can also be used for tautomerization studies. For example, 2-amino-pyridine (1) absorbs at 3500 (ν_{as}) and 3410 cm^{-1} (ν_s) in dilute carbon tetrachloride, and the value of ν_s calculated from ν_{as} is 3412 cm^{-1}; this indicates that the compound does not exist as form (2). The 1640~1560 and 900~650 cm^{-1} absorptions are also sometimes useful.

N NH_2 N NH H

(1) (2)

Aromatic amines are identified by the two νCN bonds shown in Table 7c. Although some other bands are strong, they are rather supplementary because positions and intensities are not characteristic.

The best way to determine whether or not the sample is an amine is to treat it with mineral acid and to observe the appearance of a broad strong "ammonium band" in the range of 3000~2200 cm^{-1}. **The class of the original amine** is then identified by using Table 7d; that is, it is a primary amine if the ammonium band overlaps with the νCH band, and a tertiary amine if it is clearly separated. Because the ammonium band positions of secondary amines are intermediary and cannot easily be differentiated from those of primary or tertiary amine salts, the 1600~1500 cm^{-1} absorption is also taken into account. The $>C{=}NH^+$ group can be easily identified because in addition to the νNH^+ band of the tertiary amine type, an "immonium band" appears at ≃2000 cm^{-1} (tertiary amine salts lack the ≃2000 cm^{-1} absorption). The fact that the C=N absorption is shifted to higher frequencies upon salt formation is also of practical importance (Table 7d~12). Substituted guanidines can be detected by the guanidinium bands of their hydrochlorides.

The 2800~2700 cm^{-1} bands shown in Table 7e are characteristic for the **N-methyl group** but they are somewhat confusing because *trans*-quinolizidine rings and other groups (Table 7e) also absorb in this area.

Quaternary ammonium compounds lack characteristic frequencies. It has been noted that carbon disulfide is unsuited as a solvent for amines because of the rapid reaction with primary and secondary amines to give alkyldithiocarbamates (R. N. Jones and C. Sandorfy, "Chemical Applications of Spectroscopy," p. 511, 1956).

TABLE 8 CARBONYL GROUPS (Bold figures denote νC=O)

(Non-diagnostic bands appear in italics)

The standard νC=O position (of saturated aliphatic compounds) and other characteristic frequencies are given first, and these are followed by the νC=O of conjugated unsaturated groups and others. Unless otherwise stated, the values refer to dilute solutions in non-polar solvents such as CCl_4 and CS_2. Deviations from given value are usually within ±10 cm^{-1}. When **intramolecular hydrogen-bonds** are formed, bands may be shifted lower by 50 cm^{-1} according to H-bond strength.

	Group	Band (cm^{-1})	Remarks	Figs.
1	**Ketone** -CO-	**1715**	ϵ^a 300~600. Values of $CHCl_3$ solutions and solids, lower by 10~20 cm^{-1}. Values of gases, +20 cm^{-1}. Occasionally overtone at ca. 3400 cm^{-1}.	(41)
		1100 (aliph.) } 1~ 1300 (arom.) } several	ϵ^a 50~150. Caused by C–C–C bending and C–C stretching of –C–(CO)–C–.	
	1 a) α, β-unsatd.	**1675**	νC=C shows clearly at 1650~1600 cm^{-1}; in s-*cis* forms, may be shifted below 1600 cm^{-1} and intensity becomes comparable to that of νC=O.	(41),(43)
	1 b) Ar–CO–	**1690**	Affected by *I, M,* and steric effects of substituents.	(44),(67)
	1 c) α, β-γ, δ-unsatd. 1 d) α, β-α′, β′-unsatd. 1 e) Ar–CO–Ar	**1665**		
	1 f) ▷–CO–R	**1695**	Conj. with cyclopropane lowers by ca. 20 cm^{-1}.	(42)
	1 g) 7 memb. and larger	**1705**		
	1 h) 6-membered	**1715**	Same as standard value. Shift caused by conj. with unsatd. groups is same with aliph. ketones.	(41)
	1 i) 5-membered	**1745**	Shift caused by conj. with unsatd. groups is same with aliph. ketones.	(40)
	1 j) 4-membered	**1780**		
	1 k) 3-membered	**1850**		
	1 l) α-halo- C–C– with X on first C and =O on second C X O (X: halogen)	shift of **0~+25**	Shift to higher freq. is larger the smaller the angle between C=O and C–X. No shift when angle exceeds 90°. Following values apply to satd. aliph. and cyclic ketones: –Cl, 0~+25; –Br, 0~+20; –I, 0~+10 cm^{-1}	(75)
	1 m) $-CX_2$–CO– and –CX–CO–CX (X: halogen)	shift of **0~+45**	Effect of single halogen substitution appears additively. Shift of +45 cm^{-1} results when Cl on both sides of C=O are in same plane with C=O. Co-existence of several conformers give rise to each corresp. band, the relative intensities of which are dependent upon state of measurement.	

	Group	Band (cm^{-1})	Remarks	Figs.
	1 n) –CO–CO–	**1720**	**Caused by antisym. stretching of the two C=O(a); sym. stretching is IR-inactive. When C=O groups are fixed in s-*cis* form (b), νC=O is at 1760 and 1730 cm^{-1} in 6 membered rings, and 1775 and 1760 cm^{-1} in 5 membered rings.** (a) (b) s-*cis*	(82)
	1 o) =C(OH)–CO–	**1675**	νC=C at 1650 cm^{-1}, strong.	
	1 p) –CO–CH–CO–	≅**1720**	Occasionally doublet.	
	1 q) –C(OH (or NH_2))=C–CO–	**1650** (free) **1615** (when intra. mol. H-bonded)	νC=C appears at 1605 cm^{-1} with same intensity as νC=O.	
	1 r) –C(OR)=C–CO	**1640**		
	1 s) *ortho*–CO–C_6H_4–OH (or NH_2)	**1630**	Influenced by *I, M,* and steric effects of substituent.	(45)
	1 t) 1, 4-quinone 1, 2-quinone	**1675**	Affected by *M* effect of substituent. Skeletal νC=C at ca. 1600 cm^{-1} (s). Multiple bands often appear in *p*-quinones due to Fermi resonance. When peri-OH is present, strong band is shown at 1630 cm^{-1}, and 1675 cm^{-1} band becomes much weaker. For charge transfer complexes see Ref. 8.	(80)
	1 u) extended quinone	**1645**		
	1 v) tropone	**1650**	**1600** cm^{-1} in intramolecularly H-bonded tropolones.	
2	**Aldehyde** –CHO	2820, 2720	Two bands due to Fermi resonance between νC–H and overtone of δC–H.	
		1725	Several bands also at *1400~1000* cm^{-1}, but not practical.	
	2 a) α, β-unsatd.	**1685**		
	2 b) α, β-γ, δ-unsatd.	**1675**		
	2 c) Ar–CHO	**1700**		(46),(47)
3	**Acid** –COOH		Data are mostly for dimers because H-bonding power is unusually strong, and they exist also in gas state.	(52),(53)
		3000~2500	Very characteristic. Group of small bands. Band at highest freq. is due to νOH, and others are combinations. 3550 cm^{-1} in monomers.	
		1760 (monomer) **1710** (dimer)	Considerably stronger than ketonic νC=O. ϵ^a up to 1500. Two bands may be observed with gases or solutions.	
		1420 1300~1200	Both due to coupling between in-plane O–H bending and C–O stretching of dimer. 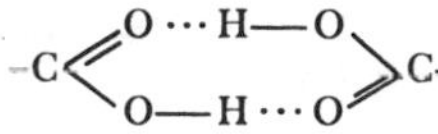	
		920	Broad, medium intensity. O–H out-of-plane bending of dimer.	

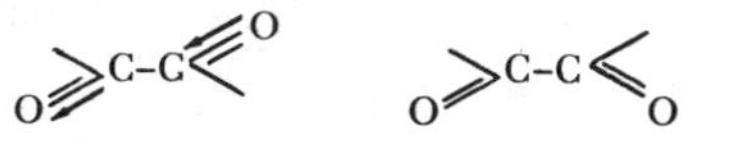

	Group	Band (cm^{-1})	Remarks	Figs.
	3 a) C=C–COOH and Ar–COOH	**1720** (monomer) **1690** (dimer)		(68)
	3 b) α-halo-	Shift of **+10~20** cm^{-1}	Values for α–Br and α–Cl. Larger for α–F, ca. +50 cm^{-1}.	
4	**Carboxylate** –COO–	**1610~1550** and **1400**	C–O antisym. and sym. stretching, resp., of –C⟨O O⟩ – When an acid is converted into its inorganic salt, the five characteristic freq. are replaced by these two. Instead of being converted to inorganic salts, they can be changed to ammonium salts by adding few drops of Et_3N to $CHCl_3$ solution; carboxylate ions are not formed in CCl_4, while addition of NH_3 or *p*-amines gives confusing N^+H_4 and N^+H_3 bands that overlap with $-COO^-$ band.	(55),(70) (64)
5	**Ester** –CO–O–	**1735**	Intensity intermediary between ketone and carboxyl, (ϵ^a 500~1000).	(23),(72)
		two at 1300~1050	Asym. and sym. stretching, resp., of ester C–O–C. The asym. stretching ("ester band") is usually stronger than νC=O and broad; also occasionally split. This higher νC–O band is fairly constant according to type of ester:[7] H–CO–OR, 1180 cm^{-1}; CH_3–CO–OR, 1240 cm^{-1}; R–CO–OR, 1190 cm^{-1}; R–CO–OCH_3, 1165 cm^{-1}	(58),(86) (38)
	5 a) –C=C–CO–O–	**1720**	Shift caused by conjugated unsaturation relatively small. νC–O at 1300~1250 (vs) and 1200~1050 cm^{-1} (s).	
	5 b) Ar–CO–O–	**1720**	νC–O at 1300~1250 (vs) and 1180~1100 cm^{-1} (s).	(59)
	5 c) –CO–O–C=C–	**1760**	νC=C at 1690~1650 (s) (may shift up to 1715 cm^{-1})[1] In CH_3–CO–O–C=C (enol and phenol acetates), one of the νC–O appears at 1210 cm^{-1} (vs).	(58)
	5 d) α-halo-	shift of **+10~+40** cm^{-1}	Shift dependent on electronegativity and number of α-halo atoms; dependent also on angle between C–X and C=O as with ketones.	
	5 e) Ar–CO–O–Ar	**1735**		
	5 f) α-keto ester	**1745**		
	5 g) β-keto ester	keto: **1735**	The β-keto group (νC=O, 1720 cm^{-1}) does not affect ester νC=O.	(51)
		enol: **1650**	Broad C=C at ca. 1630 cm^{-1} (vs).	
	5 h) (six-membered lactone ring, O, =O)	**1735**	Same as with aliph. esters.	
	5 i) (six-membered lactone ring with C=C, O, =O)	**1720**	Same as with aliph. esters.	
	5 j) (six-membered lactone ring with C=C adjacent to O, O, =O)	**1760**	Same as with aliph. esters. νC=C at 1685 cm^{-1} (s).	
	5 k) (six-membered lactone ring with two C=C, O, =O)	Table 10-4	α-Pyrones. See Table 10-a for γ-pyrones.	(79)

	Group	Band (cm^{-1})	Remarks	Figs.
	5 l) 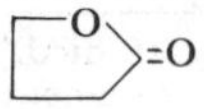	**1770**		63
	5 m) 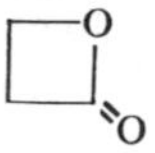	**1840**		
	5 n) 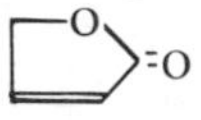	**1750** when α-H is present: **1785 and 1755** (doublet)	In non-polar solvents (CCl_4, CS_2), the higher band is stronger, while in polar solvents ($CHCl_3$, CH_3CN, CH_3OH) or liquids and solids, the lower band is stronger[2]. νC=C is weak.	
	5 o) 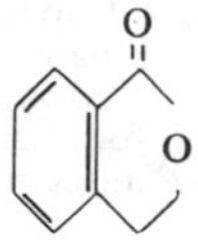	**1770**	Around 1780 in non-polar solvents, and ca. 1760 in polar solvents; affected by polarity of solvents. Influenced by *M* and other effects of substituents on arom. ring.	60
	5 p) 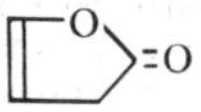	**1790**	The band is not split, contrary to α, β-unsatd.-γ-lactones. νC=C at 1660 cm^{-1} (s).	
6	**Acid anhydride** -CO-O-CO	**1820** and **1760**	Relative intensity of two bands variable. Band separation usually ca. 60 cm^{-1}, but this may differ from 35 to 90 cm^{-1} according to type[3]; lower band absorbs near νC=O of corresp. ester or lactone. Higher and lower band, resp., is stronger in acyclic and cyclic anhydrides[4].	62
		1300~1050	1~2 strong bands arising from C-O-C portion.	
	6 a) acrylic- or benzoic-	**1785** and **1725**		
	6 b) 6-membered	**1800** and **1750**	ΔC=O, 50~70 cm^{-1}[3].	
	6 c) α, β-unsatd. 6-membered	**1780** and **1735**	ΔC=O, 45 cm^{-1} when double bond is endo (75 cm^{-1} when exo)[3].	51
	6 d) 5-membered	**1865** and **1785**	ΔC=O, 80 cm^{-1}[3].	61
	6 e) maleic	**1850** and **1790**	ΔC=O, 60 cm^{-1}. 1790 cm^{-1} band is split and behaves similarly to 5 m[3].	
	6 f) phthalic	**1850** and **1770**	1770 cm^{-1} band is split and behaves similarly to 5 m[2].	61
7	**Peroxy acid**[9] -COOOH	3300~3250	γOH, narrower than in carboxylic acid	
		1745~1735	Doublet (Δν≃20 cm^{-1}) in solid state at 1740~1720 cm^{-1}	
		~1400 cm^{-1}	δOH	
		1260 cm^{-1}	νC-O	
		850 cm^{-1}	νO-O	
8	**Peroxy acid anhydride** -CO-O-O-CO			51
	alkyl	**1815** and **1790**		
	aryl	**1790** and **1770**		
9	**Acid halide** -CO-X	satd.: **1800** unsatd.: **1780~1750**	Value for -COCl, small band also at 1750~1700 cm^{-1}, due to Fermi resonance[10]. -COF, higher. -COBr and -COI, lower.	83
	Carbonate	satd.: **1750**	Absorptions also at 1250 (s) and 1000 (w) cm^{-1}.	51
	-O-CO-O-	unsatd.: **1770**	Absorptions also at 1250 (s) and 1160 (m) cm^{-1}.	

	Group	Band (cm^{-1})		Remarks	Figs.
10	**Amide** -CO-N	Free	Assoc.	Complicated but extensively studied.[11] Influenced by state of measurement as with -COOH. Data for both free and assoc. states are given.	
	primary $-CONH_2$	two at: 3500,3400	several: 3350~3200	νNH	(74)
		1690	**1650**	νC=O, "Amide I band."	
		1600	**1640**	Mainly δNH, "Amide II band." Shifted higher upon association. Solid spectra of $-CONH_2$ show two strong bands at 1650~1640, but "I band" is stronger. In concentrated solutions, all four bands arising from free and assoc. states may appear.	
	secondary -CO-NHR-	3440	3300 3070	νNH, 3300 cm^{-1} band considered to be due to *trans* assoc. form (structure I in figure shown). Band at 3070 is due to first overtone of N-H in plane bend.	
				In cyclic lactams, 3440 when free, and 3175 (dimeric, structure II) and 3070 cm^{-1} when assoc. (3300 lacking).	
		1680	**1655**	"Amide I band"	
		1530	1550	Mainly δNH, mixed with νC-N, "Amide II band." Present only in *trans* amides; in *cis* amides (e.g., small ring lactams) it is weak and shifted to 1440 cm^{-1}.	
		1260	1300	Mainly νC-N, mixed with δNH, "Amide III band."	
				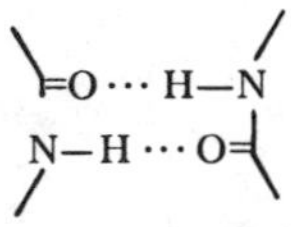 Shift in positions of free and assoc. νNH and "Amide I~III bands" is consistent with assignment. Thus contribution of limiting structure Ib is enhanced upon association. 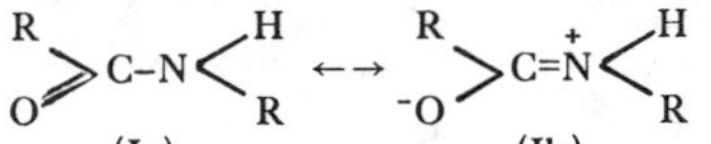 (R)(O)>C-N<(H)(R) (Ia) ←→ (R)(⁻O)>C=N⁺<(H)(R) (Ib); >C=O ··· H—N< / >N—H ··· O=C< dimeric lactam (II)	
				Solid IR of amides are, however, especially sensitive to crystal orientation. Effect of substituents may thus not be exerted correctly, and a slight complication in structure may prevent the shift of "I" and "II" bands to lower frequencies in the solid state.	
				In concentrated solutions the assoc. and free absorptions both appear.	
	tertiary -CO-N<	**1650**	**1650**	Since H-bonding with NH is absent, position is only slightly higher in the free state (similar to other C=O groups).	
	10 a) R-CO-N-C=C	shift of ca. +15 cm^{-1}		Higher shift caused by overlap of the N pi-electrons with both C=C and C=O.	(75)
	10 b) C=C-CO-N	shift of ca. +15 cm^{-1}		Extent of resonance within amide group is already large, and only *-I* effect of C=C operates (?).	

	Group	Band (cm^{-1})	Remarks	Figs.
	10 c) C=C–CO–N–C=C	shift of ca. **–15** cm^{-1}	The state of C=O becomes similar to ordinary carbonyl groups because N is already in conj. with C=C; thus *+M* effect of the other C=C is exerted.	(77)
	10 d) –CO–NH–X	**1705** (free)	X: halogen.	
	10 e) α-halo	+5~+50	Depends upon electronegativity and number of α-halo atoms; +50 cm^{-1} shift is observed in CF_3CO–N.	(74)
	10 f) 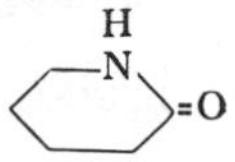 H N =O	**1670** (free)	"Amide II band" lacking; however, this is present in 9- or higher membered lactams.[12] Shifted slightly to higher freq. when α, **β-unsatd.**	
	10 g) 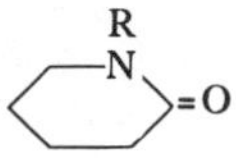 R N =O	**1640** (free)	+20 cm^{-1} when α, **β-unsatd.** Shifted higher in N-bridged systems.	
	10 h) 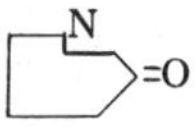 N =O	~**1700** (free)	Shifted higher in N-bridged systems. In general, the steric inhibition of conj. between N and C=O is expected to increase νC=O.	
	10 i) 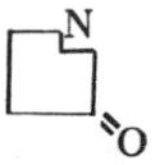 N O	~**1745** (free)		
	10 j) 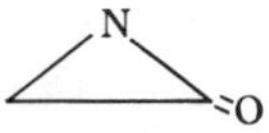 N O	~**1850**[13]		
	10 k) cyclic amide			
	6 membered	**1710** and **1700**	Usually lower band stronger.	
	α, β-unsatd. 6 membered	**1730** and **1670**		
	5 membered	**1770** and **1700**		
	α, β-unsatd. 5 membered	**1790** and **1710**		
	10 l) R–NH–CO–NH–R	**1660**		
	10 m) cyclic ureide			
	6 membered	**1640**		
	5 membered	**1720**		
	10 n) urethane R–O–CO–N	**1740~1690**	Around 1720 cm^{-1} in mono- and di-substituted derivatives, ca. 1700 cm^{-1} in non-substituted compounds.	(78)
		"Amide II band"	Non- and mono-substituted compounds absorb, resp., in the "Amide II band" region of *p*- and *s*- amides.	
		νNH	Non- and mono-substituted compounds absorb in usual NH region.	
11	R–CO–SH	**1720**		
	R–CO–SR′[5]	**1690**	cf. Chap. 4	
	Ar–CO–SR[5]	**1665**	"	
	R–CO–S–Ar[5]	**1710**	"	
	Ar–CO–S–Ar[5]	**1685**	"	

References

[1] N. J. Leonard and F. H. Owens, *J. Am. Chem., Soc.* **80**, 6039 (1958).
[2] R. N. Jones, C. L. Angell, T. Ito and R. J. D. Smith, *Can. J. Chem.* **37**, 2007 (1959).
[3] R. G. Cook, *Chem.* and *Ind.* 142 (1955).
[4] W. G. Dauben and W. W. Epstein, *J. Org. Chem.*, **24**, 1595 (1954).
[5] R. A. Nyquist and W. J. Potts, *Spectrochim. Acta*, **15**, (1959).
[6] IR of Me, Et, *n*- and *i*-Pr, *n*- and *i*-Bu esters: A. R. Katritzky, J. M. Lagowski, and J. A. T. Beard, *Spectrochim. Acta* **16**, 954, 964 (1960); for structural correlations in *far IR* of alkyl alkanoates, see E. Kroezinger, Ber. Bunsenges, *Phys. Chem.*, **78**, 1199 (1974); J. J. Lucier and F. F. Bentley, *Spectrochim. Acta*, **20**, 1 (1964).
[7] Y. Matsunga, *J. Chem. Phys.*, **41**, 1609 (1964); *idem. Nature*, **205**, 72 (1965); J. B. Chattopdhyay, M. N. Deshmukh, C. I. Jose, *J. Chem. Soc., Faraday Trans. II*, **71**, 1127 (1975).
[8] E. D. Becker, H. Ziffer and E. Charney, *Spectrochim. Acta*, **19**, 1871 (1963).
[9] R. Kavcic, B. Plesnicar and D. Hadži, *Spectrochim. Acta*, **23A**, 2483 (1967).
[10] H. N. Al-Jallo and M. G. Jahloom, *Spectrochim. Acta*, **28A**, 1655 (1972).
[11] E. A. Cutmore and H. E. Hallam, *Spectrochim. Acta*, **25A**, 1767 (1969).
[12] For the use of IR in determining lactam conformations, see references cited in: K. L. Williamson and J. D. Roberts, *J. Am. Chem. Soc.*, **98**, 5082 (1976).
[13] H. Baumgarten, R. L. Zey and U. Krolls, *J. Am. Chem. Soc.*, **83**, 4469 (1961).

ABSORPTION OF CARBONYL COMPOUNDS

The C=O stretching vibrations of various carbonyl groups absorb in the region 1900~1600 cm^{-1}; a more specific range is defined by the type of carbonyl (e.g., ketones, esters), and the position is further affected in a subtle manner by a variety of effects (cf. Chap. 5 and Table 8). Thus, the absorption will give information on various effects as well as the type of carbonyl group. This is why the region has a tendency to be overemphasized by organic chemists. The type of carbonyl group is identified by taking into account vibrations other than the C=O stretching. Rigorous discussions should be avoided unless the measurement is made in dilute non-polar solvents, since the *bands* are *shifted* to lower frequencies in more polar solvents. It should be noted that chloroform contains ethanol as a stabilizer, and this may considerably affect the νC=O position by interacting selectively with the solute molecule. In certain cases, valuable information can be obtained by changing the method of sample measurement.

TABLE 9 NITRO, NITROSO, etc.

(Non-diagnostic bands in italics)

-NO_2

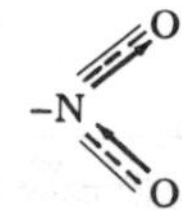

ν_{as} 1650~1500 cm^{-1}

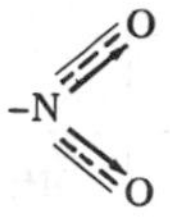

ν_s 1370~1250 cm^{-1}

Group	Band	Remarks	Figs.
1) Nitro $C-NO_2$	aliph. 1550 arom. 1525±15	ν_{as}, very stable band in nitroalkanes; aromatic compounds affected by *I*- and *M*- effects.[1]	(48), (60)
	aliph. 1370±10 arom. 1345±10	ν_s. In aromatics affected by planarity of nitro group and phenyl ring; bulky ortho subs. may increase to 1380 cm^{-1}.[1-4]	
	870	C–N stretching. In arom. nitro compounds, the out-of-plane δCH may be shifted higher and in addition this 870 cm^{-1} band is present; thus care should be taken in identifying substitution pattern.	
	610	CNO bending.	
2) Nitrate $O-NO_2$	1630±10	ν_{as}. Other bands: 870~855 (O–N stretching); 760~755 (out-of-plane bending); 710~695 (NO_2 bending).[5]	
	1275±10	ν_s. Appears as a doublet in secondary alkyl nitrates, except bicyclic compounds.[5]	
3) Nitramine $N-NO_2$	1630~1550	ν_{as} NO_2, intense.	
	1300~1250	ν_s NO_2, intense.	

-N=O

Group	Band	Remarks	Figs.
4) Nitroso C–N=O	1600~1500	Subject to *I*, *M* effects of adjacent group. Aromatic nitroso at ca. 1500, *t*-aliph. nitroso at ca. 1550 cm^{-1}.	
dimer	*trans* 1290~1190 *cis* 1425~1380	As nitroso compounds dimerize with ease, these bands may often be observed in addition to the 1600~1500 band in spectra of nitroso compounds.[6]	
5) Nitrite O–N=O	1680~1610 (two)	νN–O. Two bands, the higher one assigned to *trans* and the lower to *cis*, resp. In *t*-nitrites *trans* form content increases, and band intensity increases consequently.	
	815~ 750	νO–N	
	690~ 620 625~ 565	O–N=O bending of *cis* and *trans* forms, resp.	
6) Nitrosamine N–N=O	1460~1425	νN=O, intense. Lower than other N=O stretching, thus indicating increased contribution of limiting structure, N^+=N–O^-. In aromatics shifted to 1500~1460.[7]	
	1150~ 925	νN–N	

Group	Band	Remarks	Figs.
-N-O-			
7) Oxime C=N–OH	3600~3570	νOH; value for free OH. Shifted lower by H-bonding.	
	1680~1650	νC=N, variable intensity. Position influenced by ring strain as with C=O (cyclopentanone oxime, 1684 cm^{-1}). Aromatics may be shifted as low as 1615.	
	~1300 cm^{-1}	δOH. Syn isomers absorb at lower frequencies (10~50 cm^{-1}) than anti in the solid state or concentrated solution. [12]	
	960~ 930[8]	νN–O. In quinone mono-oximes shifted to 1075~925 cm^{-1}[4] because of contribution of =C–N=O^{+}H.	
8) Imine C=N–	1690~1650	νC=N, medium. Shifted as low as 1620 in aromatics.	
9) N-oxide N–O	aliph. [10] 970~ 950	N–O stretching, very intense.	84
	arom. [11] 1300~1200	N–O stretching, very intense. Higher than aliph. because of contribution of structure (b). Addition of methanol to organic solvents lowers it by 20~40 cm^{-1} and causes new band to appear at ca. 3360 cm^{-1}; this is explained by associated form (c) and is useful for oxide identification. N→O (a) $\overset{+}{N}$=O (b) N→O···H–O–R (c)	
10) Azoxy $N=\overset{+}{N}-\overset{-}{O}$	1310~1250	N–O stretching, strong.	

References

[1] T. Kinugasa and R. Nakajima, *Nippon Kagaku Zasshi,* **82**, 1473 (1961).
[2] A. Van Veen, P. E. Verkade and B. M. Wepster, *Rec. Trav. Chim. Pays Bas.,* **76**, 801 (1957).
[3] C. P. Conduit, *J. Chem. Soc.,* 3273 (1959).
[4] A. R. Katritzky and P. Simmons, *Rec. Trav. Chim. Pays Bas.,* **79**, 361 (1960).
[5] R. A. G. Carrington, *Spectrochim. Acta,* **16**, 1279 (1960).
[6] B. G. Gowenlock, H. Spedding, J. Trotman and D. H. Whiffen, *J. Chem. Soc.,* 3927 (1957).
[7] R. L. Williams, R. J. Pace, G. J. Jeacocke, *Spectrochim. Acta,* **20**, 225 (1964).
[8] A. Palm and H. Werbin, *Can. J. Chem.,* **32**, 858 (1954).
[9] D. Hadži, *J. Chem. Soc.,* 2725 (1956).
[10] R. Mathis-Noel, R. Wolf and F. Gallis, *Compt. Rend.,* **242**, 1873 (1956).
[11] H. Shindo, *Pharm. Bull.,* **7**, 407, 791 (1959).
[12] D. Hadži and L. Premru, *Spectrochim. Acta,* **23A**, 45 (1967).

TABLE 10a HETEROCYCLIC COMPOUNDS

(Non-diagnostic bands in italics)

Ring	Band	Remarks	Fig.
1) Furans	3165~3125	νCH. Higher than 3000 cm^{-1} as with other aromatics.	86
	1600~1300	νC=C ring stretching. Three bands appear near 1600, 1500 and 1400; their position and relative intensity vary considerably, depending on the substituent type. (All intensities are increased by electron withdrawing groups.)[1]	
(Benzofurans similar)	1030~1015		
	885~ 870	Sharp and most characteristic.	
	800~ 740	Band and region both wide, sometimes split.	
2) Pyrroles	3490	νNH. Rather sharp and characteristic. Much more intense than saturated secondary amines. Lowered by ca. 90 cm^{-1} upon H-bonding.	39
	3125~3100	νCH, weak. Higher than alkane νCH.	
(Indoles similar)	1600~1500	νC=C. Usually two bands ca. 1565 and 1500; position and intensity vary with substituent.[2]	
3) Thiophenes	3125~3050	νCH. Position characteristic, clearer than pyrrole νCH.	
	~1520 and ~1040	νC=C. Becomes clear when conjugated with other groups (lower band stronger). Mostly absent when not conjugated.	
	750~ 690	Strongest of thiophene bands, shifted to higher freq. by electron-attracting groups; for effect of substitution see ref. 3.	
4) α-Pyrones	1740~1720	Frequently split; higher band stronger in non-polar solvents (CCl_4, OS_2), and lower band stronger in polar solvents ($CHCl_3$, CH_3CN) or as liquid or solid.[4] Reason for this not well-known. Considerably shifted by substituents, but cannot be explained consistently.	79
(Coumarins, isocoumarins similar)	1650~1620 1570~1540	νC=C. Useful for supporting presence of pyrone rings.	
5) γ-Pyrones	1680~1650	Position not affected much by substituents, and splitting also rare. (Change in intensity of two bands when split is same as with α-pyrones.)[4]	79
(Chromones similar)	1650~1600 1590~1560	νC=C. Intensity variable, useful in supporting presence of pyrone ring.	
6) α-Pyridones	3200~2400[5]	Complex shape, broad and intense. Stretching of assoc. NH.	
	(3400 : free)	Sharp monomer band at 3400 cm^{-1} in CCl_4 solutions less than 0.025 mole per liter.	
(α-Quinolones similar)	1690~1650[6]	νC=O. Shifted slightly to lower freq. upon association.	
	1650~1400	νC=C. Four bands, strong to medium.[7]	
7) γ-Pyridones	3200~2300	νNH. Similar to α-pyridones.	81
	1650~1630	νC=C, intense. One of the four ring stretching modes appearing in the region 1650~1400 cm^{-1}.[8]	
(γ-Quinolones similar)	1580	νC=O.[8]	
8) γ-Thiapyrones	1610[9]	Overlapping of νC=O and νC=C. Very broad and intense.	
9) Others[10]			

References

[1] A. R. Katritzky and J. M. Lagowski, *J. Chem. Soc.*, 657 (1959).
[2] U. Eisner and R. L. Erskine, *J. Chem. Soc.*, 971 (1958).
[3] H. Rosatzin, *Spectrochim. Acta*, **19**, 1107 (1963).
[4] R. N. Jones, C. L. Angell, T. Ito and R. J. D. Smith, *Can. J. Chem.*, **37**, 2007 (1959).
[5] H. Shindo, *Chem. and Pharm. Bull.*, **7**, 407 (1959).
[6] S. F. Mason, *J. Chem. Soc.*, 4874 (1957).
[7] A. R. Katritzky and A. P. Ambler in "Physical Methods in Heterocyclic Chemistry," Vol. II, A. R. Katritzky, Ed., pp. 259–262 (Academic Press, New York, 1963).
[8] L. J. Bellamy and P. E. Rogasch, *Spectrochim. Acta*, **16**, 30 (1960).
[9] D. S. Tarbell and P. Holland, *J. Am. Chem. Soc.*, **16**, 2451 (1954).
[10] References to various heterocyclic compounds are given in:

R. N. Jones and C. Sandorfy, "Chemical Applications of Spectroscopy," pp. 534, 547 (Interscience, New York, 1956).

A. R. Katritzky and A. P. Ambler in "Physical Methods in Heterocyclic Chemistry," Vol. II, A. R. Katritzky, Ed., pp. 161–360 (Academic Press, New York, 1963).

TABLE 10b PYRIDINES

Position	Assignment	Remarks	Fig.
3075±15 3030±20	νCH νCH	In $CHCl_3$ solutions νCH of $CHCl_3$ appears at 3010~2940 cm^{-1}, being shifted to lower frequencies by *H-bonding* with pyridines and especially pyridine N-oxides. [5]	
2000~1650	overtone of δCH (out-of-plane) and combination	Bands are characteristic of substitution pattern. [1,2]	
1600±15 1570±15 1500±20 1435±20	νC=C	Ring stretch. Intensity is very variable but can be correlated with type and orientation of substitution. [2]	
920~720	δCH out-of-plane	Intense. Position and relative intensity of bands is characteristic of substitution pattern (see table below).	

TABLE 10c – MONOSUBSTITUTED PYRIDINES – CORRELATIONS IN THE OUT-OF-PLANE BENDING REGION [1-6]

Substitution	Band Position		
	Alkyl Substituents	Other Substituents	Corresp. Benzene
α	795~780 755~745	780~740	770~735
β	810~790 ~715	920~880 820~770 730~690	810~750 710~690
γ	820~795 775~710	850~790 725	860~800

The substitution of N-heteroaromatic compounds can be analyzed on the basis of the number of adjacent H atoms (as for carbocyclic aromatics) by regarding the N atom as a substituent on the phenyl ring. Theoretical calculations show excellent agreement with observed values both for methylpyridines and pyridine-N-oxides.[7] This analysis holds for polyalkyl substitution and monosubstitution by polar groups. However, as for carbocyclic compounds, the effects of strongly electron withdrawing groups and polysubstitution by polar groups is not clear.

References

[1] J. H. S. Green, W. Kynaston, H. M. Paisley, *Spectrochim. Acta,* 19, 549 (1963).
[2] A. R. Katritzky and A. P. Ambler in "Physical Methods in Heterocyclic Chemistry," Vol. II, A. R. Katritzky, ed., p. 276, Academic Press, New York, (1963).
[3] H. Shindo and N. Ikekawa, *Chem. and Pharm. Bull.,* 4, 192 (1956).
[4] H. Shindo, *ibid.,* 5, 472 (1957).
[5] A. R. Katritzky and J. N. Gardner, *J. Chem. Soc.,* 2192 (1958).
[6] D. B. Cunliffe-Jones, *Spectrochim. Acta,* 21, 747 (1965).
[7] Y. Kabuiti, H. Saito, M. Abiyama, *J. Molecular Spectroscopy,* 35, 66 (1970).

TABLE 11 SULFUR GROUPS

(Non-diagnostic bands in italics)

Bonds between S, C, H

Group	Band	Remarks	Figs.
1) -SH	2600~2550w	νSH. Weaker than OH band. Shift to lower freq. upon H-bonding is not so large.	(96)
2) C–S	*800~ 600*w	νC–S. Common to C–S of mercaptanes, sulfoxides, etc.	
3) S–S	*540~ 505*w	Two bands in linear disulfides[2]	
4) S–CH_2	2700~2630[1] ~1420	Both freq. lower than corresp. C–CH_2 bands.	
5) S–CH_3	~1325[2]	Considerably lower than δCH of C–CH_3.	(32b)
6) C=S	*1200~1050*s	Influenced by *I*- and *M*- effects of substituents, ring strain, etc., as with νC=O. Not a good group frequency.	
7) Thioureide (thiolactam) -C(=S)-N<	~3400 (soln.) ~3150 (solid) 1550~1460 vs 1300~1100 s ~1300	νNH. Corresp. to "Amide II" band. νC=S. Corresp. to "Amide I" band. Corresp. to "Amide III" band.	

Bonds between S, O

νS–O 900~ 700
νS=O 1200~1040
νSO_2 1400~1310, 1230~1120

All are intense. When C–S (above-mentioned) and C–O (ca. 1000 cm^{-1}, s) bonds are present in the sulfur-containing groups, the corresponding bands appear in addition.

Group	Band	Remarks	Figs.
8) Sulfoxide R–S(=O)–	1060~1040 s, ϵ^a 300	Shifted lower by 10~20 cm^{-1} upon conjugation and H-bonding; also large solvent shifts (20 cm^{-1} lower in CCl_4 than in $CHCl_3$). Shifted higher when bonded to halogens or oxygen.	(92)
9) Sulfinic acid R–SO–OH	~1090 s	νS=O.	
10) Sulfinate R–SO–OR′	1135~1125 s	νS=O.	
11) Sulfurous acid ester R–O–SO–O–R′	~1200 s	νS=O. Higher than (8)~(10) because two O-atoms are bonded to S=O.	
12) Sulfone R–SO_2–R′	1350~1310, ϵ^a 250~600 1160~1120, ϵ^a 500~900	$\nu_{as}SO_2$, $\nu_s SO_2$: In solids shifted lower by 10~20 cm^{-1}, and frequently split into band groups. Neither affected much by conj. or ring strain.	(90)
13) Sulfonamide R–SO_2–N<	1370~1330 1180~1160	Lower by 10~20 cm^{-1} in solids. Same position in solids. The fact that the two SO-bands are higher than in sulfones is contrary to relation between ketones and amides.	(92)

Group	Band	Remarks	Figs.
14) Sulfonic acid R–SO_2–OH	1345±5 s 1155±10 s *700~ 600*	$\nu_{as}SO_2$ } ν_sSO_2 } These values are for the anhydrous acids. Easily formed hydrates show bands at ~1200 and ~1050 cm^{-1}, similar to acid salts. νS–O	
15) Sulfonate R–SO_2–OR′	1370~1335 s 1200~1170 s	$\nu_{as}SO_2$, intense doublet of which higher freq. peak is more intense. ν_sSO_2	
16) Sulfate[1] R–O–SO_2–OR′	1415~1380 1200~1185	$\nu_{as}SO_2$. Higher than (14) or (15) because of attachment of two oxygens to SO_2. ν_sSO_2	
17) Sulfonyl chloride	1370~1365 1190~1170	$\nu_{as}SO_2$ ν_sSO_2	

References

[1] S. Detoni and D. Hadži, *Spectrochim. Acta,* **1957**, Suppl., 601.

[2] K. G. Allum, J. A. Creighton, J. H. S. Green, G. J. Minkoff, L. J. S. Prince, *Spectrochim. Acta,* **24A**, 927 (1968).

TABLE 12 PHOSPHORUS GROUPS

(Non-diagnostic bands appear in italics)

The following bands assigned to respective bonds appear additively. Frequencies given refer to solid phase spectra and are subject to large shifts with changes in phase or solvent.

Group	Band	Remarks	Fig.
1) P–H	2450~2280	νP–H, medium to weak.	
	1250~ 950	δP–H, weak	
a) phosphonates $H{-}P(=O)(OR)_2$	2450~2420	νP–H	
b) phosphinic acid esters $H_2{-}P(=O){-}OR$	2380, 2340	νP–H. Two bands present; attributed to rotational isomerism.	
c) phosphine oxides $H{-}P(=O){-}R_1R_2$	2340~2280	νP–H	
2) P=O	1300~1100	νP=O, strong. Often appears as a doublet due to rotational isomerism. Shifts up to 65 cm^{-1} result from change in phase or solvent.	
a) phosphates $P(=O){-}(OR)_3$			
i) alkyl	1285~1260		94
ii) aryl	1315~1290		
b) phosphonates $H{-}P(=O){-}(OR)_2$ (alkyl)	1265~1230		95
c) phosphinates $H_2{-}P(=O){-}OR$	1220~1180		
d) phosphine oxides			
i) alkyl	1185~1150		
ii) aryl	1145~1095		
e) pyrophosphates and $-P(=O){-}O{-}P(=O)-$	1310~1210	νP=O. Single band; unlike carboxylic acid anhydrides, there is no splitting due to non-planarity of two P=O groups.	
tetra-alkyl-pyrophosphates	1240~1210		
3) phosphorus acids $-P(=O){-}O{-}H$	2700~2200	Assoc. νOH. Free OH band at ca 3600 cm^{-1} does not appear upon dilution.	
	1240~1180	νP=O, stronger than P=O absorption in (2).	

Group	Band	Remarks	Fig.
4) phosphorus amides	1290~1155	νP=O, intense. Position of band is rather variable.	
dialkoxyphosphoramides $(RO)_2-\overset{O}{\overset{\Vert}{P}}-NR_1R_2$	1275~1200	νP=O. Frequency can be correlated with nature of amine component.[1]	
5) P–O	1100~ 950	νP–O	
P–O–C alkyl	1050	Intense; usually accompanied by a weaker band near 1190 cm^{-1}.	
P–O–C aryl	950~ 875	νP–O, intense.	
6) P=S	*800~ 650*	νP=S, weak.	

References

[1] L. C. Thomas and R. A. Chittenden, *Spectrochim. Acta,* **20**, 467 (1964).

TABLE 13 SILICON COMPOUNDS

(Non-diagnostic bands appear in italics)

Group	Band	Remarks	Fig.
1) Si–H	2360~2100	νSiH; sharp, intense.	
2) R_3SiH	2130~2100	νSiH. Substitution of halide or O on Si can shift band up to 2260 cm^{-1}.	
R_2SiH_2	2135±10	νSiH; only one band visible as ν_{as} and ν_s have almost identical frequencies. Substitution of halide or O on Si can shift band up to 2200 cm^{-1}.	
	890~ 860	δSiH, strong	
3) Si–O–H	3680±10	νO–H; confirmed by presence of Si–OH deformation band at 870~820 cm^{-1} (OH band).	
4) Si–O	1110~1000	νSi–O, very broad and intense.	
5) $Si–CH_3$	1255	δ_s CH_3. Sharp and very characteristic.	
6) $R–C(=O)–SiR_3$	1618	νC=O. Unusually low frequency is due in part to low electronegativity of Si.	

TABLE 14 HALOGENS AND MISCELLANEOUS GROUPS

Group	Band (cm^{-1})	Remarks	Fig.
1) C–F[5]	1400~1000 vs	νC–F	
single C–F	1100~1000 vs	νC–F; overtones 2600~2100 cm^{-1}	
CF_2	1250~1050 vs	Often two strong bands in this region	
	below 600 s		
CF_3	1360~1300 vs	Usually accompanied by one or two strong bands at slightly lower frequency	
ϕCF_3	1330~1310 vs	ν_s	
	1150~1130 vs	ν_{as}	
	1190~1170 vs	ν_{as}	
2) C–Cl	800~600 s		
3) C–Br	600~ 500 s		
4) C–I	500 s		
5) B–H	2640~2350		
B···H···B	2200~1540	Several bands	
RBH_2	2640~2571	δ_{as}	
	2532~2480	δ_s	
	1170~1140		
6) B–CH_3	1435	δ_{as} CH_3	
	1322±7	δ_s CH_3	
7) B–O	1350~1310	νBO.	
B–O–B	1375	ν_{as} BOB	
	1260	ν_s BOB	
8) B–N	1465~1330		
R_3B–NR_3	800~ 700	νBN	
R_2B–NR_2	1500~1350	νBN	
9) Se–H[2,6]	2300~2280	νSeH	
10) Se=O[2,6]	855~ 825 vs	νSe=O	

References

[1] L. J. Bellamy, "The Infra-red Spectra of Complex Molecules," p. 329, Methuen, London, (1958).

[2] S. Detoni and D. Hadži, *J. Chim. Phys.*, 53, 760 (1956).

[3] W. Gerrard, "The Organic Chemistry of Boron," Academic Press, New York, (1961).

[4] K. Niedenzu and J. W. Dawson, "Boron Nitrogen Compounds," Academic Press, New York, (1965).

[5] J. K. Brown, K. J. Morgan, "The Vibration Spectra of Organic Fluorine Compounds" in *Advances in Fluorine Chemistry*, v. 4, M. Stacey, J. C. Tatlow, A. G. Sharpe, eds., pp. 253–313, Butterworths, Washington, D.C., (1965).

[6] K. A. Jensen, B. M. Dahl, P. H. Nielsen and G. Borch, *Acta Chem. Scand.*, 25, 2039 (1971).

TABLE 15 INORGANIC SALTS

Group	Band (cm^{-1})
1) CO_3=	1450–1410 (vs), 880-860 (m)
2) HCO_3–	2600–2400 (w), 1000 (m), 850 (m), 700 (m), 650 (m)
3) SO_3=	1000–900 (s), 700–625 (vs)
4) SO_4=	1150–1050 (s), 650–575 (m)
5) $C10_3$–	1000–900 (m–s), 650–600 (s)
6) $C10_4$–	1100–1025 (s), 650–600 (s)
7) NO_2–	1380–1320 (w), 1250–1230 (vs), 840–800 (w)
8) NO_3–	1380–1350 (vs), 840–815 (m)
9) NH_4+	3300–3030 (vs), 1430–1390 (s)
10) PO_4–3, HPO_4=, H_2PO_4–	1100–1000 (s)
11) CN–, SCN–, OCN–	2200–2000 (s)
12) Various silicates – 1100–900 (s)	
13) Various oxysalts of phosphorus[3]	
14) CrO_4=	900–775 (s–m)
15) Cr_2O_7=	900–825 (m), 750–700 (m)
16) MnO_4=	925–875 (s)

References

[1] F. A. Miller and C. H. Wilkins, *Anal. Chem.*, 24, 1253 (1952).

[2] Characteristic frequencies for 20 ions in the 700–300 cm^{-1} region: F. A. Miller, G. L. Carlson, F. B. Bentley, and W. H. Jones, *Spectrochim. Acta*, **16**, 135 (1960).

[3] D. E. C. Corbridge and E. J. Lowe, *J. Chem. Soc.*, **1954**, 4555. D. E. C. Corbridge, *J. Appl. Chem.*, **6**, 456 (1956).

[4] R. A. Nyquist and Ronald O. Kagel "Infrared Spectra of Inorganic Compounds (3800–45 cm^{-1}), Academic Press, N. Y., 1971.

CHAPTER 3 / BAND POSITIONS AND INTENSITY

The characteristic bands so far described are subject to shifts in position. The factors that cause shifts in band positions are conveniently divided into two groups, external factors that depend on the state of measurements, and internal factors that depend on the molecular structure itself[1] .

3.1 BAND POSITIONS

External Factors (State of Measurement)

Bands will change according to experimental conditions because these will affect the shape of molecules (rotational isomerism, polymorphism, etc.) or the environments in which the molecules are placed (association of solutes, solvation, etc.). Polar groups will give rise to larger variations.

Measurement of gases:

Measurements carried out under low pressure will generally give information on isolated molecules. With the advent of Fourier Transform Infrared Spectroscopy (FTIR), gaseous samples are much more readily measured and it is now possible to measure spectra of microgram quantities. Complex mixtures of volatile organic compounds can also be analyzed efficiently by using FTIR in combination with gas chromatography (GC/IR) e.g., components of the mixture are measured in the gas phase as they leave the chromatographic column.*

Measurement of liquids:

Polar groups exist in associated states, and molecules lacking polar groups exist in a state in which they are surrounded by the species of molecules that do not electrically affect each other.

---

* Peter R. Griffiths "Chemical Infrared Fourier Transform Spectroscopy" John Wiley and Sons, New York, N.Y., 1975, Chapter 10. For other applications of FTIR, see J. L. Koenig and D. L. Tabb, *Can. Res. Dev.*, 7 25 (1974).

Measurement of solids:

If the spectra of solids measured under identical conditions turn out to be the same, the samples are identical (excepting D- and L-isomers, which cannot be differentiated). This is because the solid method reflects the molecular shape and crystal lattice. In solutions, several conformational isomers may co-exist or solvent effects may play an important role. Consequently slight differences in chemical structures are frequently overshadowed, and non-identical samples may give identical spectra. It follows that for identification purposes the solid method is most reliable.

Slight differences in preparations, such as the method of crystallization, time of mixing with nujol or KBr, or time of pressing KBr disks, will generally give identical spectra. The IR spectra measured by the nujol and KBr methods are usually the same. However, this is not necessarily the case. For example, polymorphism or tautomerism may occur while grinding with nujol[1a]. The KBr spectra are affected, in addition, by the size of grains, time of pulverization and pressure[2]. The optimum particle size gives a KBr disk a brown-purple hue (or a mauve color) when it is viewed against the light.* Anomalous spectra can also be due to adsorption of polar species (e.g., carboxylic acids) on the surface of the KBr particles[2a]. A difficulty frequently encountered with KBr spectra is adsorption of moisture during pellet preparation. This is almost unavoidable, but can be remedied, if the sample is stable, by heating the pellet at 40–50° at 1 mm Hg for several hours.

Measurement of solutions:

When making comparisons of spectra measured in the same solvent, it should be noted that concentration or temperature differences will affect the extent of intermolecular association. This change will be naturally reflected in the spectra, for the IR spectra of all species present in the solution are roughly additive. When intramolecular hydrogen-bonds** are formed, absorptions due to these bonded molecules will appear in addition. Although the equilibrium between monomeric and polymeric species is dependent on the concentration, the total amount of intramolecularly hydrogen-bonded species will remain constant, provided the temperature is unchanged.

When measured in different solvents, the stretching frequencies of polar groups such as ketones, amides, or nitriles vary with the polarity of the solvent: the more polar the solvent, the lower the frequency. Frequently, it is helpful to take the spectrum in two different solvents, or in different phases. This is particularly true in the case of poly-carbonyl compounds.

Internal Factors (Structural Factors)

Internal factors are of primary importance in the elucidation of unknown structures rather than for simple identification. In order to minimize the above-mentioned environmental effects, it is advisable to conduct measurements in non-polar dilute solutions whenever possible. The band position of the sample is then compared with standard values, and the factors that cause the shifts are

* C. F. Hammer, personal communication

** Intramolecular hydrogen-bonds associated with π-electrons have also been actively studies, e.g., M. Oki and T. Yoshida, *Bull. Chem. Soc. Jap.*, **44**, 1336 (1971), and previous papers.

considered. When the sample is insoluble in non-polar solvents, polar solvent data are used instead, but with care.

Single Bond Frequencies:

The two tables below illustrate interesting systematic variations in band frequencies.

Table 3.1 X-H Stretching Frequencies (cm^{-1})

BH 2400	CH 3000	NH 3400	OH 3600	FH 4000
AlH 1750	SiH 2150	PH 2350	SH 2570	ClH 2890
	GeH 2070	AsH 2150	SeH 2300	BrH 2650
	SnH 1850	SbH 1890		IH 2310

Table 3.2 Symmetrical Methyl Deformation Frequencies (cm^{-1})

BCH_3 1310	CCH_3 1375	NCH_3 1425	OCH_3 1450	FCH_3 1475
	$SiCH_3$ 1265	PCH_3 1295	SCH_3 1310	$ClCH_3$ 1335
	$GeCH_3$ 1235	$AsCH_3$ 1250	$SeCH_3$ 1282	$BrCH_3$ 1305
	$SnCH_3$ 1190	$SbCH_3$ 1200		ICH_3 1252
	$PbCH_3$ 1165			

Tables I and II reproduced by permission from N. B. Colthup, "Interpretation of Infrared Spectra," ACS Audio Course, American Chemical Society, Washington, 1971.

Mass effects, in general, have only a small contribution to changes in group frequencies. Application of Hooke's Law (eq. 1.1, p. 3) shows that the mass effects on stretching frequencies are small, except in the case of deuterium compounds (see below). Within any row of the periodic table the frequency of ν_{XH} (or the force constant) increases linearly with the atomic number, while on descending the table there is a regular decrease in ν_{XH} . The increase in stretching frequency parallels the increase in the electronegativity of X and the consequent increase in bond strength and decrease in bond length. It can be seen from the tables that the order of the CH_3 deformation frequencies in Table II parallels that for the X-H stretching frequencies in Table I. Thus for the symmetrical methyl deformation, the frequency increases with increasing electronegativity within a period (e.g., $N < O < F$) and decreases on descending the periodic table ($F > Cl > Br > I$).

Multiple Bond Frequencies:

The variations in the IR group frequencies can be accounted for by the combination of the inductive (I) and mesomeric (M) effects on the ground state of the molecule.* Inductive or mesomeric effects which tend to decrease the bond order of the multiple bond cause a decrease in the stretching frequency, and vice versa. Multiple bonds and atoms containing unpaired electrons exert *M* effects, but when these bonds or atoms are absent, the band position can be discussed simply in terms of the *I* effect. A few examples will be cited.

I effect only: A linear relation has been reported between the band position and the sum of Pauling electronegatives of the atoms concerned[4].

M effect and *I* effect: Numerous examples exist, e.g., Chap. 4, dimedone-(j), and isopropenyl acetate (64). The combined effects are exemplified in the following by comparing the C=O band of thiol esters[5] and ordinary esters with the reference C=O value of 1715 cm^{-1}.

(i)

R–C(=O)–R′
1715 cm^{-1} (reference: saturated aliphatic ketone)

R–C(=O)→S–R′
1690 cm^{-1}
$+M > -I$ of *S*

R–C(=O)→O–R′
1735 cm^{-1}
$+M < -I$ of *O*

The $+M$ effect of sulfur is larger than its $-I$ effect, $+M > -I$; the opposite relation holds for oxygen, $+M < -I$. This is the reason for the decreased and increased frequencies, respectively, as compared to the standard value of saturated aliphatic ketones.

(ii)

ϕ–C(=O)–S–R
1665 cm^{-1}

ϕ–C(=O)–O–R
1725 cm^{-1}

* On the other hand, in UV spectroscopy both the ground and excited electronic states have to be considered. Thus the IR and UV are not exactly parallel and there are instances where a substance behaving normally in the IR gives rise to anomalous UV absorptions. For example, the νC=O of (1) is normal and is seen at 1706 cm^{-1} (in CCl_4), whereas the UV of (1) and (2) are not identical, and the appearance of a weak absorption at 258nm suggests an interaction between the π-electron clouds of the double bond and the carbonyl group [N.J. Leonard and F.H. Owens, *J. Am. Chem. Soc.* **80**, 6038 (1958)].

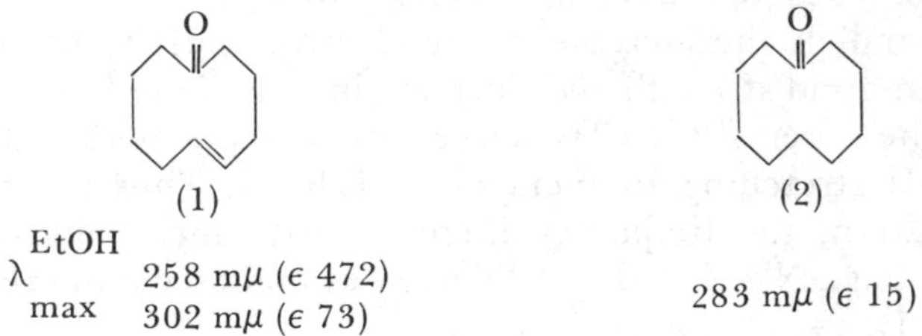

(1) (2)

λ_{max}^{EtOH} 258 mμ (ϵ 472), 302 mμ (ϵ 73) — 283 mμ (ϵ 15)

Lower than (i) because of large $+M$ effect of aryl group ($-I$ effect is small).

(iii)

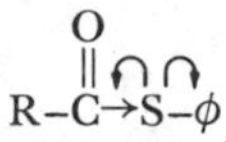

$R-\overset{O}{\overset{\|}{C}}\rightarrow S-\phi$ 1710 cm^{-1} $\qquad$ $R-\overset{O}{\overset{\|}{C}}\rightarrow O-\phi$ 1750 cm^{-1}

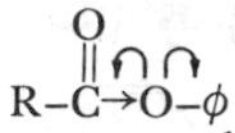

$+M$ effect of the sulfur and oxygen atoms operating in (i) is internally compensated, and consequently the $-I$ effect plays a major role.

(iv)

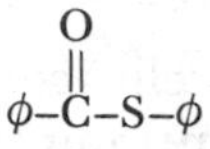

$\phi-\overset{O}{\overset{\|}{C}}-S-\phi$ 1685 cm^{-1} $\qquad$ $\phi-\overset{O}{\overset{\|}{C}}-O-\phi$ 1735 cm^{-1}

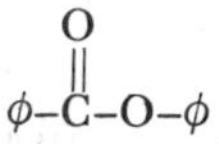

In cases such as (ii), the C=O band is located at frequencies lower than (i) (operation of so-called mesomeric effect), whereas in cases such as (iii) it is at higher frequencies than (i) (operation of so-called vinyl ester effect). Both effects are present in (iv), and thus the C=O band resumes its normal position.

The stretching frequencies of the C=O bond of a number of acids, esters and amides bearing phenyl substituents have been correlated with Hammett values.‡

Another set of data summarized in Table 3.3 are quite straight-forward.

Table 3.3* Stretching Frequencies of the Nitrile and Ester Groups, in $CHCl_3$

Compound	νC≡N	νC=O	νC=C(ϵ [a])	Remarks**
1. Me-COOEt		1733		Difference in νC=O of 1, 2 and 4 is ascribed to difference in I effect of the groups attached to COOEt; difference in νC≡N of 3 and 4 accounted for similarly.
2. Et-COOEt		1728		
3. Me-CN	2255			
4. NC-CH_2-COOEt	2267	1751		Conj. with double bond lowers νC=O and νC≡N, e.g., 5 and 6.
5. Me_2C=CH-COOEt		1705	1654(170)	
6. Me_2C=CH-CN	2221		1637 (32)	Difference in νC=C of 5 and 6 is due to difference in electronegativity of CN and COOEt.
7. Me_2C=C$(COOEt)_2$		1719	1640(115)	
8. Me_2C=$C(CN)_2$	2234		1591(100)	νC≡N of 8 is higher than 6, and νC=O of 7 is higher than 5. Thus, although the shift of C=C π-electrons in 7~9, which carry two electron-attracting groups on one side, is larger than 5 and 6, the share of these electrons per group is smaller in the former set because they have to be divided between two groups.
9. Et_2C-C(CN)COOEt	2224	1727	1629(123)	
10. NC-CH=CH-CN	2240		1612 (3)	
				Shift of νC=C in the sequence 7→9→8, $\Delta\nu$C=O of 7 and 9, and $\Delta\nu$C≡N of 8 and 9, are all due to the greater electron-attracting power of the C≡N group.
				The νC=C intensity of 10 is small because of symmetry.

‡ C.N.R. Rao and R. Venkataraghavan, *Can. J. Chem.*, **39**, 1757 (1961).

* D. G. I. Felton and S. F. D. Orr, *J. Chem. Soc.* 2170 (1955).

** Difference in mass is neglected in explanations.

Factors such as the following also affect bond orders: The tropone molecule[6] tends to maintain its state of six annular π-electrons, this increases the contribution of the limiting structure (2), and consequently the double bond character of the group is quite low.

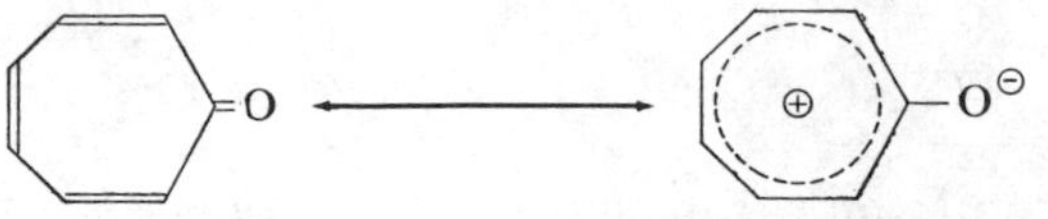

(1) (2)

tropone 1650 cm^{-1}

Steric effects:

Most typical are the ring strain effects of cyclic compounds, (succinic anhydride (61)-a), steric inhibition of resonance, and intramolecular spatial interactions* (dichloroacetamide (65)); numerous examples have been reported for all three.

The following is a common example of steric inhibition of resonance[7] :

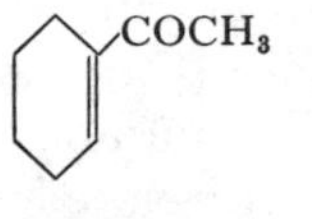

(3) 1663 cm^{-1} (4) 1686 cm^{-1} (5) 1693 cm^{-1}

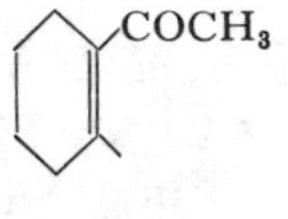

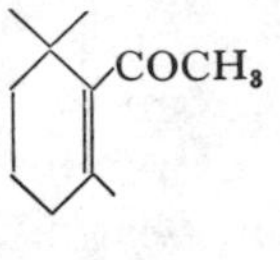

It becomes more difficult for the C=O side-chain to maintain planarity with the cyclohexene ring in the series 3→4→5, and the νC=O value approaches that of saturated compounds.

The following case has also been reported.[8] . Upon variation of *n* from 5 to 9 in compounds (6), the amide νC=O band is lowered from 1710 to 1660 cm^{-1}.

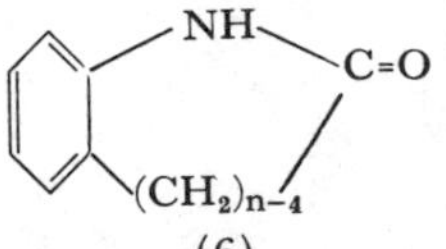

(6)

This has been interpreted by assuming that as the ring gets larger mesomerism between the amide N and benzene ring becomes hindered whereas mesomerism between NH and CO is increased, and the value approaches that of a normal chain amide. Many cases of the trans-annular effect, an intramolecular spatial effect, have been reported[9, 10] . For example, the C=O stretching of compound (7) appears at the very low wave-number of 1664 cm^{-1} (in carbon tetrachloride); in the perchlorate this band is missing and is replaced by an OH band at 3365 cm^{-1} (nujol). This has been accounted for by contribution of limiting structure (8) in the free compound, and by structure (9) for the salt.

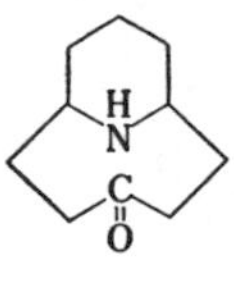

(7) 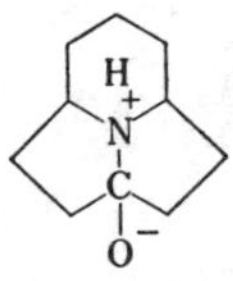 (8) 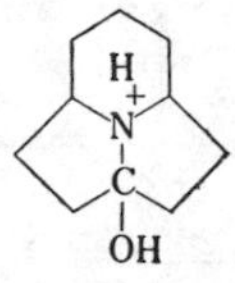 (9)

* Including intramolecular hydrogen-bonds.

3.2 DEUTERATION

If it is assumed that force constants of X–H and X–D are identical and that the mass ratio of H and D is 1:2, the wave-number ratio of stretching frequencies deduced from equation 1.1 becomes:

$$\frac{\nu X\text{–}H}{\nu X\text{–}D} = \sqrt{\frac{2+2m_x}{2+m_x}}$$

where m_x is the mass of atom X.

Values obtained using this expression are in very good agreement with experimentally derived data (see Table 3.4). Approximate calculations can be carried out using the calculation

$$\frac{\nu_{X\text{–}H}}{\nu_{X\text{–}D}} = \sqrt{2}$$

The X–H bending frequencies behave similarly, but agreement with calculated values is not as good as for stretching frequencies.

Table 3.4*

Stretching frequency, ν	X–H	$X\text{–}H_{obs}$	$X\text{–}D_{obs}$	$X\text{–}D_{calc}$	Approximation from $(X\text{–}H) \div \sqrt{2}$
	OH	3620	2630	2634	2560
	$CHCl_3$ vapor phase	3022	2258	2218	2137
	CH_4 vapor phase	3025	2206	2220	2139
	Mo . . . OH	3615	2660	2630	2556
Bending frequency, δ					
	$CHCl_3$	1220	910	895	863
	⟩C=C–H	815	678‡	598	576

Substitution of active hydrogen by deuterium can usually be carried out easily by recrystallizing from heavy water or repeated dissolution of sample in heavy water and drying. However, deuterium-containing starting materials or reagents such as $LiAlD_4$ have to be employed for preparation of compounds that contain nonactive C–D bonds. Whatever the case may be, frequencies of bands associated with D-containing bonds are at approximately $1/\sqrt{2}$ those of the corresponding H-containing bonds. This relation is of great help in making band assignments (triketohydrindene hydrate (73)).

* Additional information may be found in "Infrared Spectra of Labelled Compounds," S. Pinchus, Academic Press, 1971.

‡ B. A. Yeit, L. Bohor, S. Rubinraut, *J. Chem. Soc., Perkin II*, 253 (1976).

3.3 BAND INTENSITIES

The more polar the group is, the greater the band *intensity*. This is apparent in the following data on aliphatic and aromatic nitriles (Table 3.b).

Table 3.5 Intensity of Aliphatic[1] and Aromatic[2] Nitriles

Aliphatic	cm^{-1} (intensity*)	Aromatic	cm^{-1} (intensity*)
MeCN	2255(793.4)	Benzonitrile	2231(3742)
ICH_2CN	2248(754.8)	Hydroxybenzonitrile	
$BrCH_2CN$	2255(333.6)	ortho-	2227(6031)
$ClCH_2CN$	2259(92.4)	meta-	2234(4723)
FCH_2CN	2256(126.5)	para-	2227(9128)
$HOCH_2CN$	2257(160.2)	Nitrobenzonitrile	
NH_2CH_2CN	2242(425.7)	ortho-	2237(1189)
		meta-	2239(1752)
		para-	2237(1349)
		Picolinonitrile (2-CN)	2238(565.5)
		Nicotinonitrile (3-CN)	2236(2678)
		Isonicotinonitrile (4-CN)	2239(919.7)

The nitrile peak intensity will depend on the extent of contribution of polar structure (11) to the state of the molecule. In aliphatic compounds, only the inductive effect is exerted on the nitrile group, and it is seen that substituent groups having larger

$$\underset{(10)}{-C\equiv N} \longleftrightarrow \underset{(11)}{-\overset{+}{C}=\overset{-}{N}}$$

electron-attracting (–*I*) effects tend to suppress polarization and diminish the intensity (behavior of FCH_2CN is exceptional). The explanation for positional shifts is not so simple because the mass effect is also participating. In aromatic compounds the *I* and *M* effects are both operating, but the latter is decisive in affecting the intensity of bands. Intensities are larger than those of the aliphatic series. The intensity is increased upon substitution by the +*M* hydroxyl group, whereas it is decreased by the –*M* nitro group. Differences among the ortho-, meta-, and para-isomers are also governed by the efficiency in the transmission of the *M* effect. The α- and γ-carbon atoms of the pyridine nucleus are more positively charged than the β-carbons; this is well reflected in the data on the three pyridine derivatives.

[1] P. Sensi and G. G. Gallo, *Gazz. Chim. Ital.* **85**, 224 (1955).
[2] *idem., ibid.* **85**, 235 (1955).
* In $CHCl_3$, 0.1 cm cell. Intensities measured and expressed according to: D. A. Ramsay, *J. Am. Chem. Soc.* **74**, 72 (1952).

References

[1] L. J. Bellamy, "The Infra-red Spectra of Complex Organic Molecules," 2nd edition, pp. 377–409, Wiley, New York (1958); R. N. Jones and C. Sandorfy in W. West, "Chemical Applications of Spectroscopy," pp. 247–580, Interscience, New York (1956).

[2] M. Ogawa and S. Tanaka, "IR Spectra," Vol. 8, p. 143 (1959).

[3] A. Tolk, *Spectrochim. Acta*, **17**, 511 (1961).

[4] H. W. Thompson and D. J. Jewell, *Spectrochim. Acta* **13**, 254 (1958); Bellemy, p. 380.

[5] Many cases reported, e.g., L. Daasch, *Spectrochim. Acta* **13**, 257 (1958); R. E. Kagarise, *J. Am. Chem. Soc.* **77**, 1377 (1955).

[6] R. A. Nyquist and W. J. Potts, *Spectrochim. Acta* 514 (1959).

[7] E. A. Braude and C. J. Timmons, *J. Chem. Soc.* 3766 (1955).

[8] R. Huisgen, I. Ugi, H. Brade and E. Rauenbusch, *Ann.* **386**, 30 (1954).

[9] N. J. Leonard, D. E. Morrow, and M. T. Rogers, *J. Am. Chem. Soc.* **79**, 5576 (1957).

[10] E. W. Warnhoff and W. C. Wildman, *ibid.* **82**, 1472 (1960).

CHAPTER 4 / EXAMPLE OF ABSORPTION BAND SHIFTS[1]

The information which can be derived from the infrared spectrum can be considerably expanded by measuring the spectrum under varying conditions. The dimedone spectra which follow are an example of the type of information which can be obtained.

Dimedone is a cyclic β-diketone capable of intermolecular conjugate chelation. The β-diketone moiety is found in many natural products and although its very characteristic IR spectra have been studied quite widely[2] there still remain some points to be clarified. However, measuring the IR spectra of dimedone and derivatives under varying conditions provides a means of interpreting rather consistently the complicated band shifts and demonstrates some of the factors that cause these shifts.

The forms that are conceivable for dimedone (5, 5–dimethylcyclohexanedione-1, 3) are the keto, the enol, the dimeric enol, and the polymeric enol forms as shown in Fig. 4. 1. The following assignments are mainly for the carbonyl region; assignments for the lower frequency region are omitted since more detailed studies such as comparison with carboxylic acid dimers are necessary. In any case, considerable coupling of vibrational frequencies will invariably occur in a compact conjugated system such as is present in the associated enol form of dimedone and a one-to-one correspondence between the frequencies and vibrations cannot be obtained.

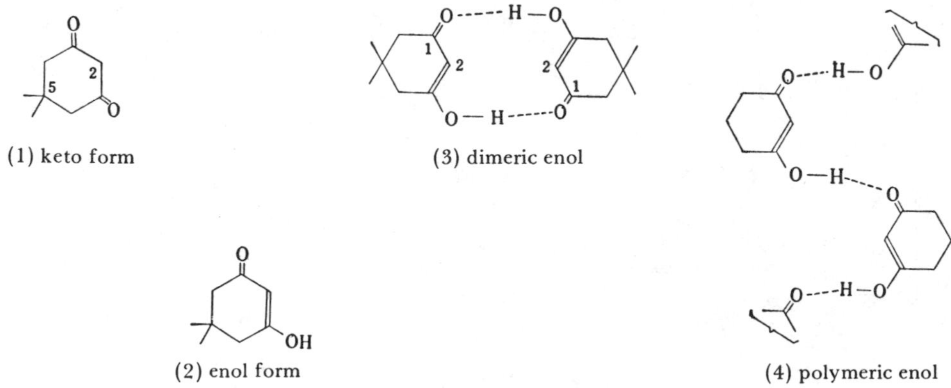

(1) keto form

(3) dimeric enol

(2) enol form

(4) polymeric enol

Fig. 4. 1. Dimedone

(a) 2 mg/600 mg KBr

Dimedone exists as either form 3 or 4 in the solid state, and the results described under (e) show that in solution at least it exists in the dimeric enol form (3).

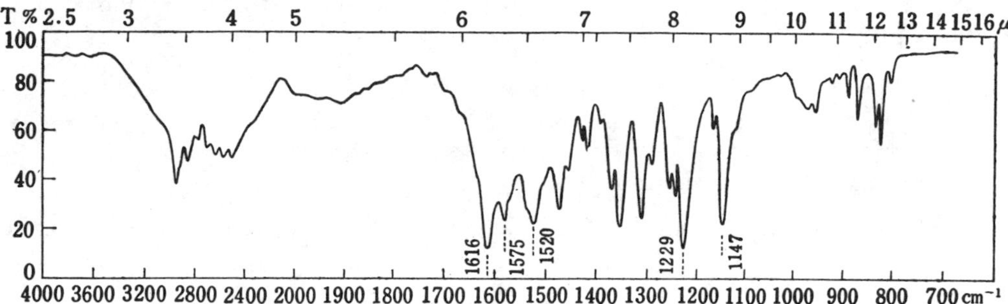

Fig. 4. 2. Dimedone, 2 mg/600 mg KBr

$\simeq$2500 cm^{-1}: The usual 3400~3200 cm^{-1} associated hydroxyl band is shifted to around 2500 cm^{-1} upon formation of extremely strong hydrogen-bonds. When this absorption overlaps with the 3000 cm^{-1} νCH band it may not be detected. With deuterodimedone, which was prepared by dessicating twice a solution of dimedone in heavy water, the $\simeq$2500 cm^{-1} band was replaced by a group of small bands at 2317, 2129, 2046, 1957, and 1916 cm^{-1}.
1616 cm^{-1}: νC=O
1575 cm^{-1}: νC=C
1520 cm^{-1}: Stretching of C_1–C_2?
The three bands 1616, 1575, and 1520 cm^{-1} were not shifted upon deuteration.

(b) **Dimedone in $CHCl_3$, 20mg/ml**

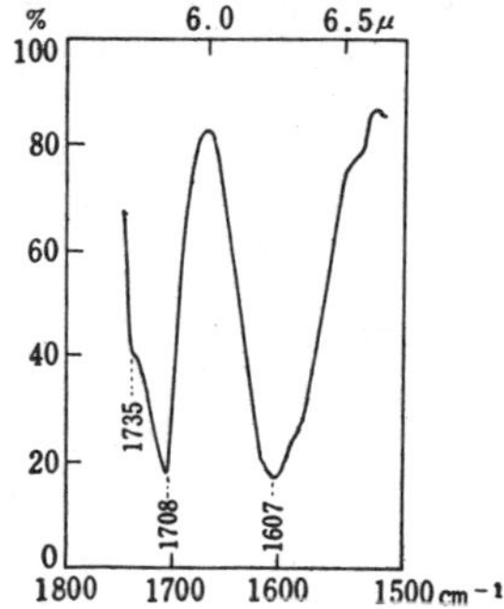

An equilibrium between the keto form (1) and enolic dimer (3) exists in this medium. The chloroform should be freed from ethanol to avoid solvation of the type described in (g). 1735 and 1708 cm^{-1}: νC=O of keto form. 1607 cm: νC=O and νC=C of dimeric enol.

These assignments are consistent with those given in (c) and others.

(c) 2, 2, 5, 5-Tetramethylcyclohexanedione-1, 3, in $CHCl_3$, 20mg/ml

This is fixed in the keto form; the band is doubled due to coupling of the carbonyls to give in-phase and out-of-phase vibrations.

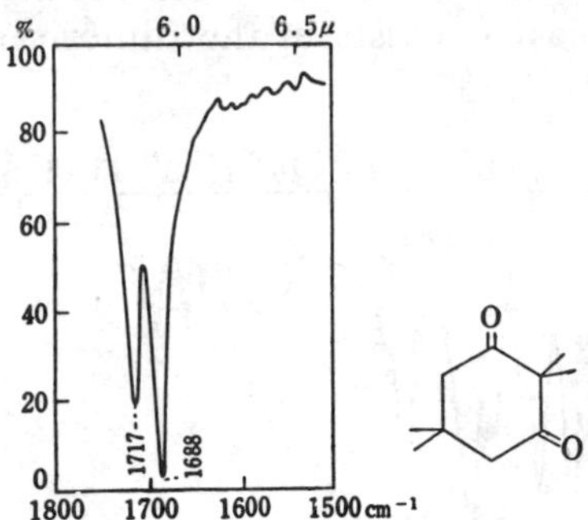

1717 and 1688 cm^{-1}: $\nu C{=}O$ of keto form.

(d) Concentration variation of dimedone in $CHCl_3$

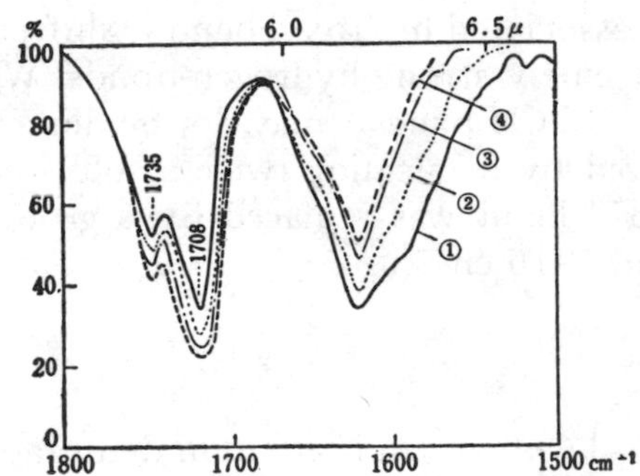

curve 1.	20 mg/ml	0.125 mm cell
curve 2.	10 "	0.250 "
curve 3.	5 "	0.500 "
curve 4.	2.5 "	1.000 "

An isosbestic point is obtained when the concentrations and cell lengths are varied so as to keep the number of light-absorbing molecules constant. The presence of the isosbestic point shows that an equilibrium between only two forms is involved within this range of concentration variation. The fact that the 1735 and 1708 cm^{-1} bands become stronger upon dilution shows that they can be assigned to the less polar form. Thus:

$$\text{(1) keto form} \underset{\text{concn. decrease}}{\overset{\text{concn. increase}}{\rightleftharpoons}} \text{enol form (dimeric or polymeric)}$$

1735 cm^{-1}	1607 cm^{-1}
1708 cm^{-1}	
less polar	more polar

(e) Changes caused by increasing size of C-2 substituent, 20 mg/ml, 0.125 mm cell.

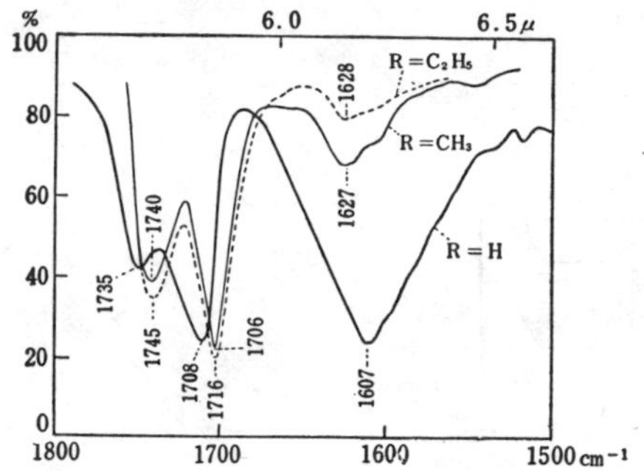

C-2 substituent	keto form	enol form
–H (dimedone)	1735, 1708	1607
$-CH_3$	1740, 1706	1627
$-C_2H_5$	1745, 1716	1628
$-n\text{-}C_3H_7$	1740, 1708	1622

An increase in the size of the C–2 substituent weakens the enolic band. This demonstrates that the associated form of the enol is the dimer (3); that is, an increase in the substituent size causes the equilibrium (1)$\rightleftharpoons$(3) to be shifted towards (1).

(f) Dimedone in tetrahydrofuran, 20 mg/ml, 0.200 mm cell

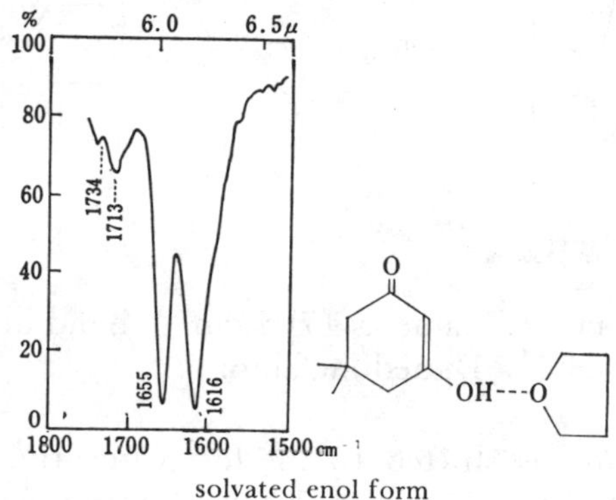

solvated enol form

1734 and 1713 cm^{-1}: νC=O of form (1)
1655 and 1616 cm^{-1}: νC=O and νC=C of solvated enol form (2).

(g) Dimedone, in 20% alcohol and 80% chloroform mixture

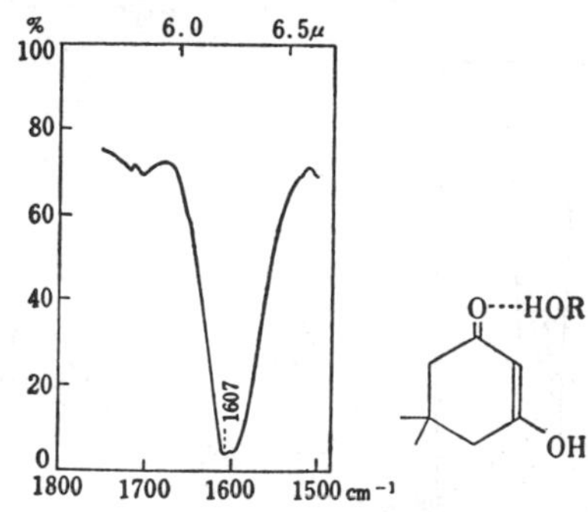

1607 cm^{-1}: νC=O and νC=C, same as in associated enol form described in (b).

(h) Dimedone sodium salt, nujol

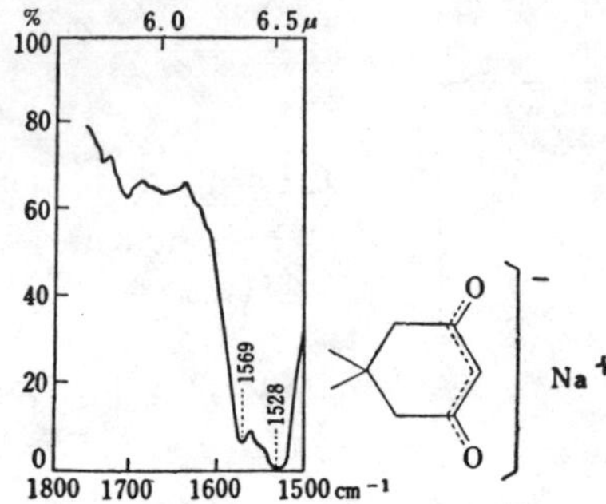

(i) Triethylamine added to 2 mg/ml chloroform solution of dimedone

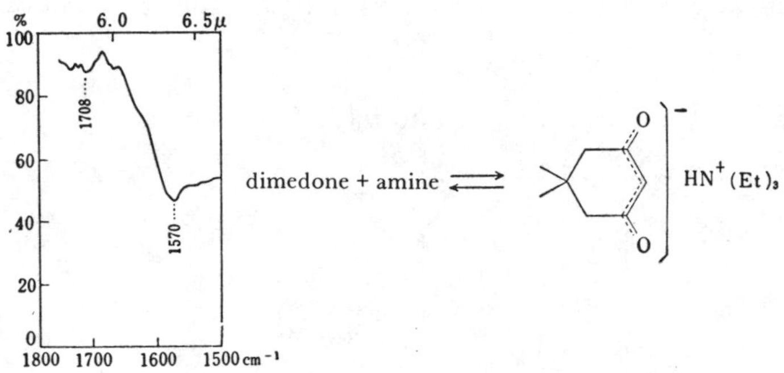

1708 cm^{-1}: Same as 1708 cm^{-1} band in (b).
1570 cm^{-1}: Dimedone anion.

The equilibrium was shifted to the left when the dimedone concentration was increased; thus, the 1570 cm^{-1} band became weaker, the 1708 cm^{-1} band stronger, and two bands at 1735 and 1607 cm^{-1} (corresponding to those in Fig. b) started to appear.

(j) Dimedone ethyl ether in $CHCl_3$, 20 mg/ml, 0.200 mm cell

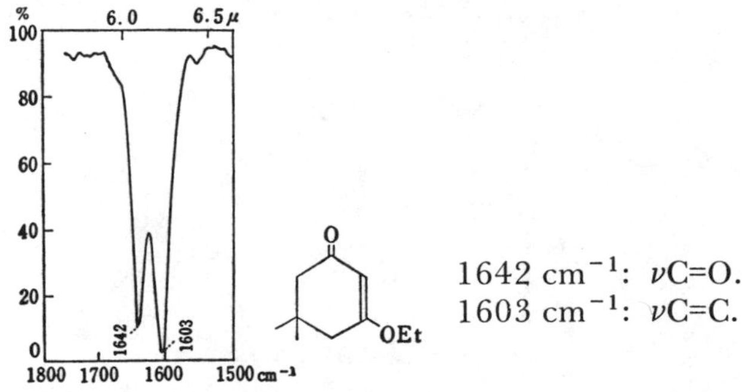

1642 cm^{-1}: νC=O.
1603 cm^{-1}: νC=C.

The absorptions for saturated esters and ketones are at 1735 and 1715 cm^{-1}, respectively. This indicates that the ester carbonyl has a greater double bond character and can be explained by assuming that the $+M$ effect of the oxygen atom is smaller than its $-I$ effect:

$$-\mathrm{O} \leftarrow \mathrm{C}=\mathrm{O}$$
$$-I > +M$$

Dimedone ethyl ether is a vinylogous ester and the fact that the carbonyl absorption is at 1642 cm^{-1} shows that only the $+M$ effect is important, in the present case:

$$-\mathrm{O}-\mathrm{C}=\mathrm{C}-\mathrm{C}=\mathrm{O}$$
only $+M$

Both electric effects, the M and I, should be considered when the static state of the sample is involved, as in IR spectroscopy and other physico-chemical measurements, but the present example makes it clear that the I effect becomes negligibly small as compared to the M effect in ordinary vinylogous systems. Thus a difference exists between the carbonyl absorptions of an ester and a β-alkoxy-α, β-unsaturated carbonyl group.

(k) Formalin dimedone in $CHCl_3$,

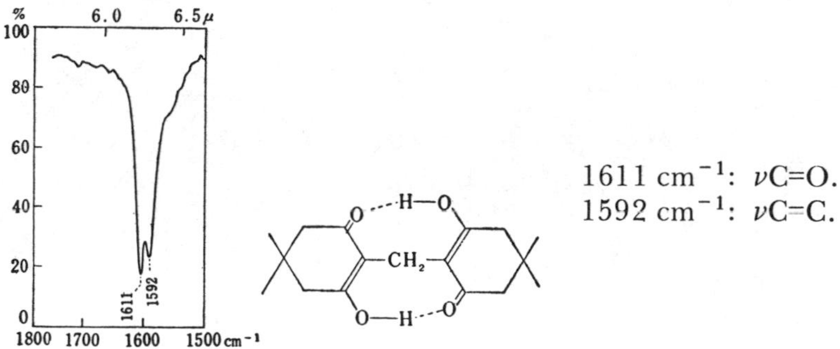

Positions are comparable to those of the νC=O and νC=C bands in (a) and (b).

(l) Propionaldehyde dimedone anhydride (nujol)

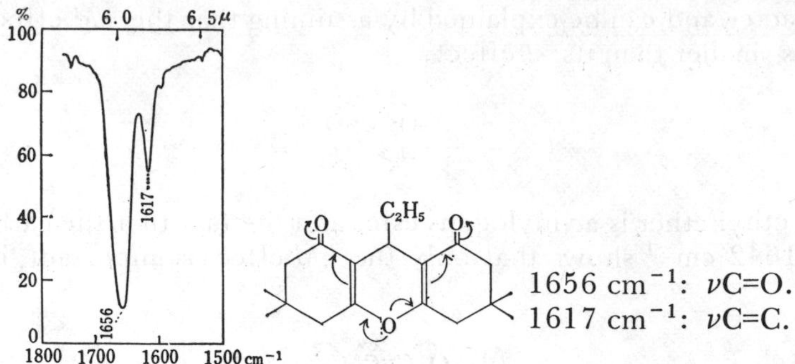

The lone pair electrons of the oxygen are now shared by two α, β-unsaturated ketone systems, and this would cause the νC=O to be at a higher frequency than in (j).

References

[1] K. Nakanishi, T. Goto, and S. Eguchi, 10th Annual Meeting of Chem. Soc. Japan, April, 1957.

[2] Bellamy, p. 142.

CHAPTER 5 LASAR RAMAN SPECTROSCOPY

5.1 INTRODUCTION

The Raman effect is due to a light scattering process resulting from collision of photons of energy $h\nu_0$ with a molecule. Of the various molecular vibrations, those which strongly absorb the infrared light and hence give rise to strong infrared bands give rise to only weak Raman bands; in contrast, molecular vibrations which produce strong Raman bands give rise to only weak infrared bands. The two methods are thus complementary, and a complete picture of vibrational frequencies can be secured only by use of both spectroscopic methods.

Most of the early data on molecular vibrations was provided by Raman spectroscopy, which was theoretically predicted in 1923 by Smekal[1] and experimentally observed in 1928 by Raman.[2] Until the mid-1960's when the laser became the common light source, the conventional exciting line for "classical" Raman spectroscopy was the 439 nm Hg emission. This limited the sample to those which do not absorb in this region, since otherwise both the incident and scattered light are absorbed. Also, because of the extremely weak intensity of the scattered light, the sample had to be non-fluorescent; fluorescence, which is an emission process, interferes with measurements of the scattered light. In addition, due to low intensities, gram quantities of samples either as a liquid or solution were required. Thus when commercial infrared spectrophotometers were introduced in the late 1940's, Raman spectroscopy became a largely neglected tool by organic chemists. This picture was rapidly changed, however, when laser sources became available for Raman studies in the mid-1960's, and it is now possible to record Raman spectra as rapidly as infrared spectra and with less than mg quantities.

A molecule can be considered as being an assembly of positively charged nuclei and negatively charged electrons. When a monochromatic laser beam of high frequency (of the order of 10^{15} sec^{-1}) strikes a molecule, it interacts much more strongly with the electrons since, being much lighter, the force of the laser field can modify their motions more easily. As a result, the electrons start oscillating somewhat at the frequency of the incident laser field, but the heavier nuclei do not. The incident light thus leads to a fluctuating change in the position of the electron cloud relative to the nucleus, i.e., an oscillating dipole is induced in the molecule, which is now polarized. This assembly of oscillating dipoles scatters light in all directions at the laser frequency and this process is called Rayleigh Scattering. Rayleigh Scattering is viewed as an **elastic collision** between the molecule and the light photon, since the photon neither loses nor gains energy and hence its energy is still $h\nu_0$. This is by far the strongest of scattered light phenomenon from molecular systems (Fig. 5.1).

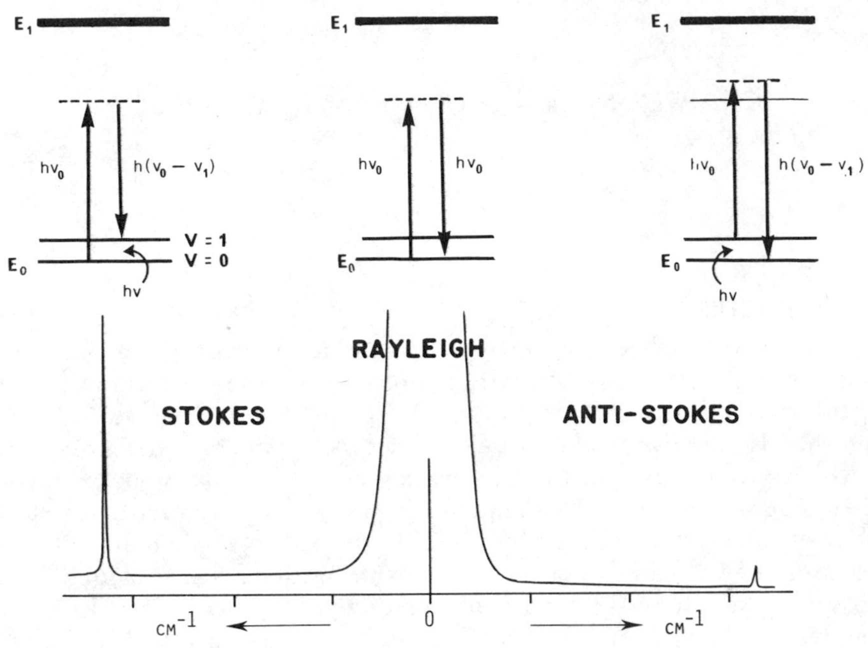

Fig. 5.1 Rayleigh, Stokes and Anti-Stokes Scattering.

The second type of collision is **inelastic collision** where the photon either loses energy to or gains energy from the molecule. The energy of the scattered light is $h(\nu_0 - \nu_1)$ or $h(\nu_0 + \nu_1)$ (Fig. 5.1), the lost or gained energy $h\nu_1$ corresponds to the vibrational energy. Thus the energy of scattered light depends on the frequency of incident light, but the displacement $h\nu_1$ from the Rayleigh line is a constant corresponding to the vibrational level. This type of scattering process arises from the fact that the electrons are coupled to the nuclei as well as to the incident laser field so that the vibrational frequency of the nuclei is superimposed upon the electronic motions. Thus the molecular electrons are oscillating to some degree at the laser frequency plus and minus the vibrational frequencies, again creating an oscillating dipole moment.

A record of vibrational levels as measured by displacements from the incident frequency, namely the record of the inelastic collisions, is the **Raman spectrum**. The Stokes bands* on the low frequency side of the Rayleigh line have intensities typically of the order of 10^{-5} of the Rayleigh line because the electronic-nuclei interaction is weak; the Stokes bands are of higher intensity than the higher-frequency anti-Stokes bands. The higher intensity of the Stokes line is due to the higher population of molecules in the ground vibrational state

* The Raman bands are frequently referred to as "lines." This is a reflection of earlier days when instead of using photomultipliers, the spectra were recorded photographically.

ν_0 as compared to those in the excited ν_1 level (Boltzmann distribution).

The intensity of an infrared band is proportional to the change in the **dipole moment** as atoms pass through their equilibrium positions. On the other hand, the **intensity of a Raman band** is governed by the change in **polarizability** as atoms pass through their equilibrium positions. The polarizability can be considered to be a measure of how readily electron clouds are displaced by nuclear motion. Therefore, the Raman intensity is dependent on how different the shapes of electron clouds are before and after equilibrium.

This is exemplified by the linear triatomic molecule, carbon disulfide, which has 3n–5 = 4 modes of vibrations, ν_1, ν_2 and ν_3 (doubly degenerate), shown in Fig. 5.2. All vibrations of a CS_2 molecule can be regarded as a com-

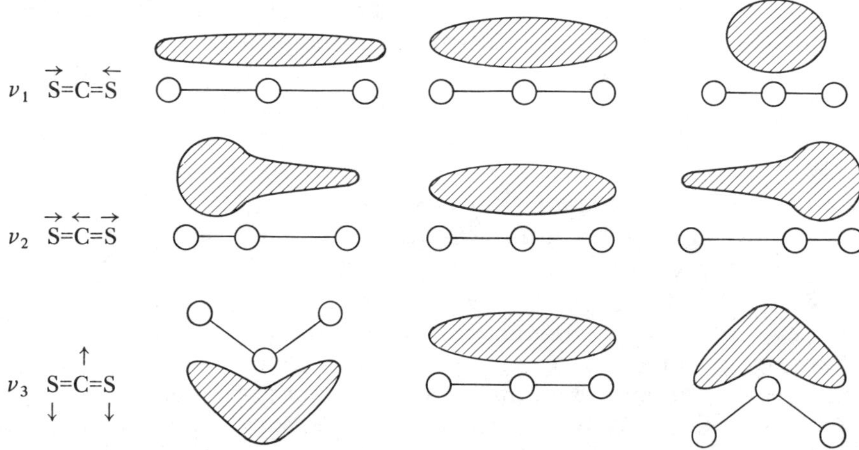

Fig. 5.2 Vibrations of carbon disulfide and accompanying changes in polarizability. Bending vibration ν_3 is degenerate, the other ν_4 being perpendicular to the paper plane.

posite of these four fundamental vibrations. There is no change in the dipole moment in the symmetric stretch ν_1 and hence this is infrared inactive. However, the directions of dipole moments are reversed depending on whether the electron cloud is to the right or left of the central carbon (asymmetric stretch ν_2), or whether it is bent upwards or downwards (bend ν_3), or above or below the paper plane (bend ν_4); vibrations ν_2 and ν_3 are accordingly infrared active.

This picture is different in Raman spectroscopy. In ν_1, the shapes of electron clouds are different depending on whether the molecule is stretched or shrunken; ν_1 therefore gives a Raman line. In ν_2, ν_3 and ν_4, the electron cloud shapes are identical before and after the vibrations pass through their equilibrium state. They are thus Raman inactive; that is, the infrared and Raman bands are complementary. Similar tendencies are displayed in the more common non-centrosymmetric molecules although not as rigidly as in the simple case of carbon disulfide.

In measurements of Raman spectra, the laser beam is introduced into the sample cell and the scattered light is usually observed at an angle of 90° to the laser beam. If an analyzer is inserted between the cell and monochromator slit, since the laser beam is polarized, the intensity of the recorded Raman bands may differ depending on whether the analyzer is oriented parallel or perpendicular to the direction of polarization of the laser beam. The ratio of the light intensity when the analyzer is perpendicular to that when it is parallel ($I\perp/I/\!/$) is called the depolarization ratio ρ, the maximum value of which is 0.75. Vibrations for which ρ is 0–0.75 are called polarized and are caused by symmetric vibrations; on the other hand, those for which ρ is 0.75 are called depolarized and are caused by nontotally symmetric and degenerate vibrations. Measurements of depolarization ratios are useful for assigning bands to particular vibrational modes and in separation of overlapping bands.

Some of the general characteristics of Raman spectroscopy are listed below.

1. As mentioned above, symmetric and pseudosymmetric vibrations give rise to strong Raman bands, e.g., the symmetrically substituted S–S, C=C, N=N, C≡C. The intensity increases in going from single to double to triple bonds because the bonds contain more electrons that can be modulated.

2. The usually weak or variable intensity bands in the infrared spectrum due to C≡N, C=S, S–H stretch are strong in the Raman. The symmetric C–O–O–C stretch at ca. 880 cm^{-1} is also strong.

3. The symmetric breathing vibration of cyclic compounds in which all bonds forming the cyclic skeleton stretch simultaneously are frequently the strongest Raman band.

4. The symmetric stretches of bonds such as X=Y=Z, C=N=C and $O=C=O^-$ are strong in the Raman but weak in the infrared. Conversely, the asymmetric stretches are weak in the Raman but strong in the infrared.

5. The C–C stretches are strong in the Raman, but as may be anticipated, they are extensively coupled.

6. The Raman spectra of alcohols and alkanes are similar. This is due to the fact that: (i) the force constant or bond strength of the C–O bond does not differ much from the C–C bond; (ii) the mass of the hydroxyl group only differs by two from the methyl group; and (iii) the O–H band is weak compared to the C–H and N–H bands.

Some of the advantageous points of Raman over infrared spectroscopy are listed below.

1. Unlike infrared spectra, the range 40~4,000 cm^{-1} can be covered routinely.

2. Samples contained in bottles, capillaries, etc., can be measured directly because glass is completely transparent to Raman scattering.

3. Water is an excellent solvent because its Raman spectrum is generally very weak (unlike infrared).

4. Solids can be measured directly and need not be mulled or pressed into KBr disks. However, scorching of samples may occur due to the high intensity of the laser beam; hence, samples should be examined for deterioration.

5. The light from the laser beam and the Raman scattering are polarized so that one can carry out depolarization measurements by having the analyzer placed before the monochromator entrance slit horizontally or perpendicularly.

The intensity of a Raman band due to a totally symmetric vibration has drastically different values, depending on the polarizer orientation, whereas that of non-totally symmetric vibrations does not.

6. The Raman spectrum of a chromophore can be recorded in the presence of a nonchromophoric moiety by the technique of **resonance laser Raman spectroscopy**. Here the incident laser field is strongly coupled to the electrons of the chromophore since the laser frequency and the characteristic frequency of the electronic motion of the chromophore are equal (or nearly so), i.e., the resonance condition. Thus the laser can very strongly drive the electronic motion, and the Raman scattering process is strongly increased (by 10^5 more); in fact, the laser field and the electrons are so strongly coupled that significant energy can be lost from the laser to the molecule, i.e., the chromphore is absorbing. Since the nonchromophoric moiety is not in resonance, Raman scattering from this is normally small; the Raman spectrum of the chromophore is therefore dominant.

Spectra of some common solvents appear below, followed by annotated spectra of some representative compounds.

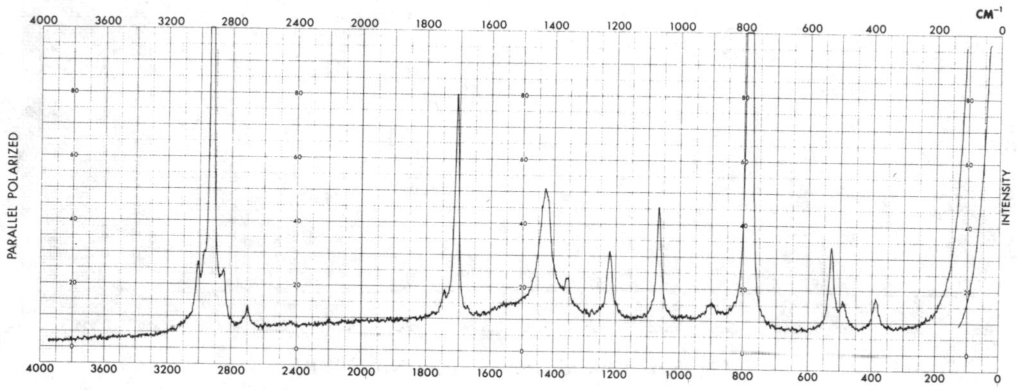

Fig. 5.3 Acetone (Sadtler No. 162R)

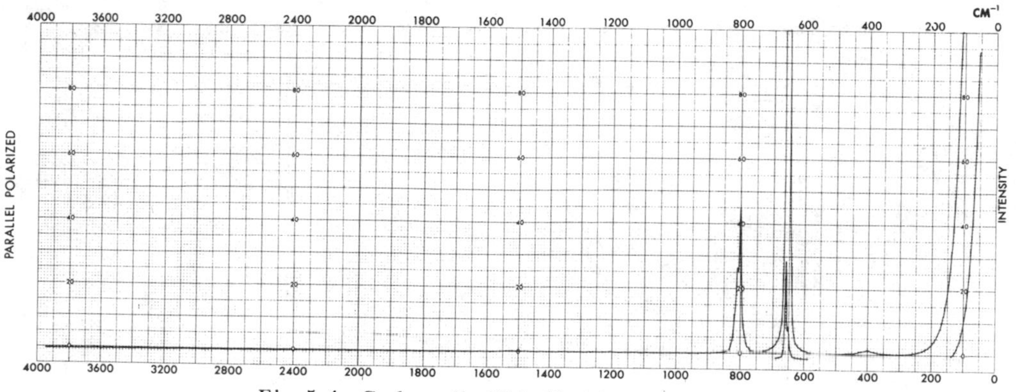

Fig. 5.4 Carbon disulfide (Sadtler No. 142R)

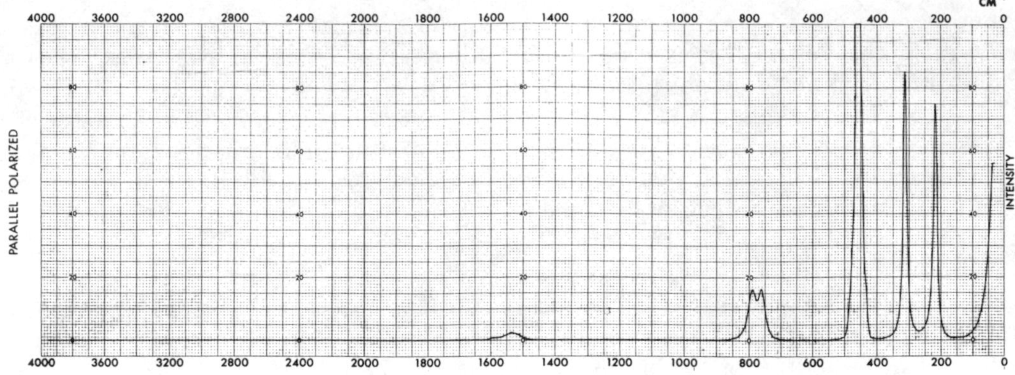

Fig. 5.5 Carbon tetrachloride (Sadtler No. 307R)

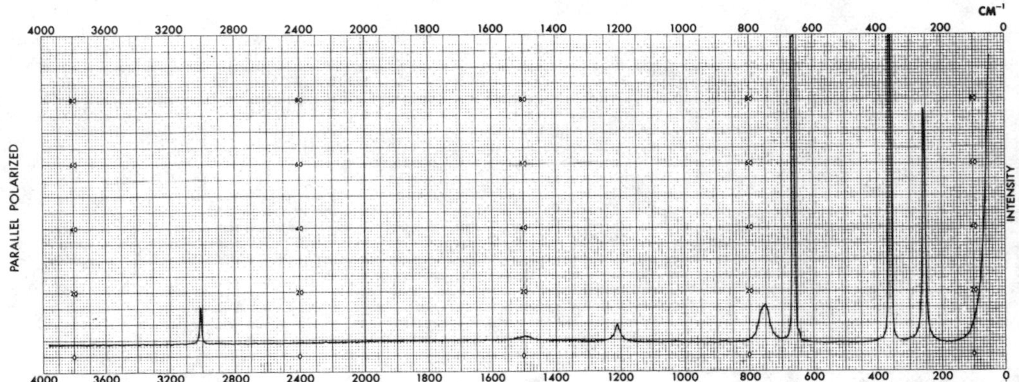

Fig. 5.6 Chloroform (Sadtler No. 308R)

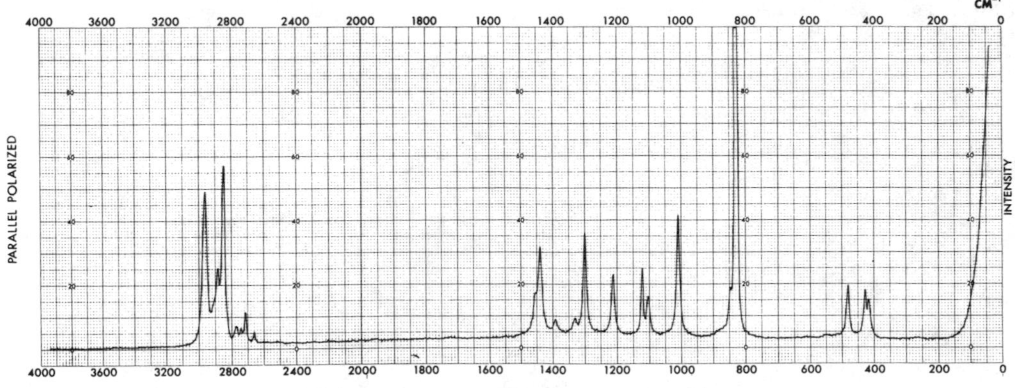

Fig. 5.7 Dioxane (Sadtler No. 704R)

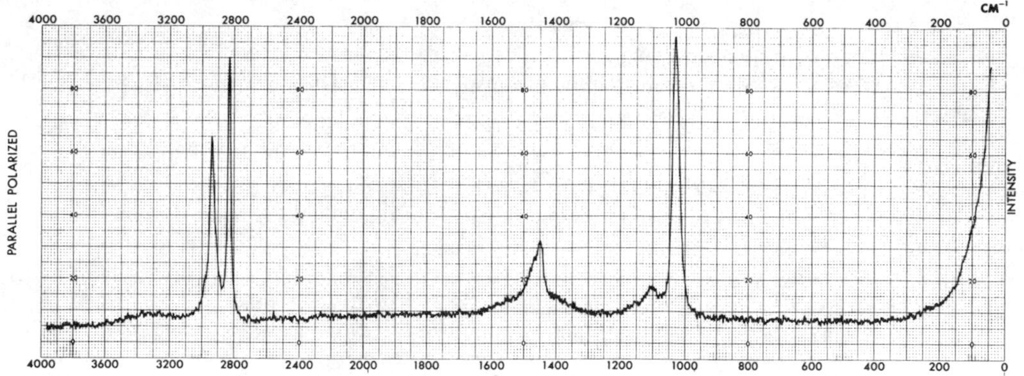

Fig. 5.8 Methyl Alcohol (Sadtler No. 176R)

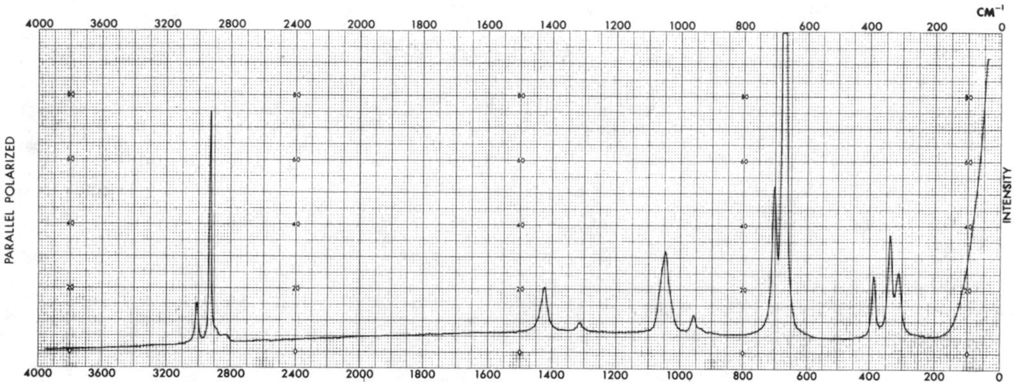

Fig. 5.9 Dimethyl Sulfoxide (Sadtler No. 2649R)

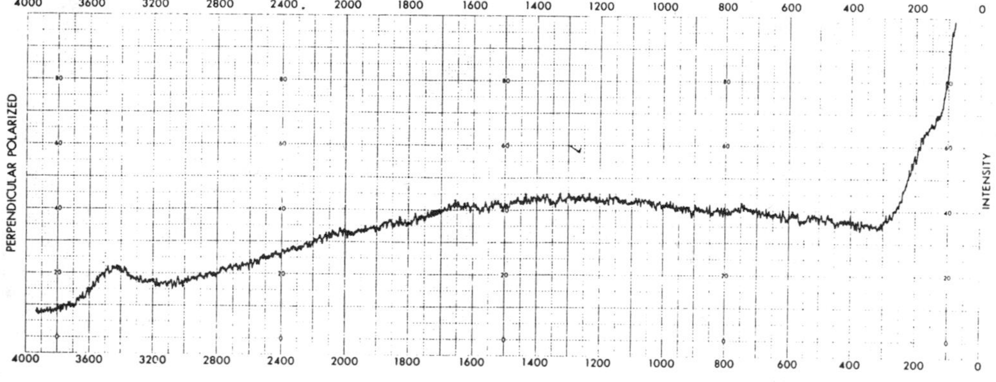

Fig. 5.10 Water (Sadtler No. 143R)

References

[1] A. Smekal, Naturwiss., *11*, 873 (1923).

[2] C. V. Raman and K. S. Krishnan, Nature, *121*, 501 (1928); Raman was awarded the 1930 Nobel Prize in Physics for this discovery.

General References

Most references which deal with infrared spectroscopy also describe Raman spectroscopy. However, some of the recent references which deal particularly with Raman spectroscopy are the following.

[1] R. S. Tobias, "Raman Spectroscopy in Inorganic Chemistry," *J. Chem. Educ.*, 44, 2, 71 (1967).

[2] H. A. Szymanski, ed., "Raman Spectroscopy," Plenum Press, New York, 1967.

[3] P. J. Hendra and P. M. Stratton, "Laser-Raman Spectroscopy," *Chem. Rev.*, 69, 325 (1969).

[4] J. Loader, "Basic Laser Raman Spectroscopy," Heyden and Sons Ltd., London, and Sadtler Research Laboratories, Philadelphia, 1970. A concise book mainly dealing with experimental aspects.

[5] T. R. Gilson and P. J. Hendra, "Laser Raman Spectroscopy," Wiley-Interscience, New York, 1970.

[6] M. C. Tobin, "Laser Raman Spectroscopy," Vol. 35 in "Chemical Analysis," P. J. Elving and I. M. Kolthoff, ed., Wiley-Interscience, New York, 1971.

[7] F. R. Dollish, W. G. Fateley and F. F. Bentley, "Characteristic Raman Frequencies of Organic Compounds," Wiley-Interscience, New York, 1974.

[8] J. B. Lambert, H. F. Shurvell, L. Verbit, R. G. Cooks and G. H. Stout, "Organic Structural Analysis," Macmillan Publishing Co., Inc., New York, 1976.

5.2 EXAMPLES OF RAMAN SPECTRA

Example 1/ Cyclohexane, cyclohexanol, cyclohexyl amine

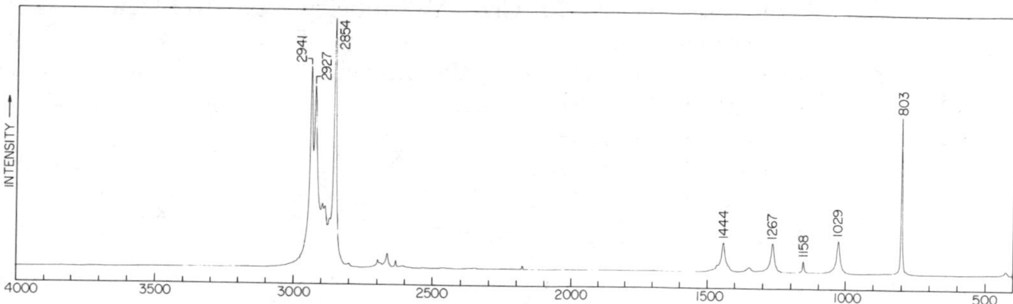

Fig. 5.11 Cyclohexane, neat liquid

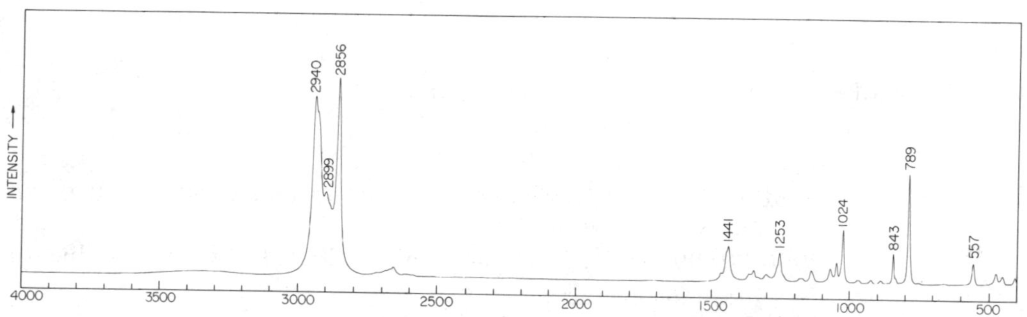

Fig. 5.12 Cyclohexanol, neat liquid

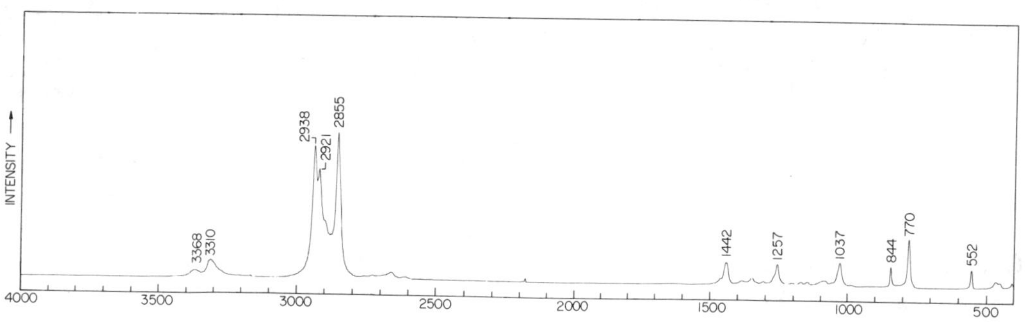

Fig. 5.13 Cyclohexyl amine, neat liquid

The Raman spectra of aliphatic hydrocarbons and their hydroxy and amino substituted derivatives are all very similar. This is due to the fact that although the Raman spectra show characteristic frequencies for hydroxyl and amino groups, the intensities are low, especially in comparison to the strong bands arising from the symmetrical ring breathing and other skeletal vibrations.

Cyclohexane

2941, 2927 cm^{-1}: ν_{as} CH_2. Multiplicity is due to lifted degeneracy.
2854 cm^{-1}: ν_s CH_2. Note the intensity.
1444 cm^{-1}: CH_2 scissoring.
1267 cm^{-1}: CH_2 twist.
1158 cm^{-1}: CH_2 rock.
1029 cm^{-1}: ν C–C.
803 cm^{-1}: Ring breathing.

Cyclohexanol

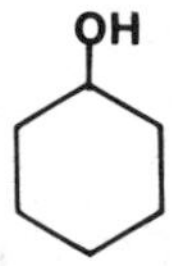

~3500 cm^{-1}: ν OH; band is so broad and weak that it often escapes detection.
2940, 2899 cm^{-1}: ν_{as} CH_2 and ν CH. Multiplicity is due to lifted degeneracy.
2856 cm^{-1}: ν_s CH_2. Note the intensity.
1441 cm^{-1}: CH_2 scissoring.
1024 cm^{-1}: ν C–C.
789 cm^{-1}: Ring breathing.

Cyclohexyl amine

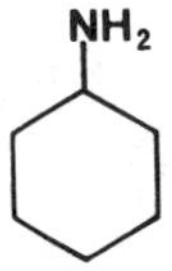

3368, 3310 cm^{-1}: ν_{as}, ν_s, NH_2.
2938, 2921 cm^{-1}: ν_{as} CH_2 and ν CH. Multiplicity is due to lifted degeneracy.
2855 cm^{-1}: ν_s CH_2. Note the intensity.
1442 cm^{-1}: CH_2 scissoring.
1037 cm^{-1}: ν C–C.
770 cm^{-1}: Ring breathing.

Example 2/ 2,3-Dimethyl-2-butene

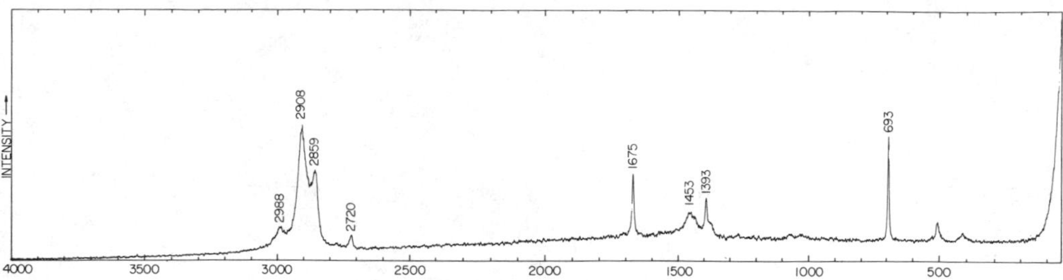

Fig. 5.14 2,3-Dimethyl-2-butene, neat liquid

CH_3 CH_3 / CH_3 CH_3

2908, 2859 cm^{-1}: Methyl ν CH.

1675 cm^{-1}: ν C=C. Tetrasubstituted double bond is clearly visible, unlike the infrared case.

1453 cm^{-1}: δ_{as} CH_3.

1393 cm^{-1}: $\delta_s CH_3$.

693 cm^{-1}: ν C–C.

Example 3/ Anisole, Thioanisole, N,N-Dimethylaniline

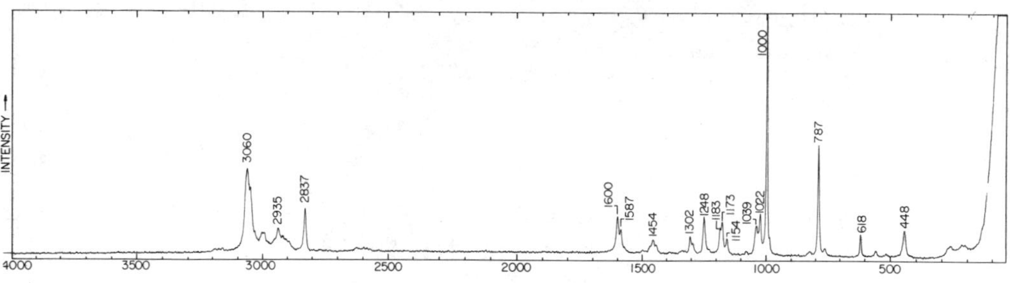

Fig. 5.15 Anisole, neat liquid

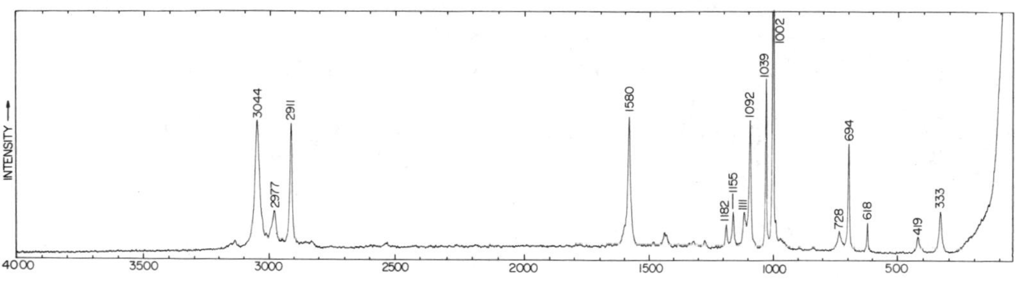

Fig. 5.16 Thioanisole, neat liquid

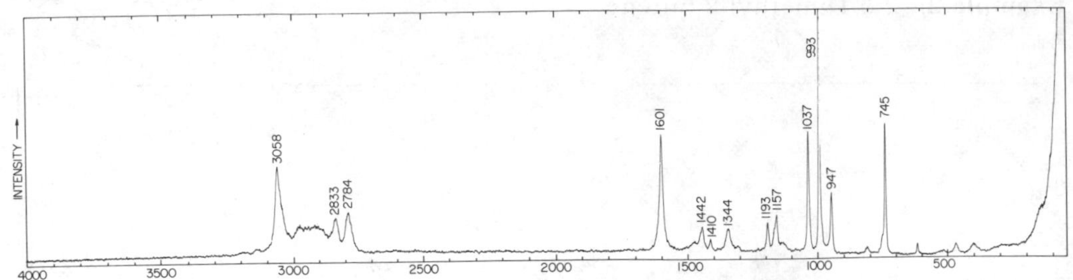

Fig. 5.17 N,N-Dimethylaniline, neat liquid

Characteristic frequencies of aromatic compounds are quite similar in Raman and infrared spectra. Raman spectra typically show absorptions:

1) Above 3000 cm^{-1} (3100~3030 cm^{-1}): ν CH.
2) 1610~1590 cm^{-1}: Phenyl ring, often appearing as a doublet.
3) 1200~1000 cm^{-1}: A number of intense bands due to in-plane δ CH and ring breathing vibrations.
4) 1000~450 cm^{-1}: A number of bands due to out-of-plane δ CH; usually more variable and less intense than infrared bands.

For monosubstituted benzene derivatives, the very intense band at 1010~990 cm^{-1} due to trigonal ring breathing ("Star of David" vibration) is usually the strongest band in the spectrum. This band near 1000 cm^{-1} is also intense in the spectra of 1,3 and 1,3,5-substituted phenyl compounds. A medium to strong absorption at 1030~1015 cm^{-1} due to in-plane CH deformation serves to distinguish mono- from 1,3-disubstituted derivatives, and further corroboration of the substitution pattern is supplied by a medium intensity band of rather variable frequency at 825~675 cm^{-1}. The δ CH_3 band is usually very weak in Raman spectra (see Thioanisole).

Anisole

OCH_3

Note the absence of ν C–O which is a prominent feature of the infrared spectrum (1251 cm^{-1}; see p. 189).

2837 cm^{-1}: ν_s CH_3; shifted to lower frequency by attachment to O–Ph.

1600, 1587 cm^{-1}: Phenyl ring; this doublet is frequently seen in substituted aromatics.

1454 cm^{-1}: δ_s CH_3, shifted by attachment to oxygen (1460~1430 cm^{-1}).

1183, 1173, 1039, 1022 cm^{-1} in-plane δ CH; 1039, 1022 cm^{-1} typical of monosubstituted aromatic (1030~1015 cm^{-1}).

1000 cm^{-1}: Trigonal ring breathing; highly diagnostic.

787 cm^{-1}: Ring deformation (825~675 cm^{-1}).

Thioanisole

While ν C–S (735~590 cm^{-1}) is usually prominent in the Raman spectrum, it is absent in the spectra of aromatic compounds, due to coupling of ν C–S with ring stretching modes.

2911 cm^{-1}: ν_s CH_3; slightly shifted to lower frequency, due to S–Ph attachment.
1580 cm^{-1}: Phenyl ring.
1092 cm^{-1}: In-plane δ CH; usually very weak.
1039 cm^{-1}: In-plane δ CH; typical of monosubstituted phenyl.
1002 cm^{-1}: Trigonal ring breathing.
694 cm^{-1}: Ring deformation mode (825~675 cm^{-1}).

N,N-Dimethylaniline

Note the absence of ν C–N which has a very intense absorption in the infrared spectrum (1350 cm^{-1}, see p. 189).

2784 cm^{-1}: ν_s CH_3; shifted to lower frequency due to N–Ph.
1601 cm^{-1}: Phenyl ring.
1442 cm^{-1}: δ_s NCH_3 (1440~1410 cm^{-1}).
1193 cm^{-1}: In-plane δ CH.
1037 cm^{-1}: In-plane δ CH; characteristic of monosubstituted phenyl.
993 cm^{-1}: Trigonal ring breathing; intense and polarized.
745 cm^{-1}: Ring deformation (825~675 cm^{-1}).

Example 4/ *o*-, *m*-, *p*-Xylene

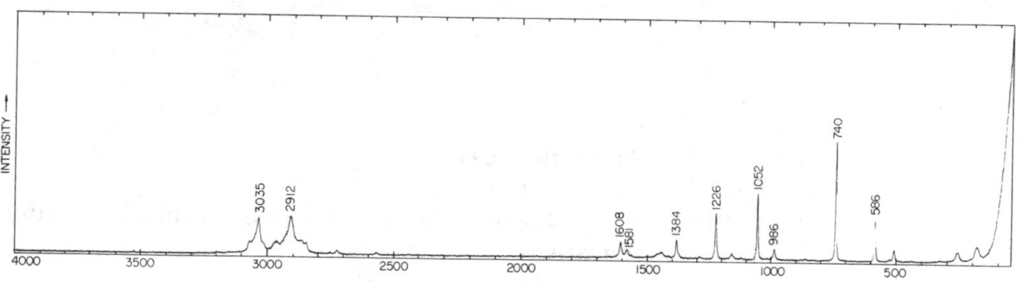

Fig. 5.18 *o*-Xylene, neat liquid

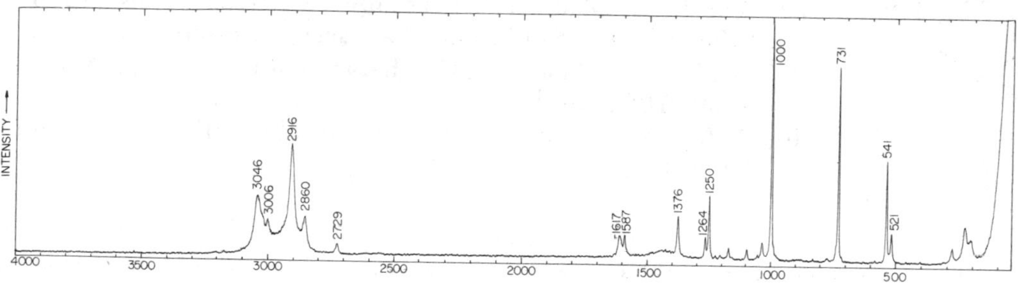

Fig. 5.19 *m*-Xylene, neat liquid

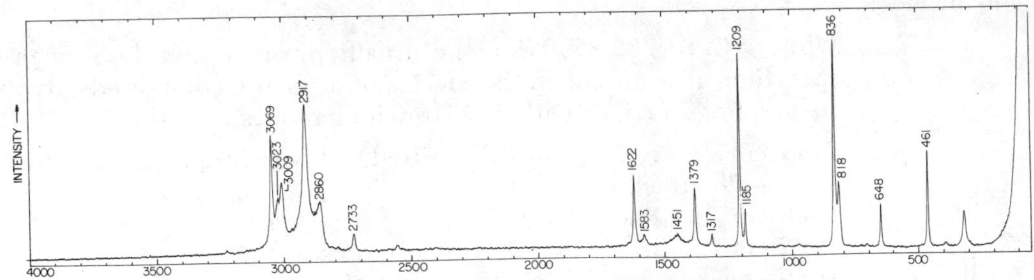

Fig. 5.20 *p*-Xylene, neat liquid

In addition to the characteristic bands above 3000 cm^{-1} (aromatic ν CH) and at 1610~1590 cm^{-1} (phenyl ring), the spectra of substituted benzene derivatives show absorption in the lower frequency range which is typical of the substitution type.* The bands for *o*-, *m*-, *p*-disubstition are shown in Fig. 5.21.

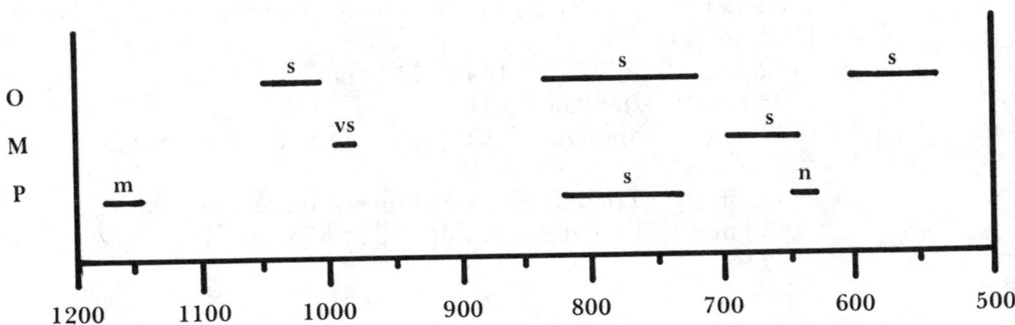

Fig. 5.21 Characteristic Raman bands for disubstituted benzene derivatives.

o-Xylene

3035 cm^{-1}: Aromatic ν CH.
2912 cm^{-1}: Methyl ν CH.
1608, 1581 cm^{-1}: Phenyl; shows the doublet typical of many benzene derivatives.
1451 cm^{-1}: Phenyl ring and δ_{as} CH_3.
1384 cm^{-1}: δ_s CH_3.
1226 cm^{-1}: In-plane δ CH; usually two weak bands (1295~1250 and 1170~1150 cm^{-1}) are seen for *o*-substitution.
1052 cm^{-1}: In-plane δ CH; characteristic of *o*-substitution (1060~1020 cm^{-1}).
740, 586 cm^{-1}: Characteristic of *o*-substitution (740~650, 600~540 cm^{-1}).

* For a detailed treatment of substituted benzenes see G. Varsanyi, "Vibrational Spectra of Benzene Derivatives," Academic Press, New York, 1969.

m-Xylene

3046, 3006 cm^{-1}: Aromatic ν CH.
2916, 2860 cm^{-1}: Methyl ν CH.
1617, 1587 cm^{-1}: Phenyl ring.
1376 cm^{-1}: δ_s CH_3.
1264, 1250 cm^{-1}: In-plane δ CH; characteristic of *m*-substitution.
1000 cm^{-1}: Trigonal ring breathing; always very intense and polarized (1010~990 cm^{-1}).
737 cm^{-1}: Characteristic of *m*-substitution (740~640 cm^{-1}).
541 cm^{-1}: Common to *m*-dialkylbenzenes.

p-Xylene

3069, 3023, 3009 cm^{-1}: Aromatic ν CH.
2917, 2860 cm^{-1}: Methyl ν CH.
1622, 1583 cm^{-1}: Phenyl ring.
1379 cm^{-1}: δ_s CH_3.
1209 cm^{-1}: This unusually strong band is characteristic of *p*-dialkyl benzenes.
1185 cm^{-1}: In-plane δ CH; characteristic of *p*-substitution (1180~1150 cm^{-1}).
836, 648 cm^{-1}: Characteristic of *p*-substitution (830~730, 650~630 cm^{-1}). The high frequency band is usually very intense.

Example 5/ L-Alanine

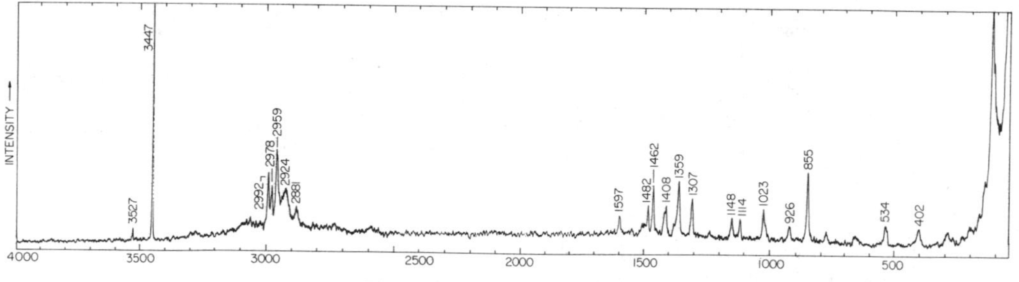

Fig. 5.22 L-Alanine, crystal

NH_3^+
$CH_3CH_2CH\text{-}COO^-$

The molecule exists in the zwitterionic form.

3447 cm^{-1}: Possibly due to water of hydration.
2992, 2978, 2959, 2924, 2881 cm^{-1}: ν_{as} and ν_s of CH_3, CH and NH_3^+. Intensity of NH_3^+ stretch is generally less than that of CH_3 etc.
1597 cm^{-1}: ν_{as} COO^-.
1462 cm^{-1}: δ_{as} CH_3.
1408 cm^{-1}: ν_s COO^-.
1359 cm^{-1}: δ_s CH_3.
1023 cm^{-1}: ν C–N.
855 cm^{-1}: ν_s C–C–N.

Example 6/ n-Butyl disulfide

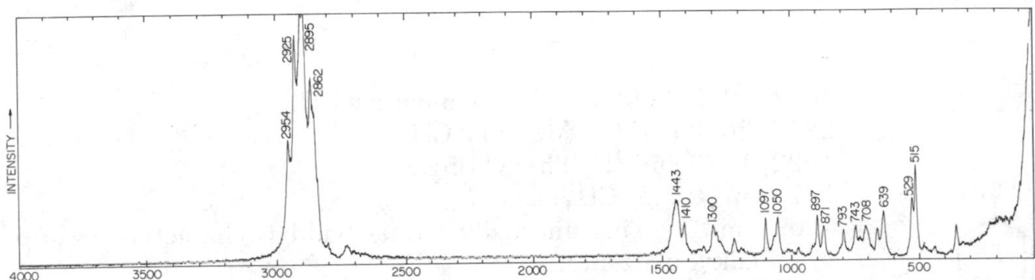

Fig. 5.23 n-Butyl disulfide, neat liquid

The Raman spectra of sulfur containing compounds are generally more informative than infrared spectra. In this case, in addition to the usual alkyl vibrations the C–S and S–S stretch are clearly visible.

$(CH_3CH_2CH_2CH_2S\text{-})_2$

708, 639 cm^{-1}: ν C–S. Generally one or more polarized bands appear in the region 705~570 cm^{-1} (medium to strong).
529, 515 cm^{-1}: ν S–S. Two polarized bands generally appear in the narrow range 525~510 cm^{-1} (medium intensity).

Example 7/ Molecular flow resonance Raman spectra of 11-cis retinal and rhodopsin[1]

It is well-known that in the visual pigment rhodopsin, the chromophoric moiety is 11-cis retinal which is linked via a protonated Schiff base to the ϵ-amino group of lysine in opsin, and that action of light triggers a series of tranformations, the end products being all-trans retinal (λ_{max} 380 nm) and the apoprotein opsin (λ_{max} 270 nm, molecular weight ca. 40,000) (Fig. 5.24).

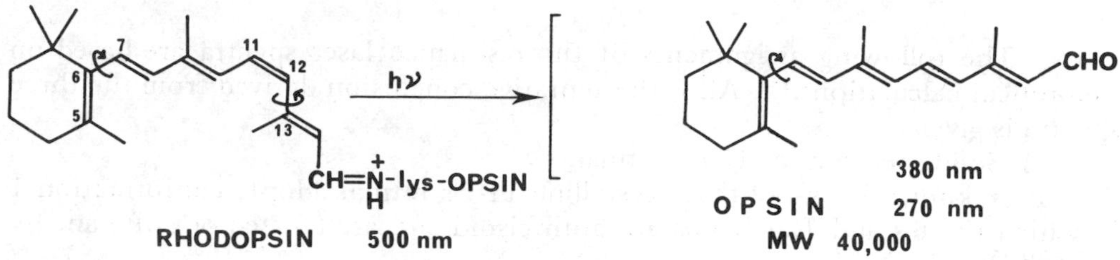

Fig. 5.24 Rhodopsin and its bleaching to trans retinal and opsin. The bent arrows denote that the single bonds are nonplanar and twisted.

One of the many unanswered questions in the area of visual pigments concerns the conformation of the 12-single bond in the 11-cis retinal moiety in rhodopsin, i.e., is it 12-s-cisoid 1 or 12-s-transoid 2? Through studies of rhodopsins derived from the model 14-methylretinal, it was previously suggested that the conformation is 12-s-trans.[2] The Raman data shown in Fig. 5.25 corroborate this view.

The spectra were obtained with a Spex 1401 attached to computers and under the following conditions of a krypton ion laser: crystals, 647.1 nm line at a power level of 10 mW; carbon tetrachloride solution, 520.8 nm line, 1 mW power level; cetyltrimethylammonium bromide (CTAB) solution, 568.2 nm line 10 mW power level. Since it became evident that a portion of these very photolabile molecules were photolysed by the intense laser beam during measurement, a special "molecular flow" device was employed for measurements of the solution. In this molecular flow apparatus the solution was passed through a capillary, which intersected the exciting laser beam, at a velocity of typically 500 cm/sec. In this manner the photobleaching of retinals was less than 2%. A similar flow system has also been used for rhodopsin measurements by Mathies, et al.[3]

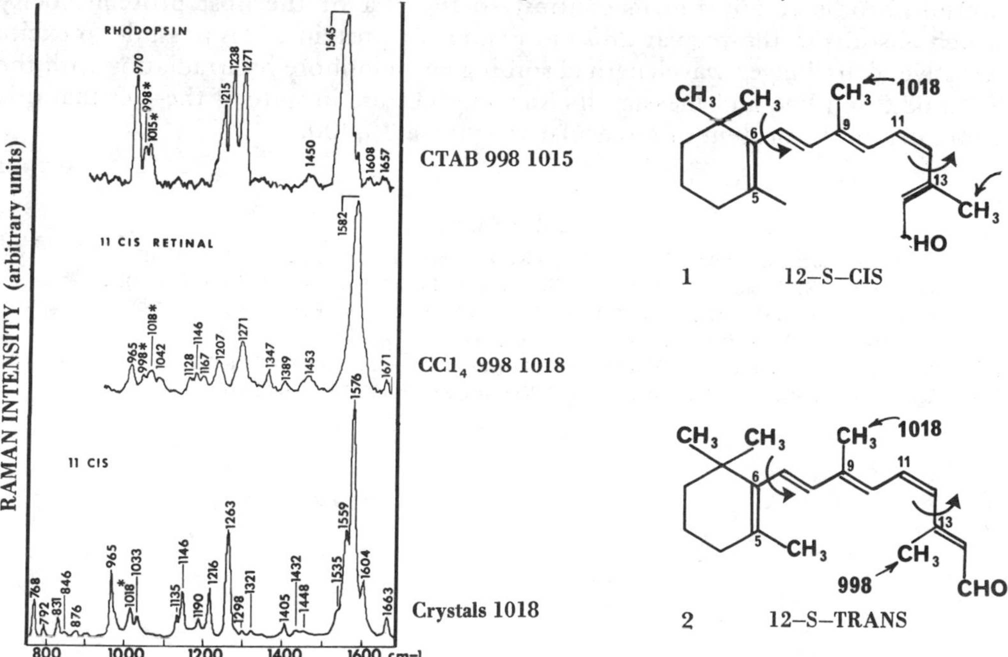

Fig. 5.25 Resonance laser Raman spectra of 11=cis retinal and rhodopsin.

The following assignments of the resonance laser spectra are based on theoretical calculations.[4] Also, the tentative conclusion derived from the three spectra is given.

a) Solid spectrum of 11-cis retinal.

It is known by x-ray that crystalline 11-cis retinal adopts conformation 1 in which the 6-s and 12-s bands are both cisoid and are twisted out of plane by ca. 40°.[5]

1663 cm^{-1}: Aldehyde.

1604–1535 cm^{-1}: Double bond.

1033 cm^{-1}: C(5)–Me stretching vibration.

1018 cm^{-1}: Degenerate C(9) and C(13)–Me stretching vibration; namely, these are assigned to the stretches of the two "outside" methyls attached to C–9 and C–13 in 1.

b) Carbon tetrachloride solution of 11-cis retinal.

Note the appearance of a weak additional band at 998 cm^{-1} in the C–Me stretching region. It is tempting to assign this to the "inside" methyl attached to C–13 in the 12-s-shape 2, and to assume that there is an equilibrium between the two conformers 1 and 2. Presence of this equilibrium is in agreement with results of NMR nuclear Overhauser experiments by Rowan, et al.[6]

c) CTAB solution of rhodopsin.

Note now that the intensity ratio of the 998 and 1015 cm^{-1} bands are roughly 1:1. Namely, if the above assignments are correct, this suggests that the shape of 11-cis retinal in rhodopsin is closer to 12-s-trans 2 rather than 12-s-cis 1.

The assignments mentioned can be checked by making the 9-, 13- and 9,13-di-trideuteriomethyl analogs of retinals. This example shows the advantage of the resonance laser Raman technique. Namely, since the chromophore in rhodopsin absorbs at 500 nm in contrast to the rest of the host protein moeity which absorbs in the regular 280 nm region for proteins, it is possible to excite selectively the longer wavelength absorbing chromophore by irradiating with the Kr^+ 568.2 nm line and measure its Raman spectrum in spite of the fact that it is contained in the protein of molecular weight ca. 40,000.

References

[1] R. H. Callender, A. Doukas, R. Crouch and K. Nakanishi, *Biochem.*, **15**, 1621 (1976).

[2] W. K. Chan, K. Nakanishi, T. Ebrey and B. Honig, *J. Am. Chem. Soc.*, **96**, 3642 (1974).

[3] R. Mathies, A. R. Oscroff, and L. Streyer, *Proc. Natl. Acad. Sco., U.S.A.*, **73**, 1 (1976).

[4] A. Warshel and M. Karplus, *J. Am. Chem. Soc.*, **96**, 5677 (1974).

[5] R. D. Gilardi, I. L. Karle and J. Karle, *Acta Crystallogr. Sect. B*, **28**, 2605 (1972).

[6] R. Rowan, A. Warshel, B. D. Sykes and M. Karplus, *Biochem.*, **13**, 970 (1974).

PROBLEMS

Ordinate: Percent transmission
Abscissa: Wave-number (cm^{-1}) and wavelength (μ)

The actual curves were measured by a Japan Spectroscopic Company (JASCO) Koken DS 301 spectrophotometer and by JASCO IRA-1 and Perkin Elmer 621 grating infrared spectrophotometers.

PROBLEM 1 / Calculate the thickness of the three cells from their interference fringes.

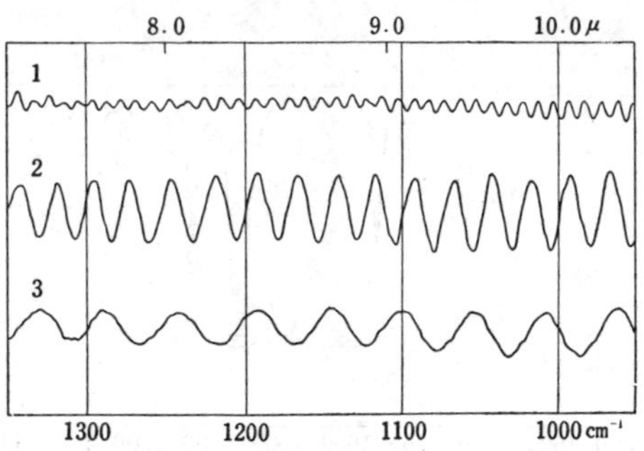

Fig. 1 Interference fringes of 3 cells

PROBLEM 2 / Deduce the structure of compound C_6H_{14}.

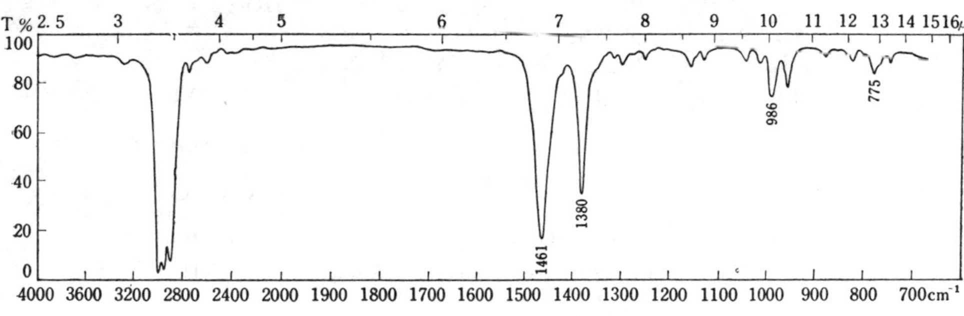

Fig. 2 Liquid film (IRDC 256)

PROBLEM 3 / Deduce the structure of an optically active monoterpenoid essential oil $C_{10}H_{16}$. No conspicuous UV maximum.

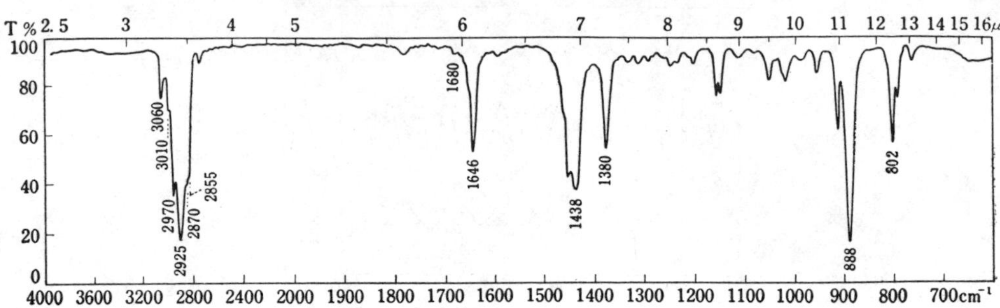

Fig. 3 Liquid film (IRDC 1155)

PROBLEM 4 / Assign out-of-plane CH bending bands (a)~(g) to the dienes (I)~(VII).

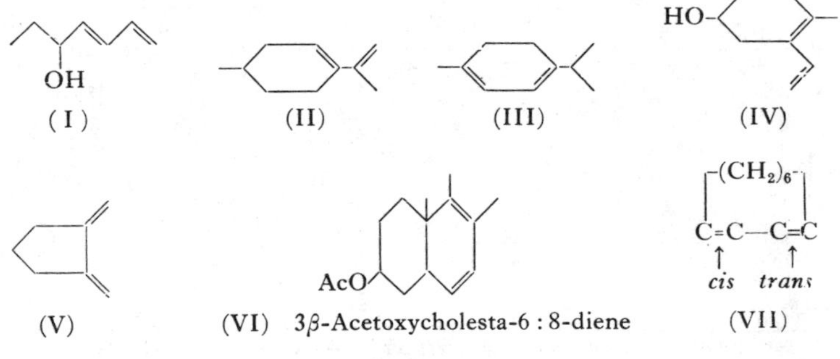

Out-of-plane δCH bands

(a) 720 cm^{-1}, ε^a 130

(b) 833 cm^{-1}

(c) 980 and 885 cm^{-1}

(d) 976 cm^{-1} s and 836 m, 815 m, 772 m, 720 s, 700 cm^{-1} s.

(e) 890 and 826 cm^{-1}

(f) 1002, 958 (doublet) and 900 cm^{-1}

(g) 880 cm^{-1}

PROBLEM 5 / Deduce the structure of C_7H_{10}.
λ_{max} at 273 mμ (log ε 4.569) and 263 mμ (log ε 4.671)

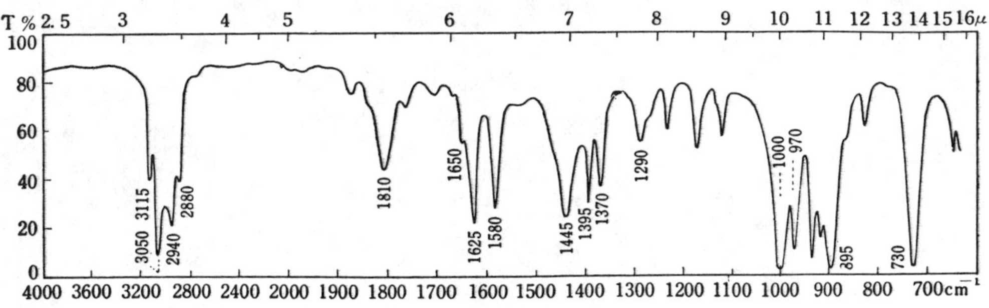

Fig. 5 Liquid film

PROBLEM 6 / Correlate bands with groups.

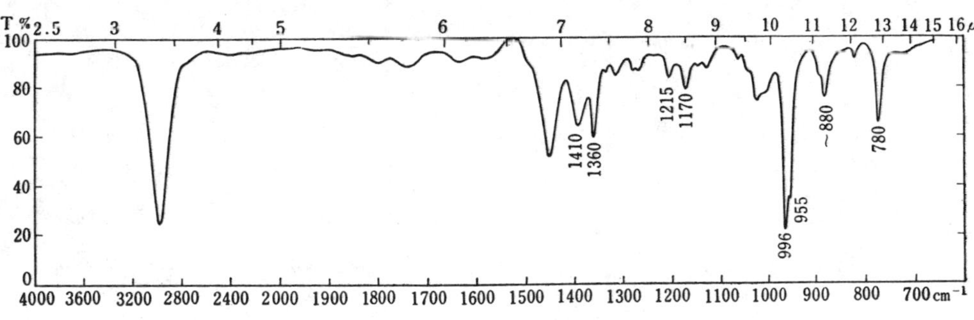

Fig. 6 (in CCl_4 soln)

15 15′

PROBLEM 7 / The following three figures are the IR spectra of *o*–, *m*,– and *p*–xylene. Correlate respective figures with the three isomers.

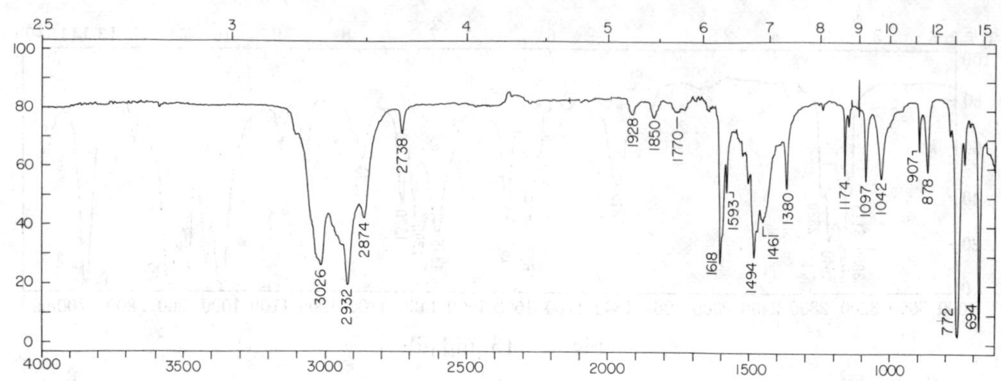

Fig. 7-a Liquid film

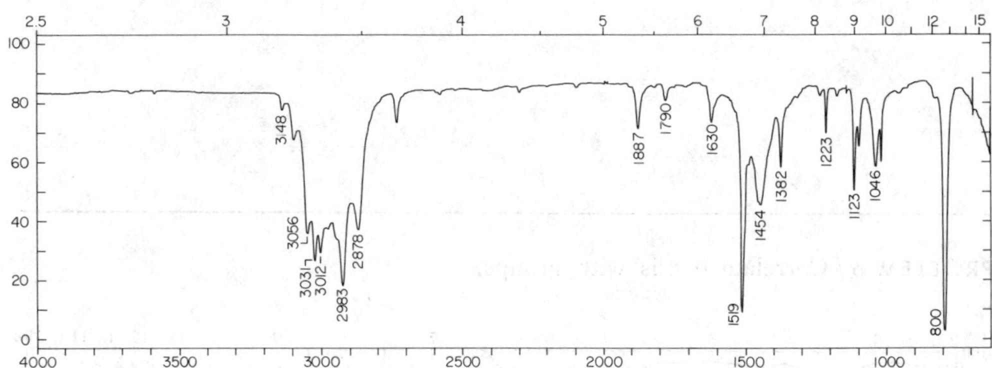

Fig. 7-b Liquid film

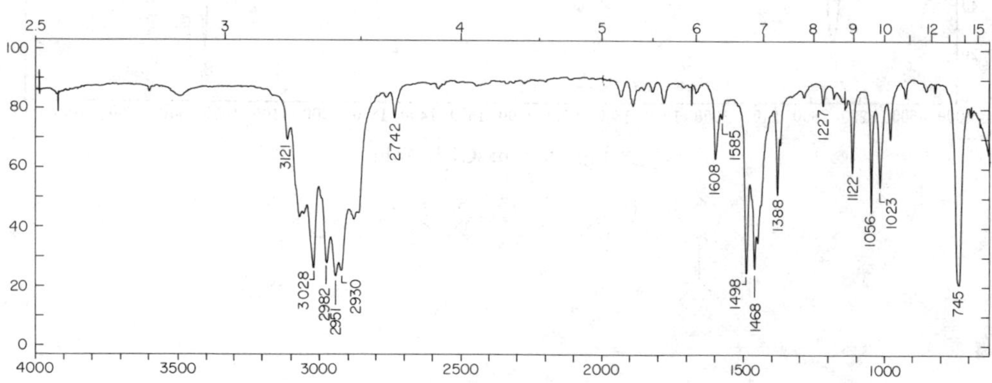

Fig. 7-c Liquid film

PROBLEM 8 / Deduce the structure of C_7H_7Br, m.p. 28.5°.

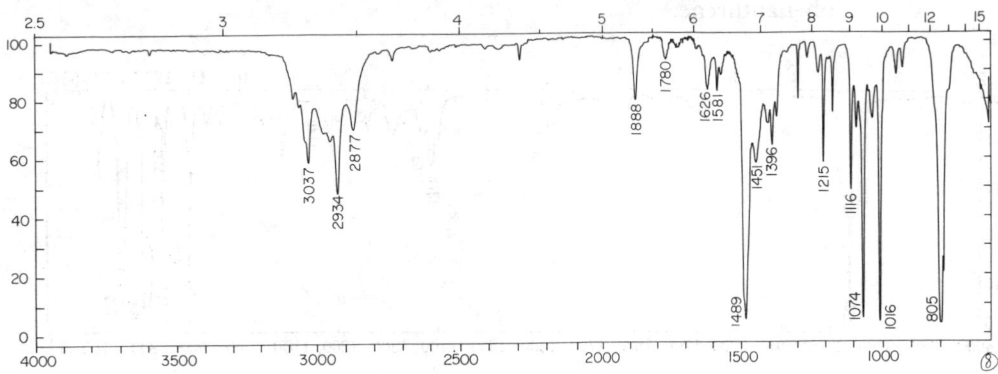

Fig. 8 Liquid film

PROBLEM 9 / Deduce the structure of the solid $C_{14}H_{14}$, m.p. 51.8~52°.

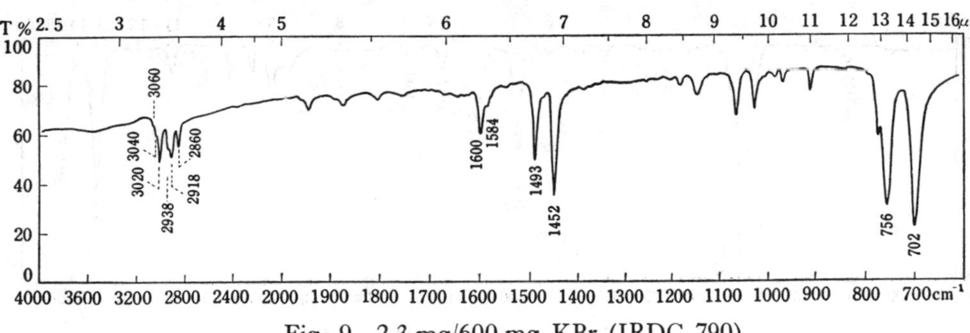

Fig. 9 2.3 mg/600 mg KBr (IRDC 790)

PROBLEM 10 / Deduce the structure of the selenium dehydrogenation product, $C_{17}H_{16}$, of a natural substance. The UV spectrum is that of an alkyl phenanthrene.

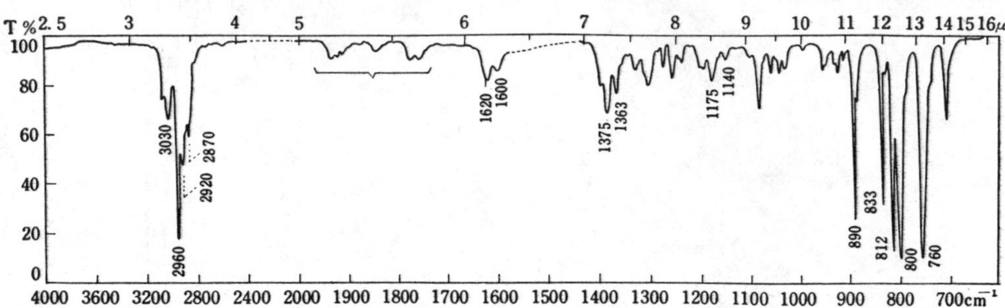

Fig. 10 CS_2 solution in 0.2 mm cell

PROBLEM 11 / Deduce the structure of C_7H_5NO.

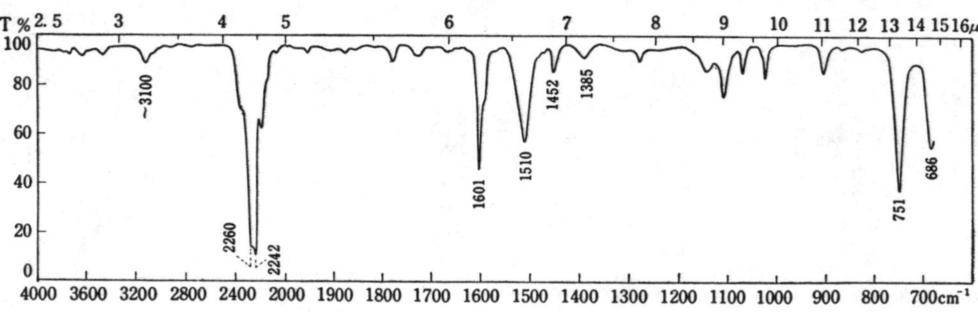

Fig. 11 Liquid film (IRDC 4)

PROBLEM 12 / What is the nature of the two nitrogen atoms in $C_{13}H_{22}N_2$? The compound adds water and is converted into a urea derivative.

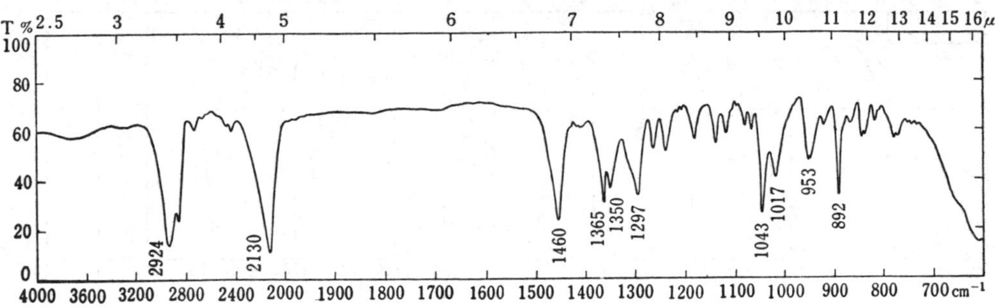

Fig. 12 Liquid film

PROBLEM 13 / Deduce the structure of C_8H_7N, m.p. 29.5°.

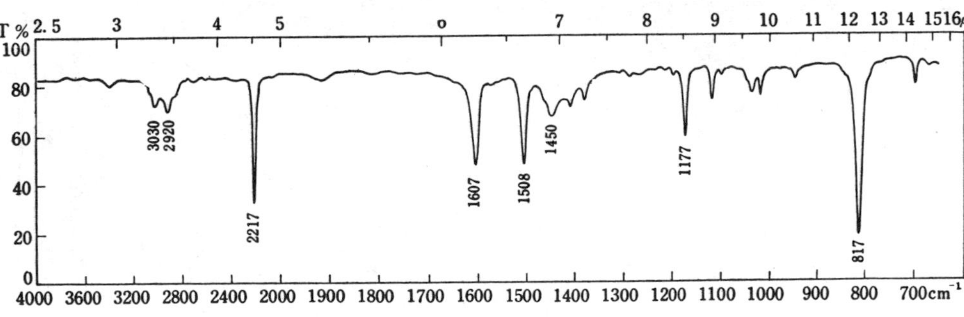

Fig. 13 Liquid film (IRDC 655)

PROBLEM 14 / Deduce the structure of C_7H_5NS.

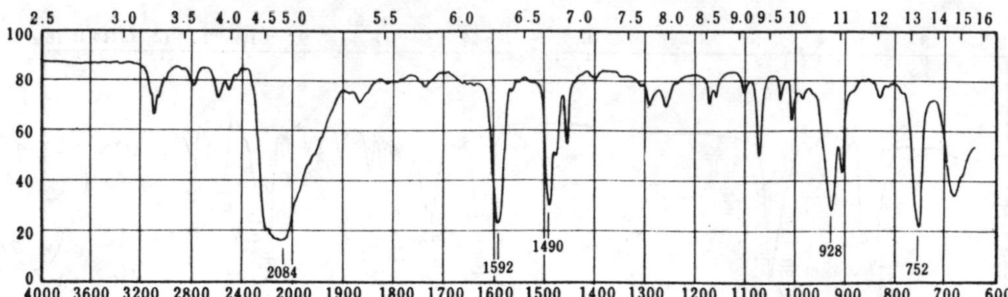

Fig. 14 Liquid film (IRDC 5638)

PROBLEM 15 / What are the substituents on the phenyl derivative $C_6H_3FN_3O_2$?

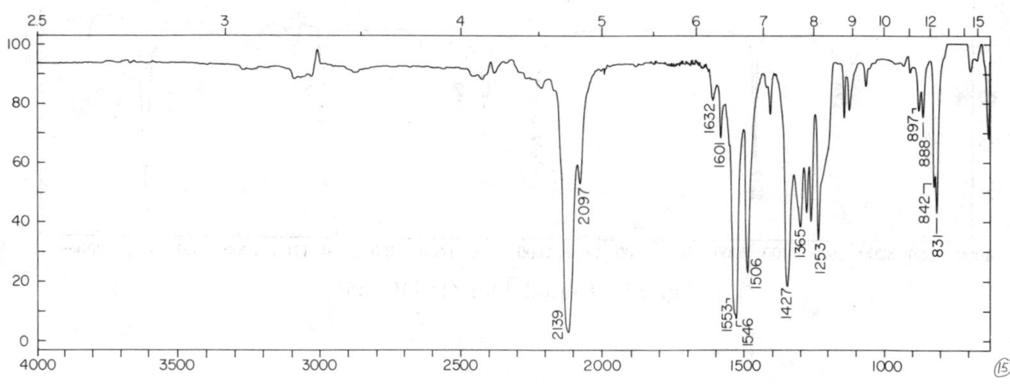

Fig. 15 CHCl₃ solution

PROBLEM 16 / The following is the infrared spectrum of the crude product (C_9) obtained when $CH_3(CH_2)_3C{\equiv}C{-}MgBr$ is treated with $BrCH_2C{\equiv}CH$. Associate bands with main product and infer nature of by-product.

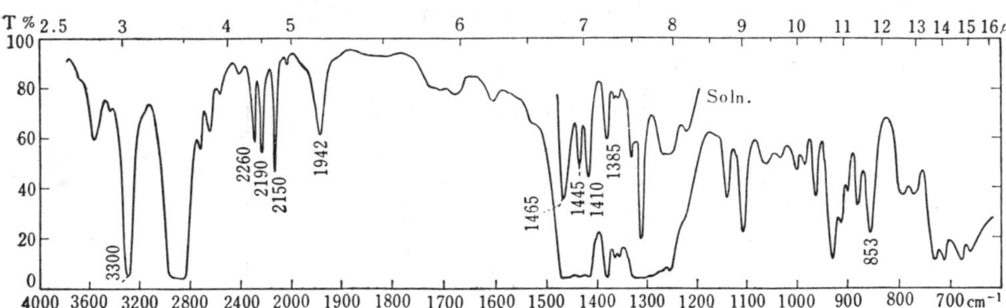

Fig. 16 Liquid film and 0.15 mm cell, CCl_4 solution

PROBLEM 17 / Deduce the structure of an essential oil $C_{12}H_{10}$.
λ_{max}^{EtOH} 239 mμ (ε 537), 253 mμ (ε 340)

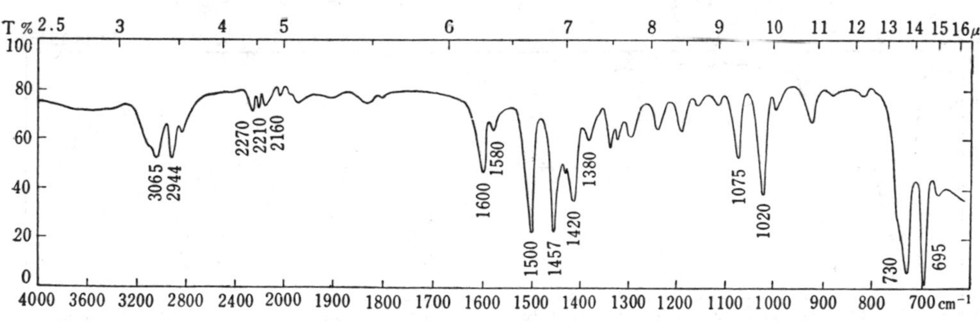

Fig. 17 Liquid film.

PROBLEM 18 / The figure shows the concentration variation of ethanol in carbon tetrachloride. What facts can be deduced from it?

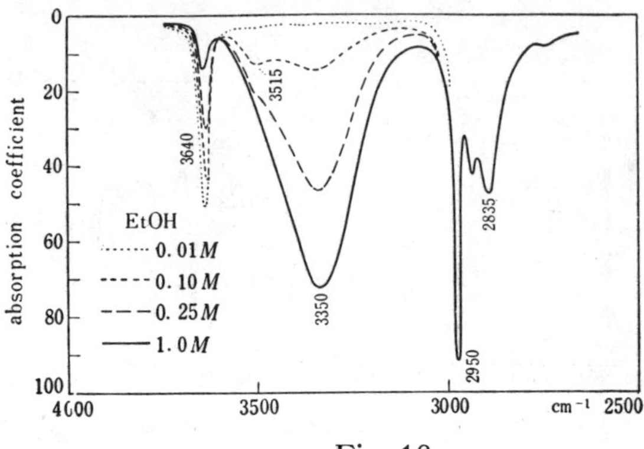

Fig. 18

PROBLEM 19 / Deduce the structure of the monoterpene $C_{10}H_{18}O$. No conspicuous λ_{max}, and $[\alpha]_D^{25}$ $-18°$.

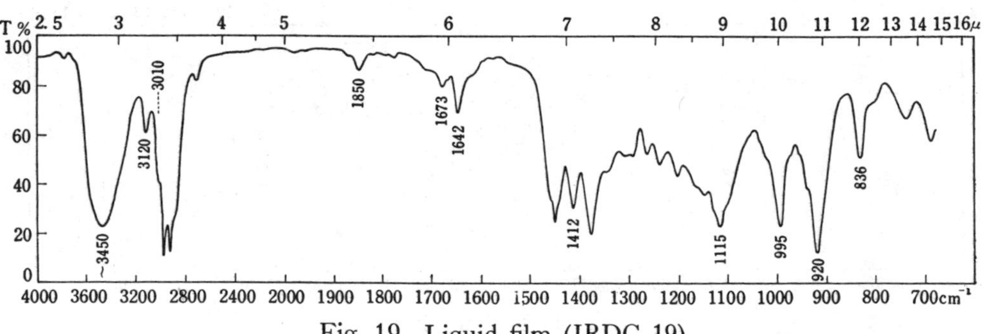

Fig. 19 Liquid film (IRDC 19)

PROBLEM 20 / What is the structure of the product $C_{13}H_{22}O$ obtained when dihydro-α-ionone is allowed to react with hydrogen chloride in *sec*-butanol at $-13°C$ for ten days? When ethanol was employed as solvent the ethyl ether of the product was obtained instead.

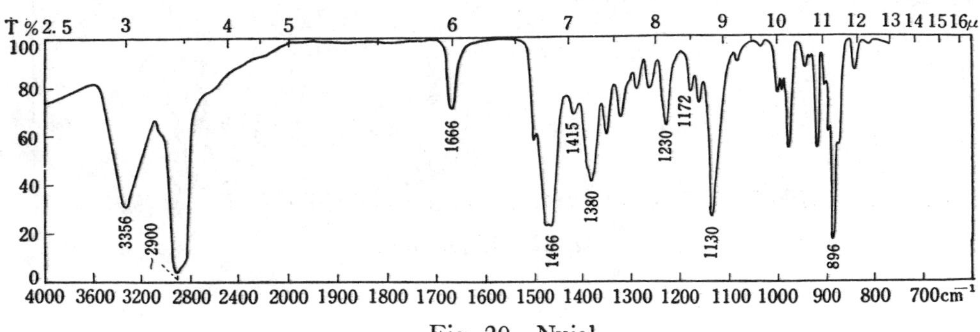

Fig. 20 Nujol

$\xrightarrow[\text{-13°C, 10 days}]{\textit{sec}\text{-BuOH-HCl}}$ $C_{13}H_{22}O$

PROBLEM 21 / Deduce the structure of C_7H_9NO. An ammonium band well separated from the C–H stretching bands appears in the IR spectrum of its hydrochloride.

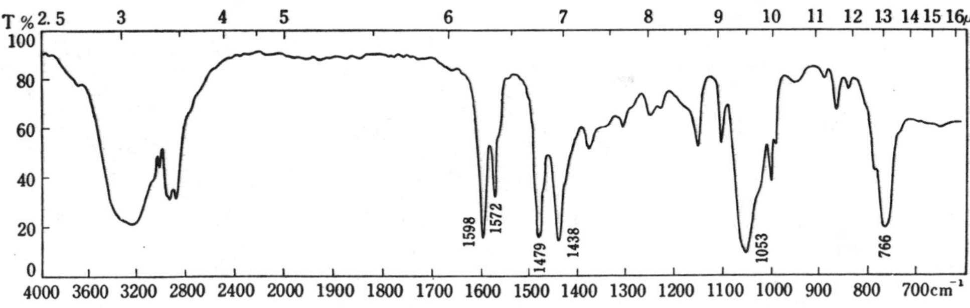

Fig. 21 Liquid film (IRDC 1111)

PROBLEM 22 / Deduce the structure of C_3H_6O.

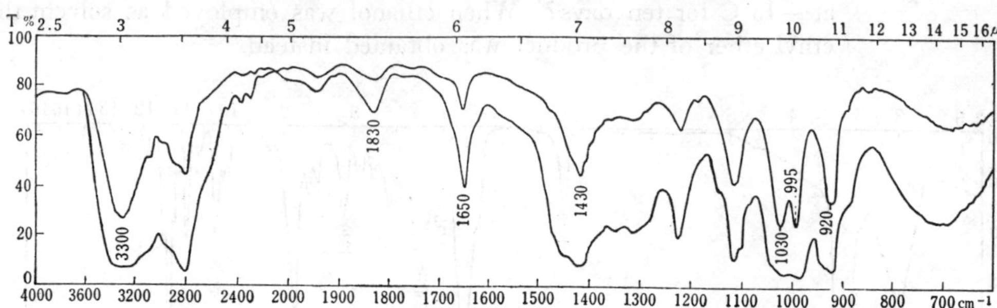

Fig. 22 Liquid film

PROBLEM 23 / The following infrared spectrum is that of a simple derivative of allyl alcohol. What derivative is it? Also compare spectrum with previous Fig. 21, and summarize the difference in the IR spectrum of the present derivative.

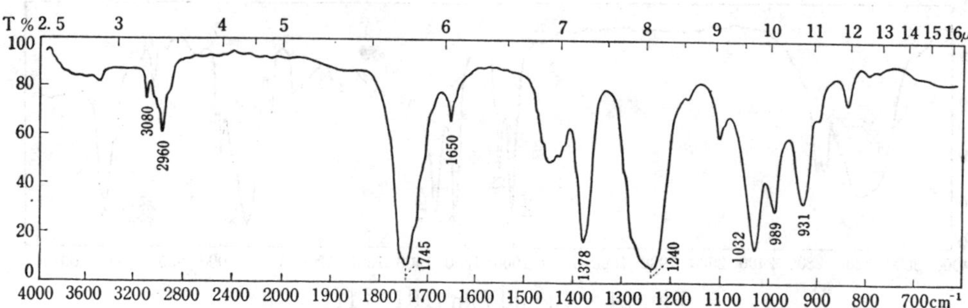

Fig. 23 Liquid film

PROBLEM 24 / Deduce the structure of C_4H_5N.

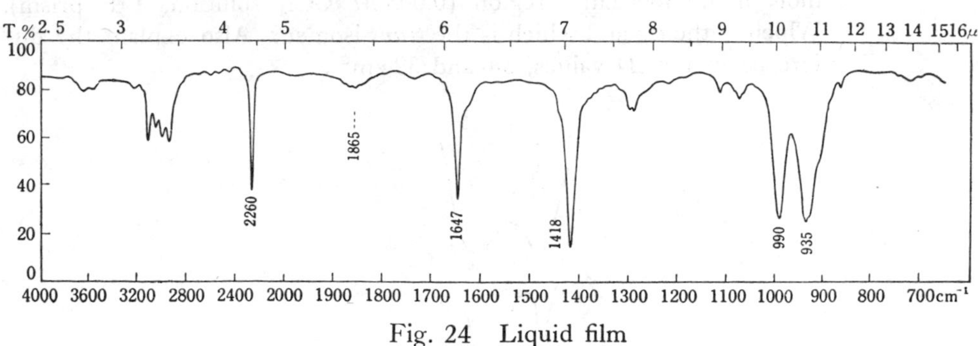

Fig. 24 Liquid film

PROBLEM 25 / What is the nature of the two oxygen atoms contained in $C_{18}H_{24}O_2$? Figs. 25-a and -b are the IR spectra of the same compound.

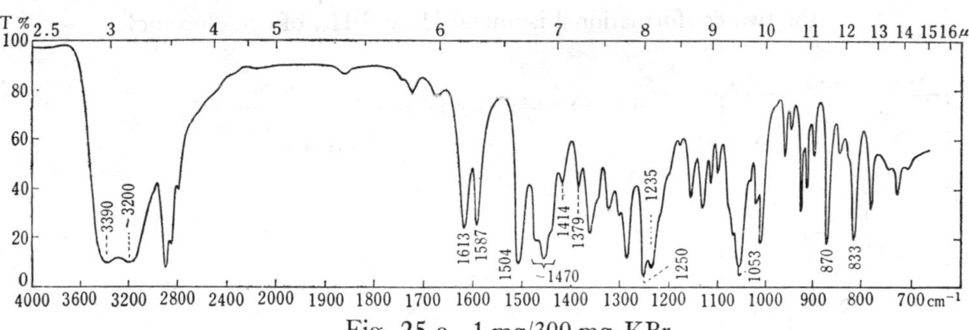

Fig. 25-a 1 mg/300 mg KBr

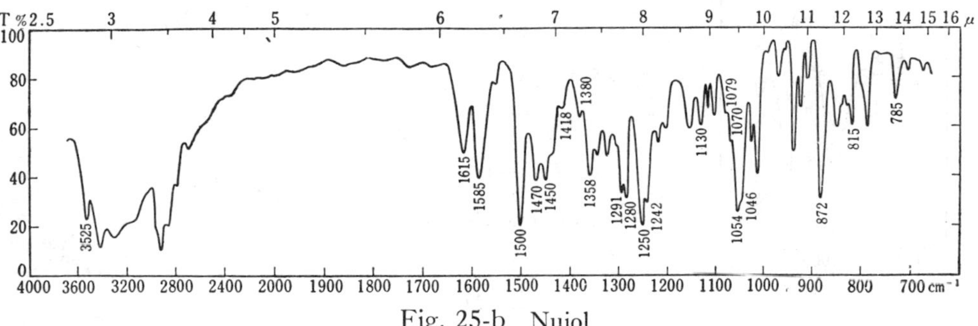

Fig. 25-b Nujol

PROBLEM 26 / The figure shows the absorption of *cis*- and *trans*-cyclohexane–1, 2–diols in the 3600 cm^{-1} region (0.005 M CCl_4 solution, LiF prism). Which is the *cis* and which is the *trans* isomer? Also explain the difference in the $\Delta\nu$ values, 38 and 32 cm^{-1}.

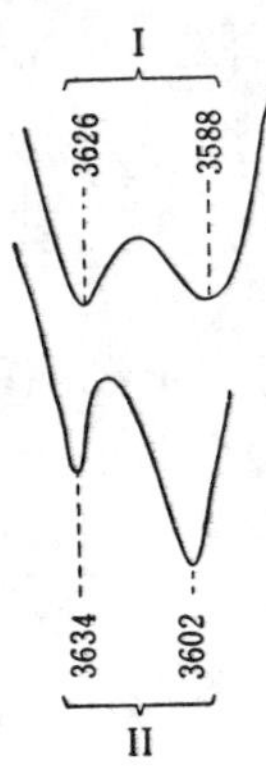

PROBLEM 27 / The 1069 cm^{-1} band intensity of a 30.05 mg/g CS_2 solution of cyclohexanol (M.W. 100.16) was identical with the 1062 cm^{-1} band intensity of a 31.04 mg/g CS_2 solution of *trans*–4–*t*–butylcyclohexanol (M.W. 156.26) (20°). From this, calculate the equilibrium constant between the two conformational isomers, II_e and II_a, of cyclohexanol.

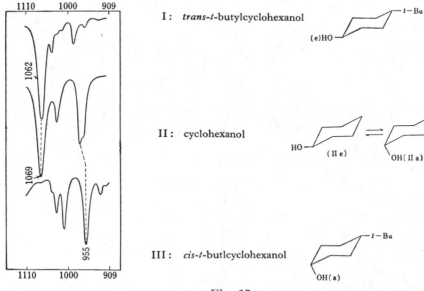

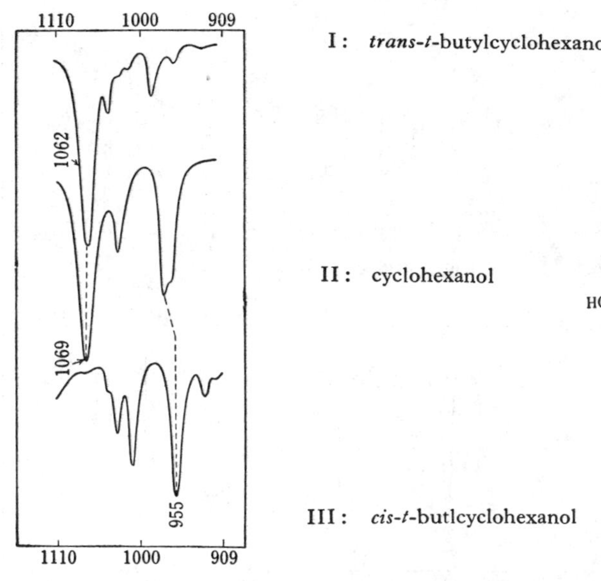

Fig. 27

PROBLEM 28 / Deduce the structure of $C_{10}H_{10}O$. A three proton methyl singlet is present in the nuclear magnetic resonance spectrum.

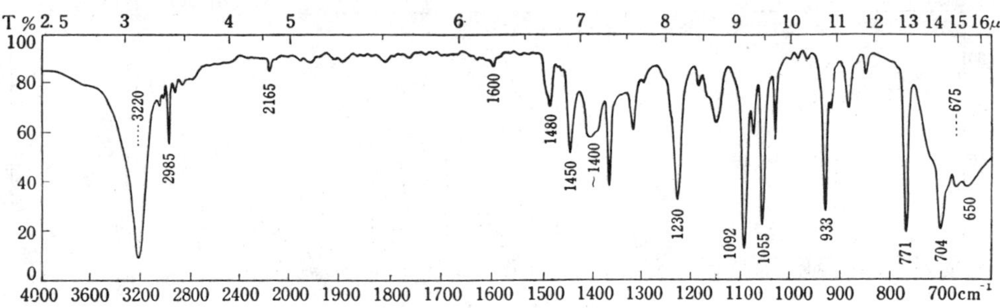

Fig. 28 2.8 mg/600 mg KBr (IRDC 1345)

PROBLEM 29 / To what type of compound does the following belong?

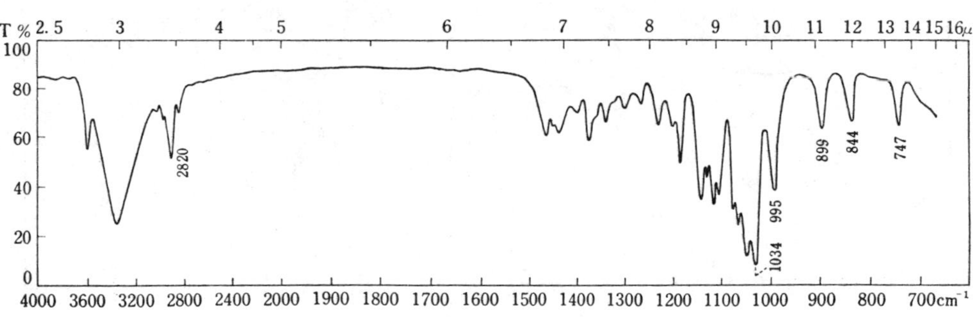

Fig. 29 1.2 mg/600 mg KBr (IRDC 53)

PROBLEM 30 / Deduce the structure of $C_6H_{12}O$. The nmr spectrum shows the presence of one methyl group.

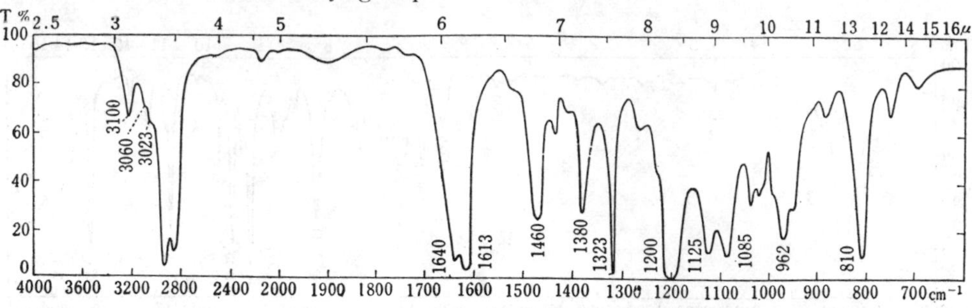

Fig. 30 Liquid film

PROBLEM 31 / Deduce the structure of $C_{11}H_{10}O$.

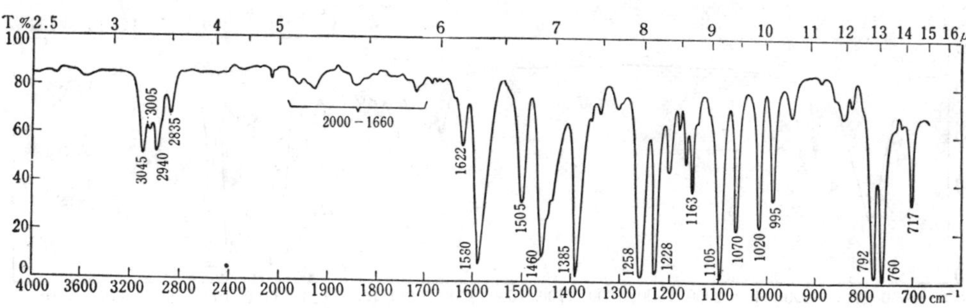

Fig. 31 Liquid film

PROBLEM 32 / Three compounds with molecular formulas C_7H_8O, C_7H_8S, and $C_8H_{11}N$ have the spectra shown below. Deduce the structures.

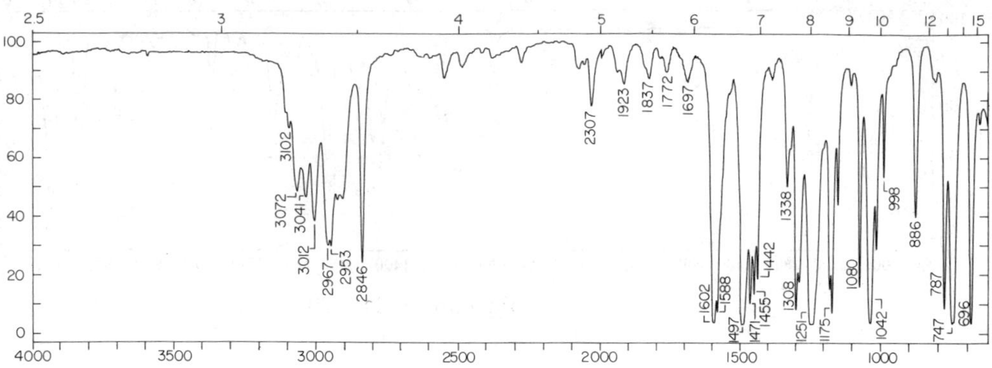

Fig. 32-a Liquid film

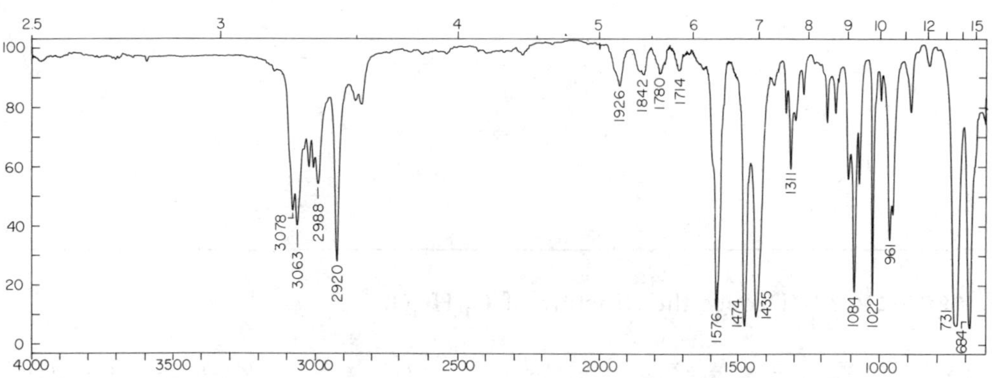

Fig. 32-b Liquid film

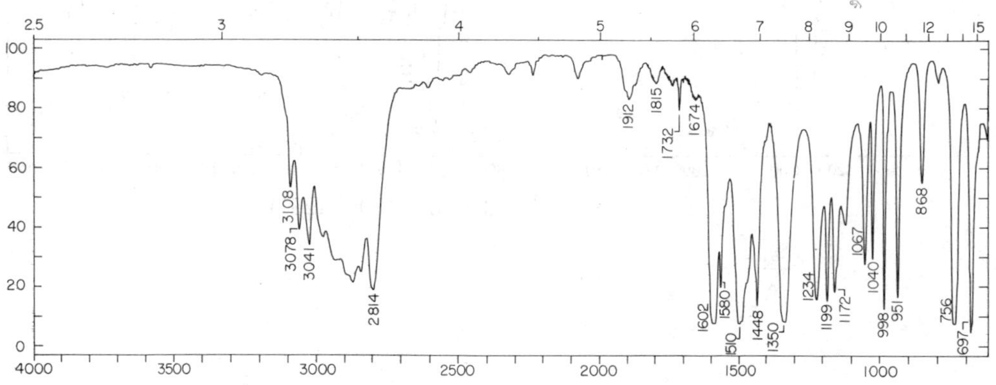

Fig. 32-c Liquid film

PROBLEM 33 / Deduce the structure of $C_9H_{12}O_2$.

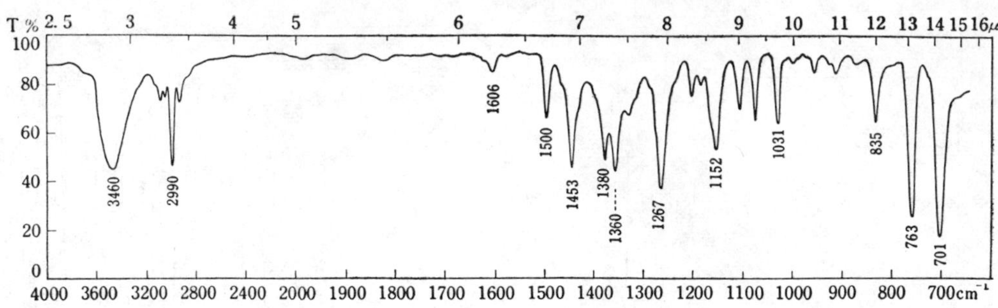

Fig. 33 Liquid film (IRDC 207)

PROBLEM 34 / Deduce the structure of $C_{12}H_{24}O$.

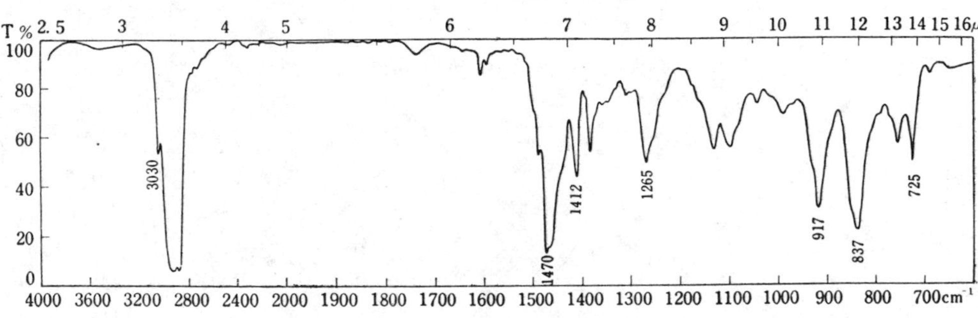

Fig. 34 Liquid film (IRDC 1251)

PROBLEM 35 / What is the structure of $C_6H_{15}N$ (Fig. 35-a)? Fig. 35-b shows the IR of its simple derivative. Rationalize the peak changes.

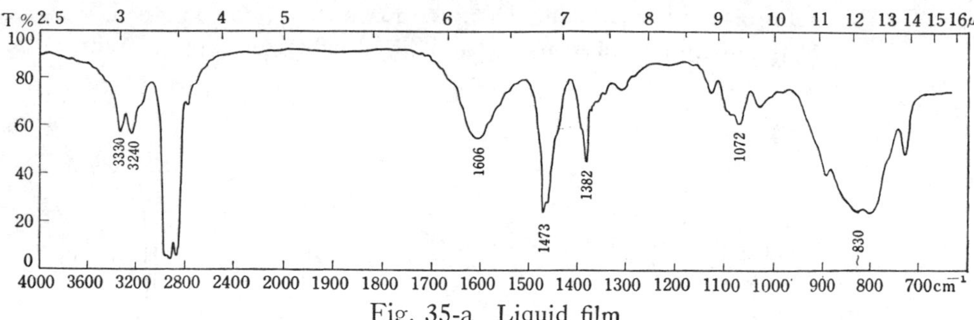

Fig. 35-a Liquid film

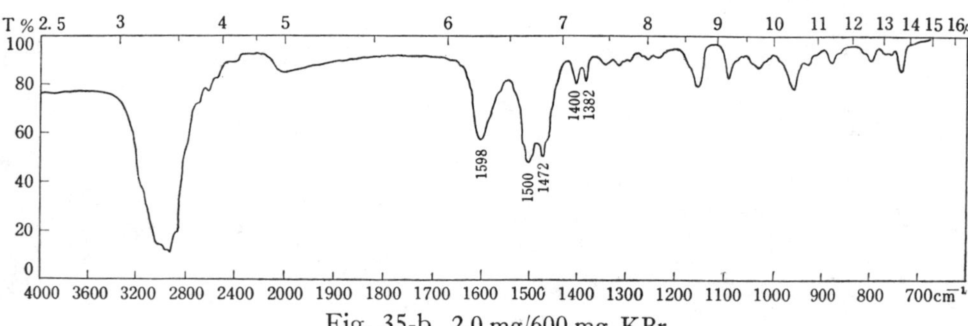

Fig. 35-b 2.0 mg/600 mg KBr

PROBLEM 36 / Deduce the structure of C_7H_9N.

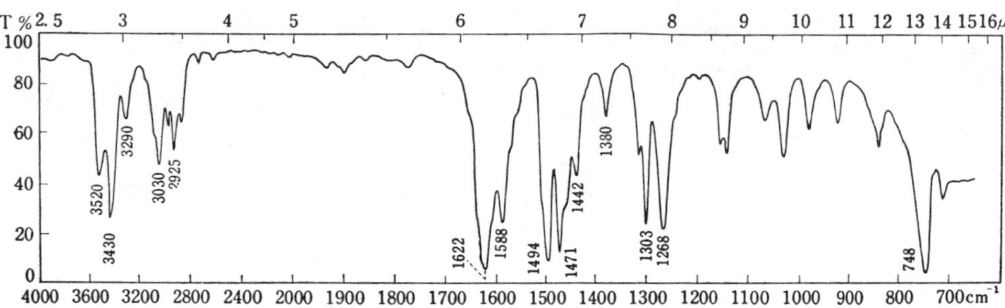

Fig. 36 Liquid film (IRDC 562)

PROBLEM 37 / The IR spectrum of $C_{10}H_{16}N_2 \cdot HCl$ is shown. The amine also forms a dihydrochloride. When alkali is added to the dihydrochloride solution the UV spectrum changes in two steps; namely, the UV spectra of the diamine, and its mono- and dihydrochlorides are different. What is the structure of the diamine?

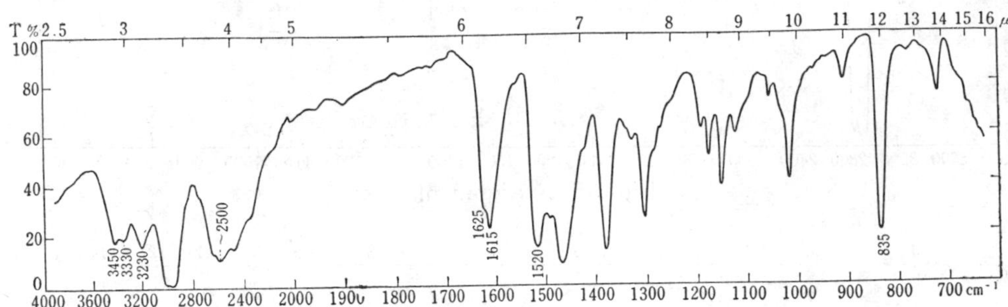

Fig. 37 Nujol

PROBLEM 38 / Correlate peaks with structure.

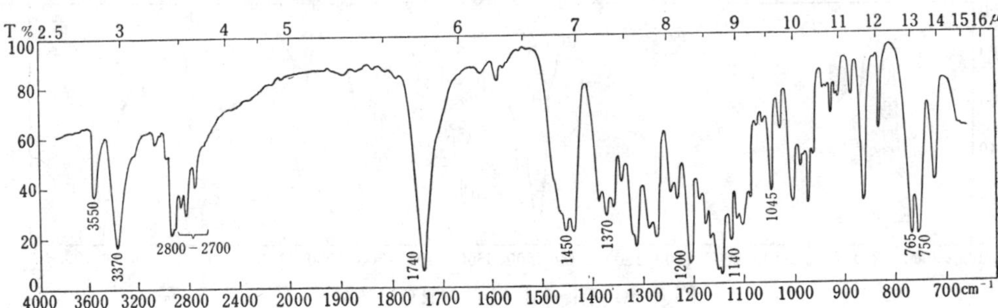

Fig. 38 Yohimbine (KBr)

C
N
NH
D
CH_3OOC
OH

PROBLEM 39 / Correlate IR peaks with the structure of yohimbine hydrochloride (cf. previous problem).

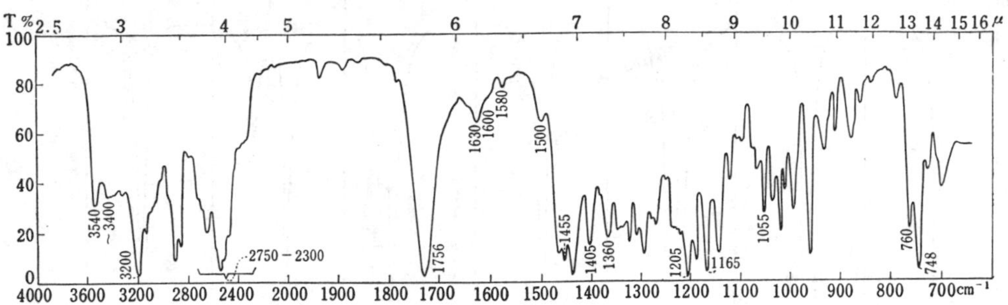

Fig. 39 Yohimbine hydrochloride (KBr)

N H
+N H
CH_3OOC
OH

PROBLEM 40 / Deduce the structure of the monoterpene, $C_{10}H_{16}O$. Three methyl singlets are apparent in the nuclear magnetic resonance spectrum.

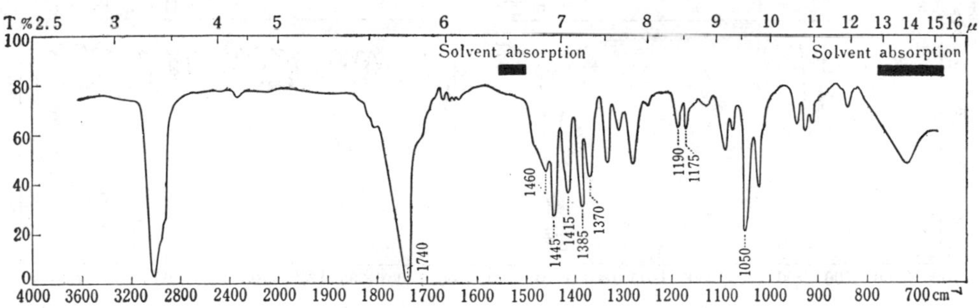

Fig. 40 CCl_4 solution, 0.5 mm cell

PROBLEM 41 / Correlate steroids with figures.

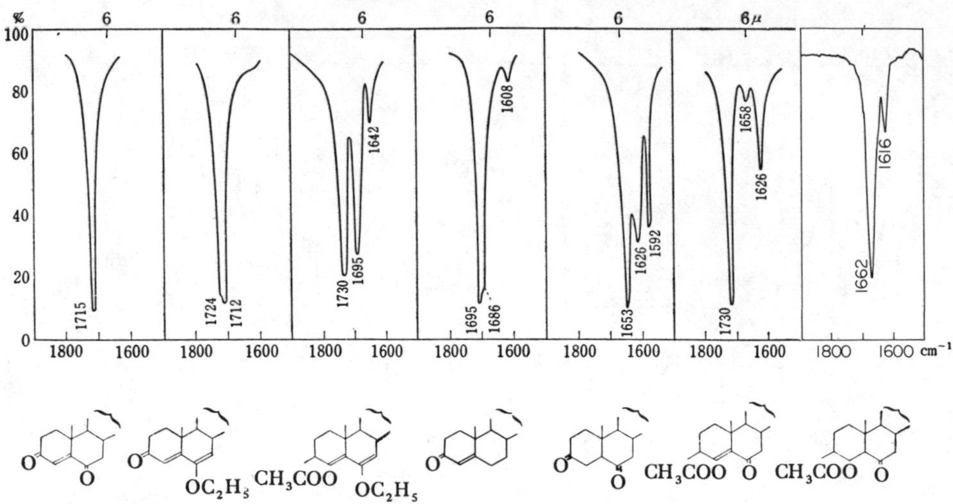

Fig. 41 Cholestane derivatives with unsaturated A, B rings ($CHCl_3$ solutions)

PROBLEM 42' / What is the structure of the thujopsene derivative? The molecular formula is $C_{14}H_{22}O$, and it lacks olefinic protons, m.p. 105.0~105.5°.

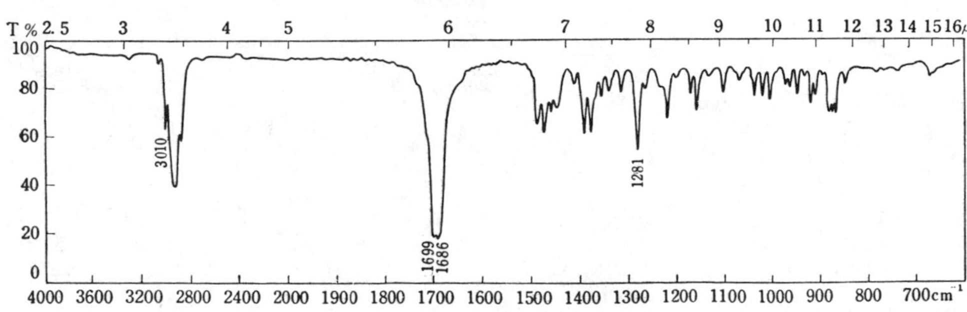

Fig. 42 1.8 mg/600 mg KBr (IRDC 856)

thujopsene

PROBLEM 43 / Deduce the structure of $C_6H_{10}O$.

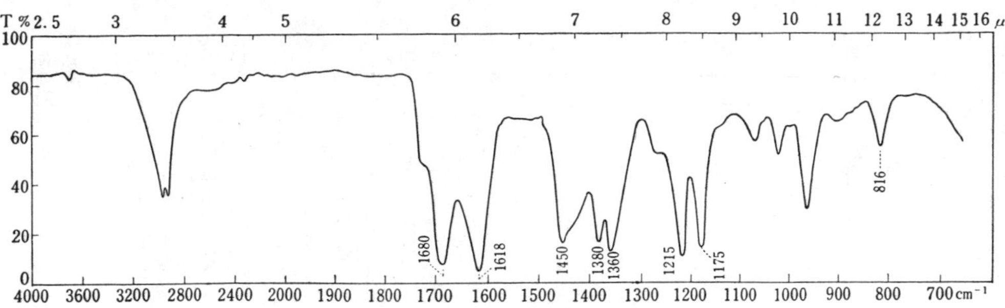

Fig. 43 Liquid film

PROBLEM 44 / Deduce the structure of C_8H_8O.

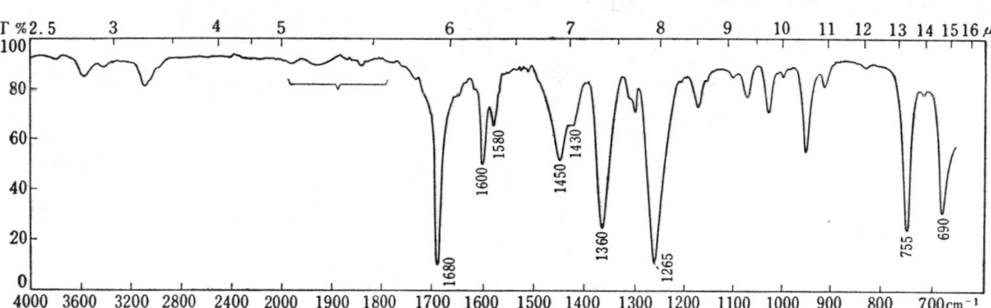

Fig. 44 Liquid film

PROBLEM 45 / Deduce the structure of $C_9H_{10}O_3$, m.p. 50°. The compound forms a crystalline derivative with 2,4-dinitrophenylhydrazine.

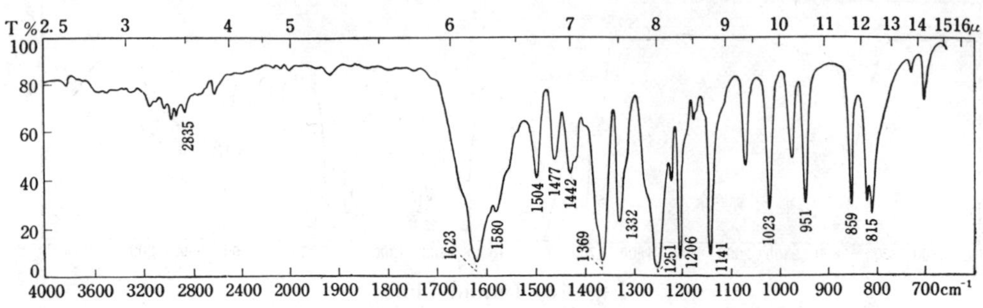

Fig. 45 1.6 mg/600 mg KBr (IRDC 307)

PROBLEM 46 / Deduce the structure of $C_8H_8O_2$.

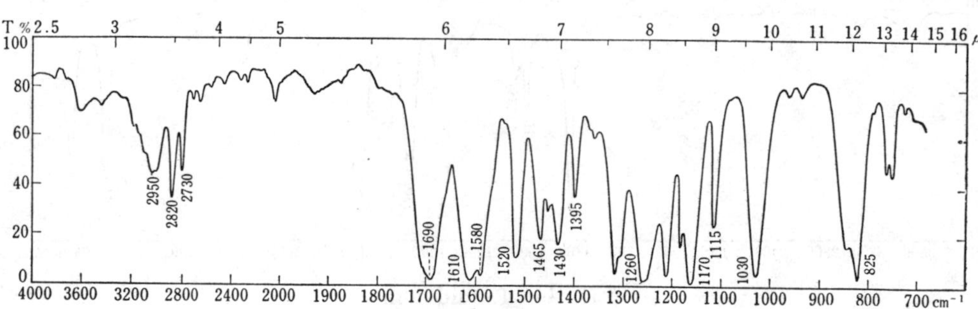

Fig. 46 Liquid film

PROBLEM 47 / Deduce the structure of $C_8H_6O_3$, m.p. 37 °C.

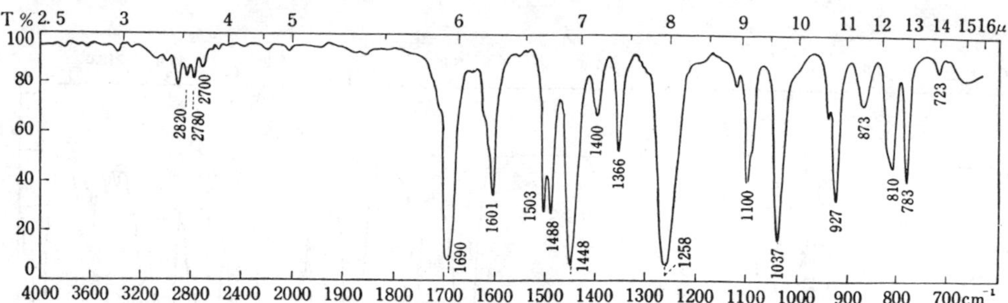

Fig. 47 Molten solid between plates (IRDC 102)

PROBLEM 48 / Deduce the structure of $C_7H_5NO_3$, m.p. 106°.

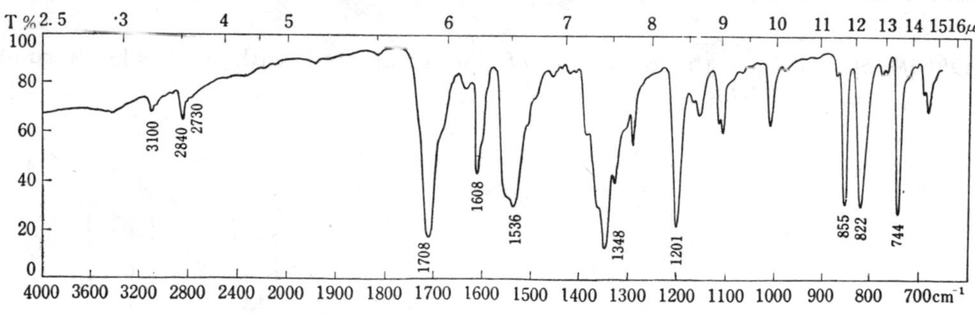

Fig. 48 1.0 mg/300 mg KBr

PROBLEM 49 / The following spectrum is that of a simple derivative of cholesterol. What is the derivative?

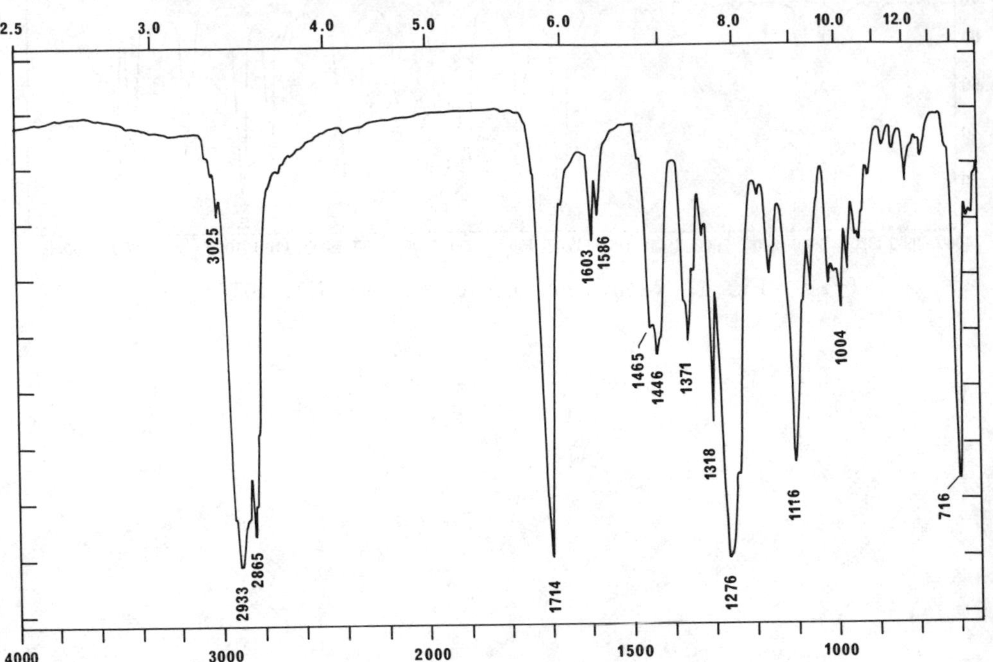

Fig. 49 3 mg/200 mg KBr

PROBLEM 50 / Deduce the structure of the compound with molecular formula $C_9H_9NO_3$.

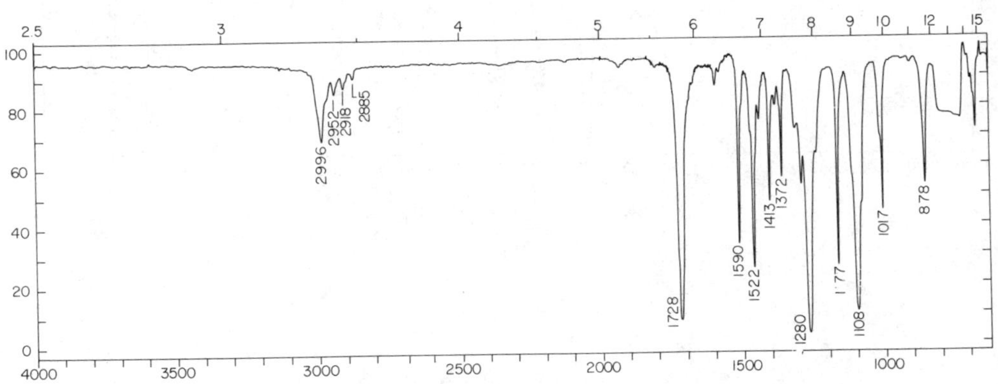

Fig. 50 5% solution in CCl_4

PROBLEM 51 / Correlate structures with IR spectra.

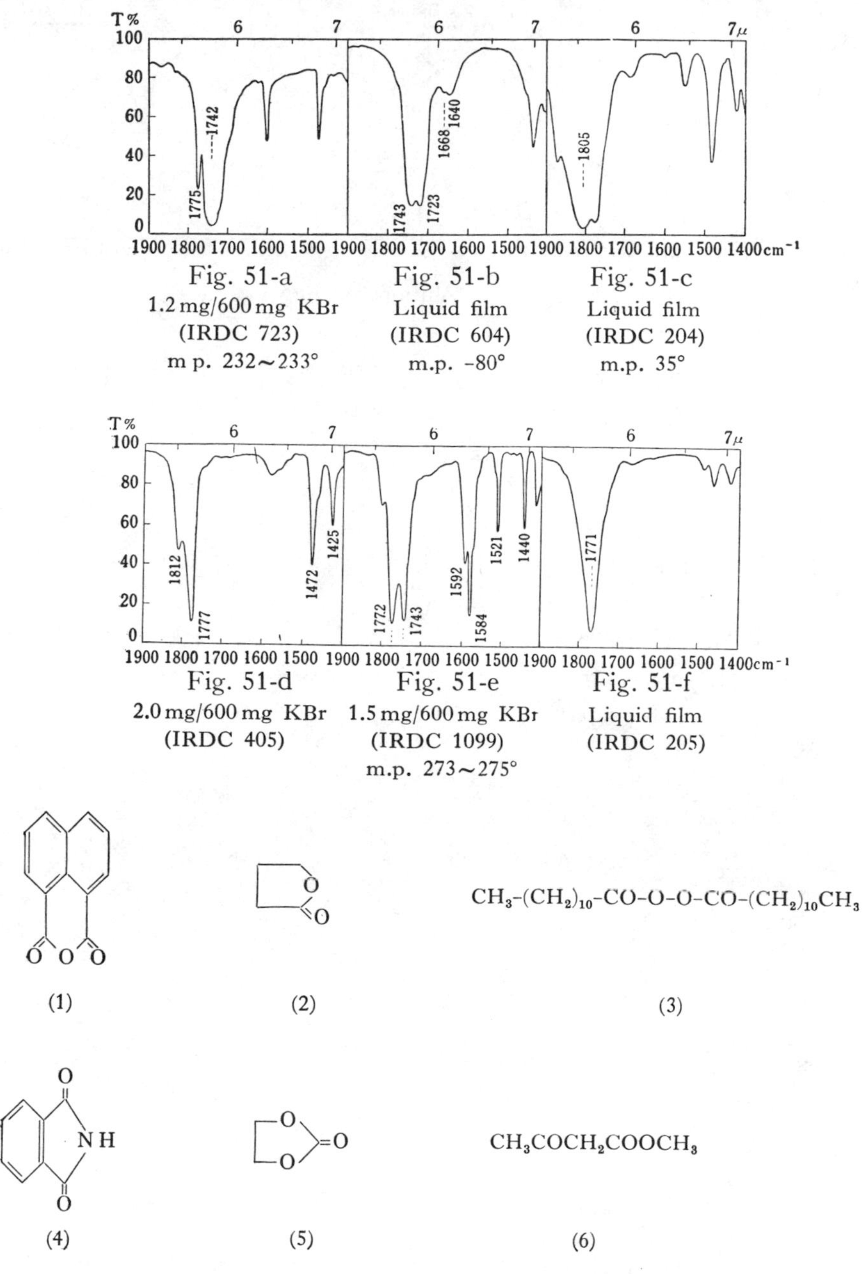

Fig. 51-a
1.2 mg/600 mg KBr
(IRDC 723)
m p. 232～233°

Fig. 51-b
Liquid film
(IRDC 604)
m.p. -80°

Fig. 51-c
Liquid film
(IRDC 204)
m.p. 35°

Fig. 51-d
2.0 mg/600 mg KBr
(IRDC 405)

Fig. 51-e
1.5 mg/600 mg KBr
(IRDC 1099)
m.p. 273～275°

Fig. 51-f
Liquid film
(IRDC 205)

PROBLEM 52 / Deduce the structure of $C_3H_6O_2$.

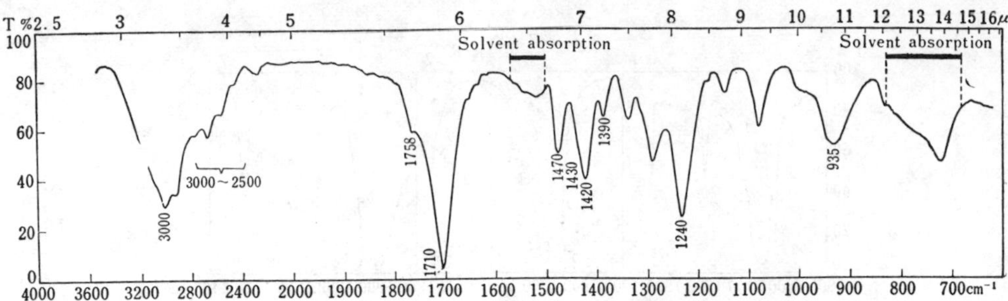

Fig. 52 CCl_4 1% solution, 0.5 mm cell

PROBLEM 53 / Figs. 53 a～d are the IR spectra of propionic acid measured as liquid film, gas, carbon tetrachloride solution, and dioxane solution. Correlate the spectra with the state of measurement.

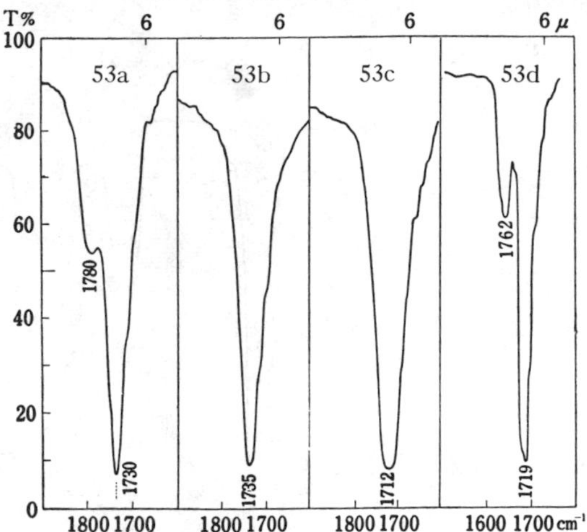

Fig. 53 C=O bands of propionic acid

PROBLEM 54 / Deduce the structure of $C_{16}H_{34}O_2$.

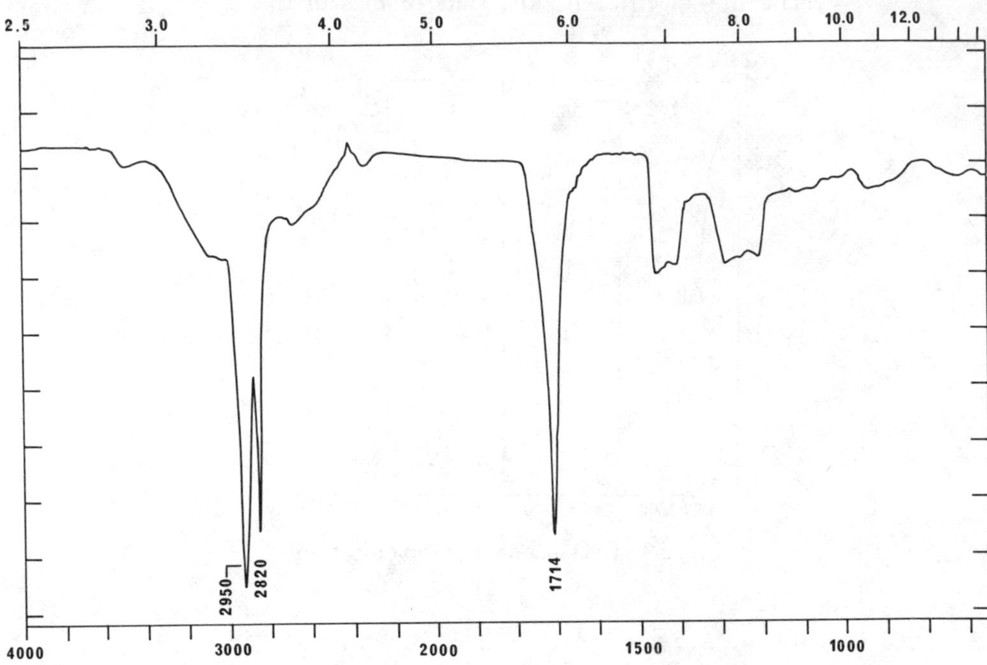

Fig. 54-a 5% solution in $CHCl_3$, 0.1 mm cell

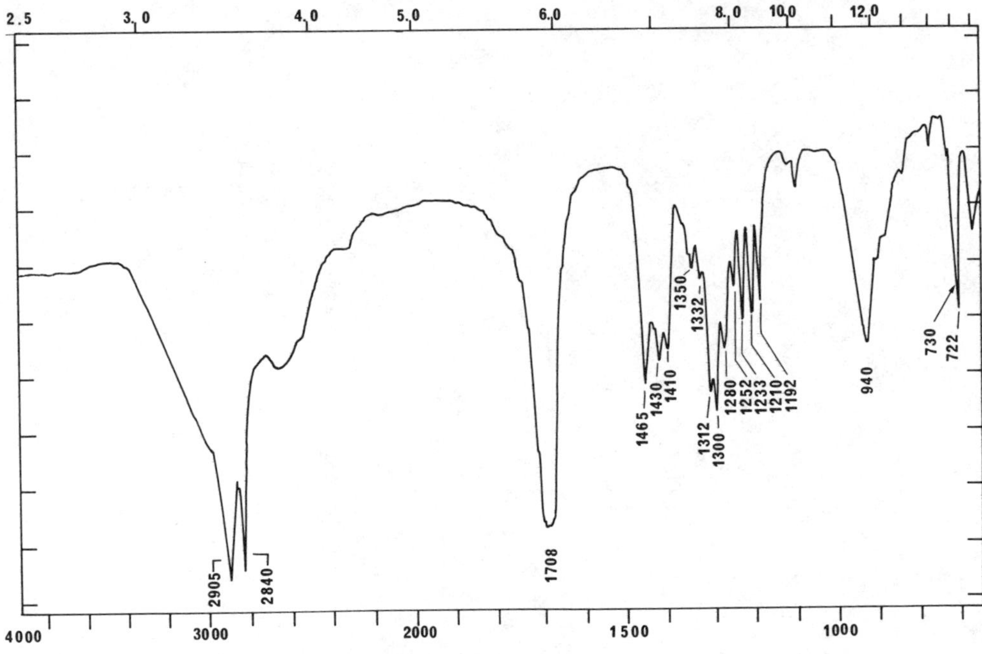

Fig. 54-b 3mg/200 mg KBr

PROBLEM 55 / What is the following compound that contains sodium?

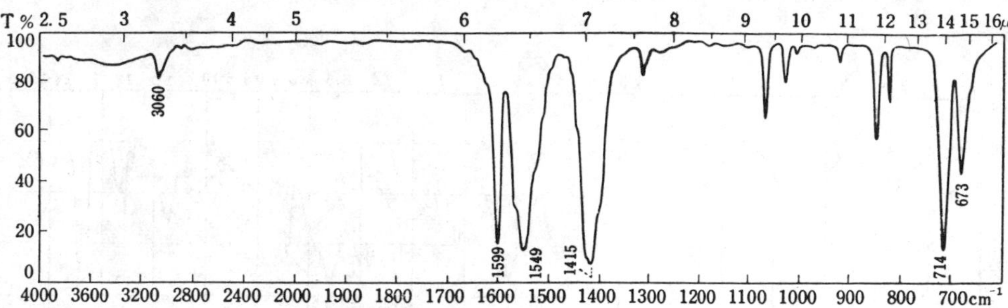

Fig. 55 1.5 mg/600 mg KBr (IRDC 1004)

PROBLEM 56 / Identify the functional groups present in the gibberellin derivative whose molecular formula is $C_{20}H_{22}O_6$.

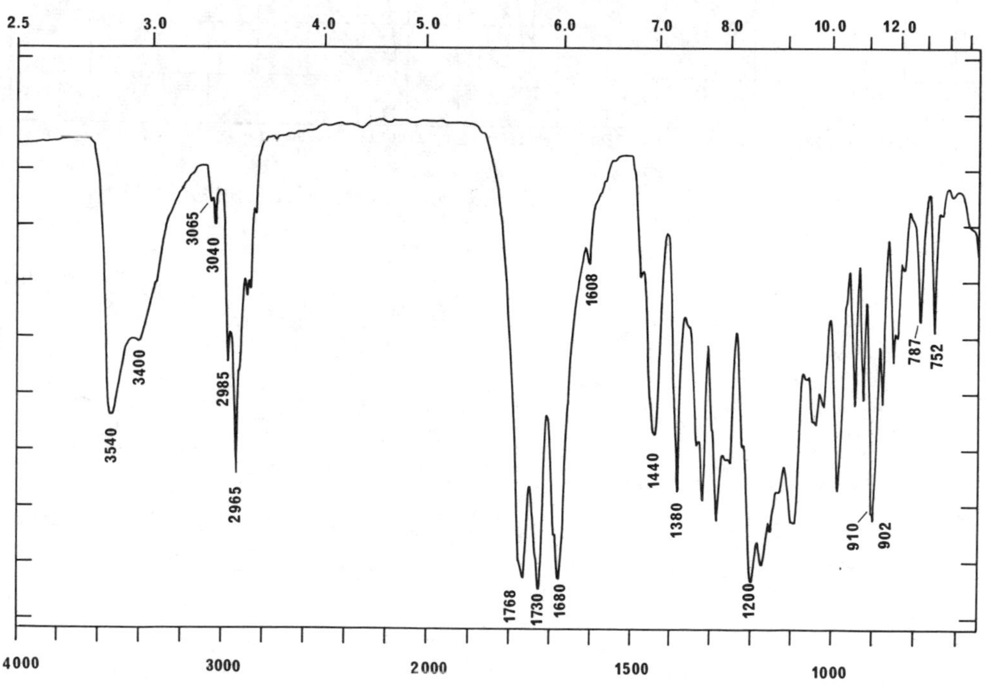

Fig. 56 3 mg/200 mg KBr

PROBLEM 57 / Below are the spectra for *o*-aminobenzoic acid and *m*-aminobenzoic acid. The spectra appear quite different. Explain.

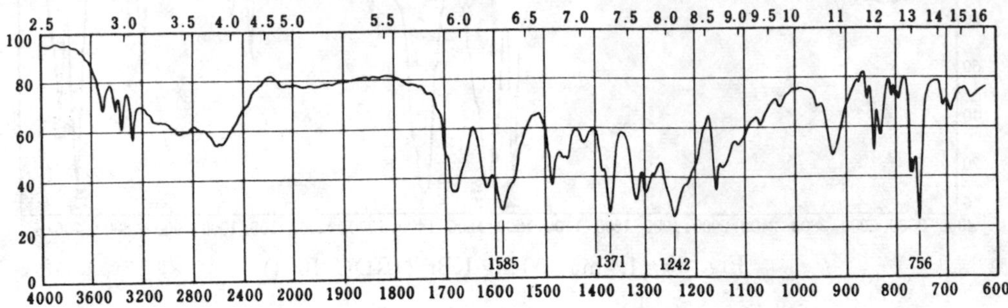

Fig. 57-a *o*-aminobenzoic acid, 1.3 mg/600 mg KBr (IRDC 3490)

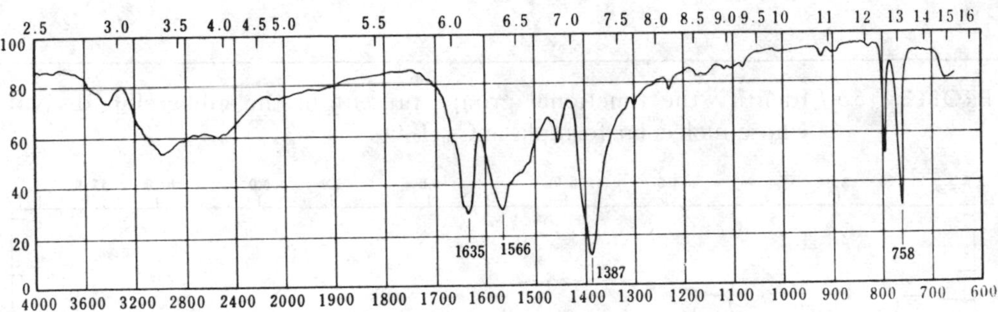

Fig. 57-b *m*-aminobenzoic acid, 0.8 mg/600 mg KBr (IRDC 7452)

PROBLEM 58 / Deduce the structure of $C_5H_8O_2$.

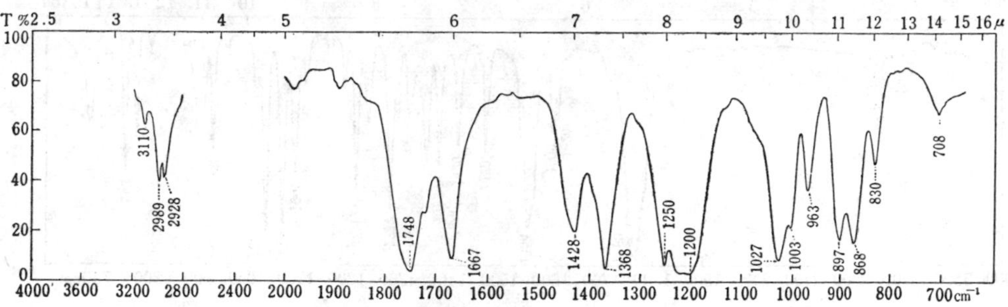

Fig. 58 Liquid film

PROBLEM 59 / Deduce the structure of $C_{12}H_{14}O_4$.

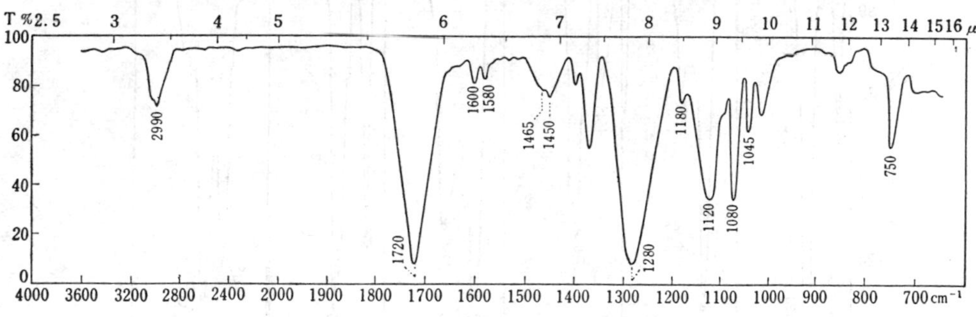

Fig. 59 Liquid film

PROBLEM 60 / Deduce the structure of $C_9H_7NO_4$, m.p. 130~131°.

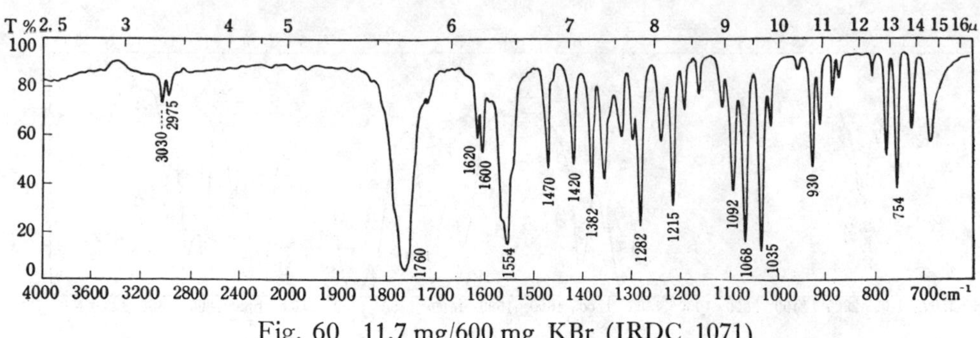

Fig. 60 11.7 mg/600 mg KBr (IRDC 1071)

PROBLEM 61 / Correlate figures with compounds. Figs. a, c, d, and f lack bands between 1600~1500 cm^{-1}.

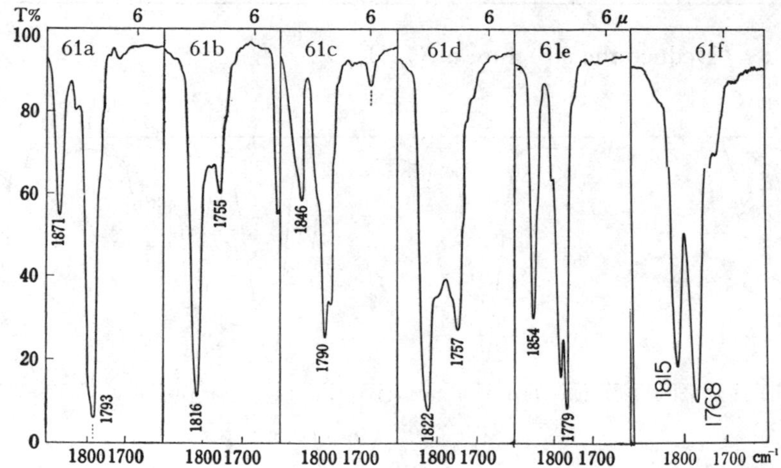

Fig. 61 νC=O of various anhydrides in $CHCl_3$ (0.1 mm cell).

acetic	succinic	*o*-nitrobenzoic	phthalic	glutaric	itaconic

PROBLEM 62 / Deduce the structure of $C_8H_{14}O_3$.

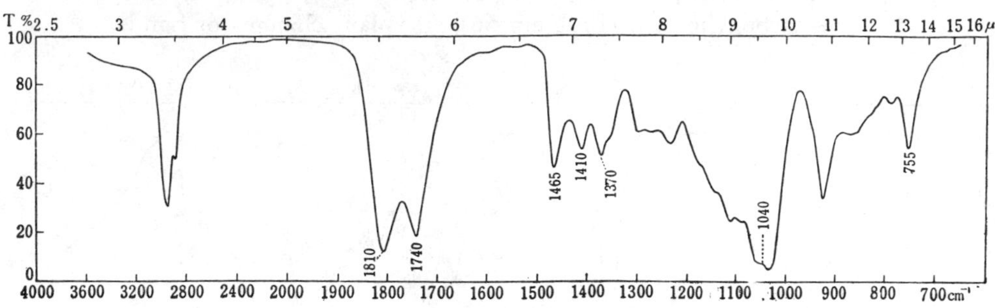

Fig. 62 Liquid film

PROBLEM 63 / What functional groups can be guessed from the spectrum of a compound $C_8H_{12}O_4$?

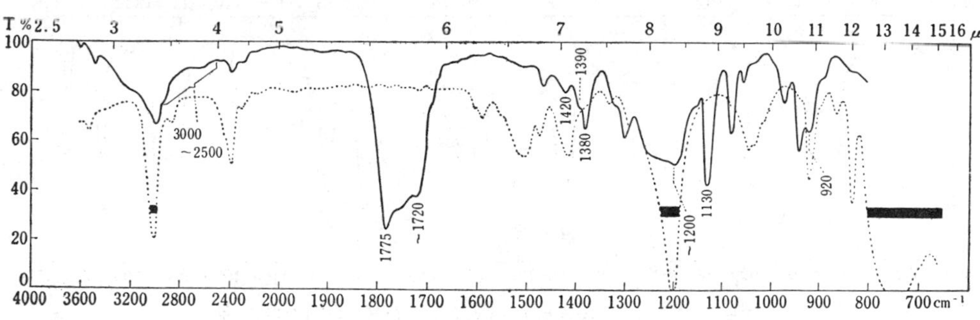

Fig. 63

——— : Sample in $CHCl_3$, 0.5 mm cell.

- - - - : $CHCl_3$, 0.5 mm cell. The region indicated by broad black lines cannot be used because of solvent absorption.

PROBLEM 64 / The following chart was obtained when a drop of diethylamine was added to the previous $CHCl_3$-solution of 3, 5, 5-trimethylbutyrolactone-4-carboxylic acid (Problem 56). Explain changes of bands.

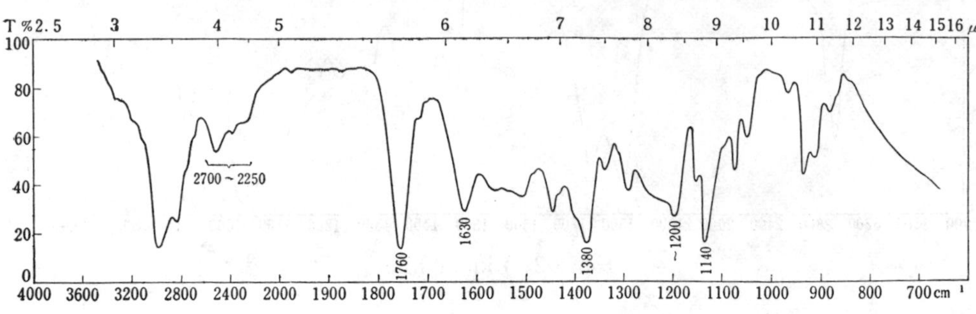

Fig. 64 ($CHCl_3$, 0.5 mm cell)

PROBLEM 65 / Deduce the structure of $C_9H_5NO_4$, m.p. 183~184° (dec.).

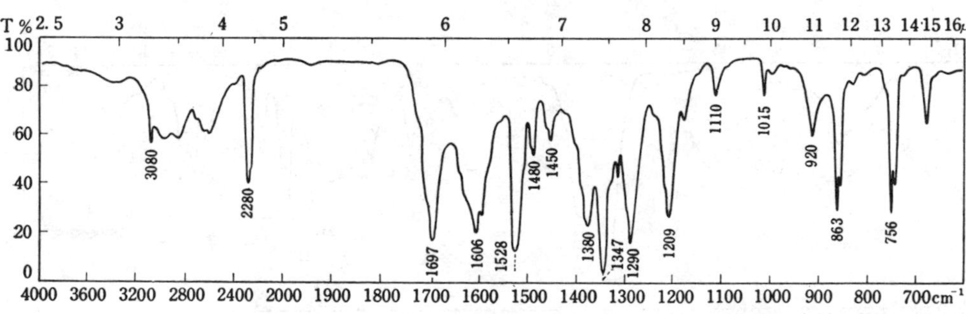

Fig. 65 1.5 mg/600 mg KBr (IRDC 1109)

PROBLEM 66 / Correlate main bands with vibrations.

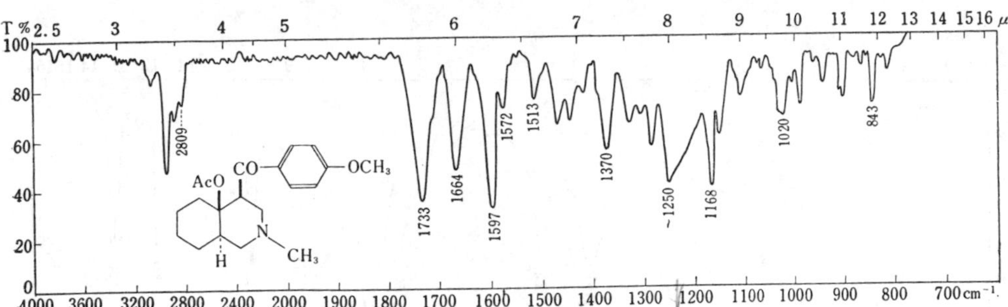

Fig. 66 DL–N–methyl-4–anisoyl–10–acetoxy–decahydroisoquinoline (4,10–*cis*) ($20\,mg/ml$ $CHCl_3$, $0.2\,mm$ cell)

PROBLEM 67 / The hydrochloride of an epimer of compound (1) has the following IR spectrum. When this hydrochloride is converted into the free form the $\nu C{=}O$ is shifted from 1639 to 1672 cm^{-1}. What is the structure of this epimer?

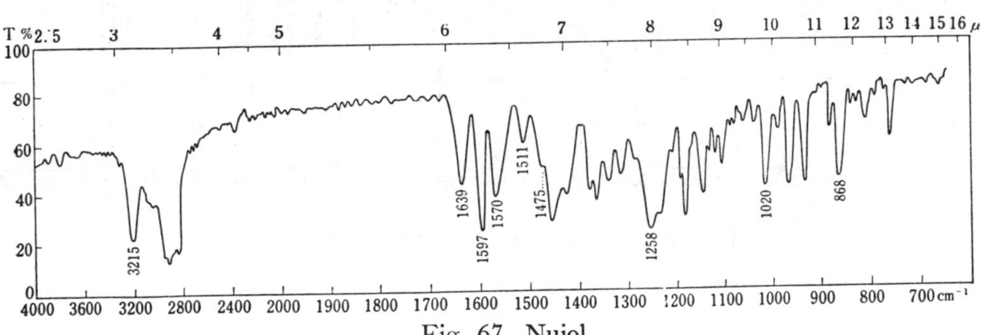

Fig. 67 Nujol

$CO-C_6H_4-OCH_3$
OH
N
CH_3
H

(1)

Bonded $\nu C{=}O$ at $1645\,cm^{-1}$
Free $\nu C{=}O$ at $1656\,cm^{-1}$ (data in $CHCl_3$)

PROBLEM 68 / Deduce the structure of $C_7H_7NO_2$, having pK′a values of ≃5.0 and 10.5.

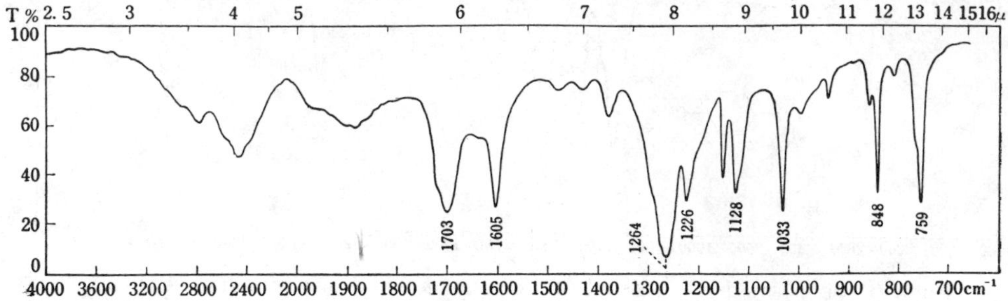

Fig. 68 1.1 mg/600 mg KBr (IRDC 147)

PROBLEM 69 / Correlate the IR peaks with structure of 2-cyano-6-methoxyquinoline.

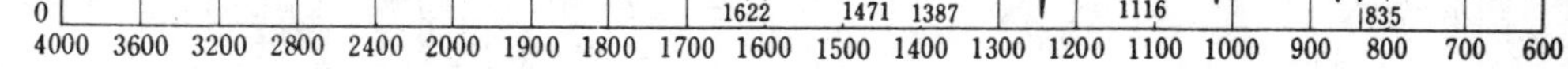

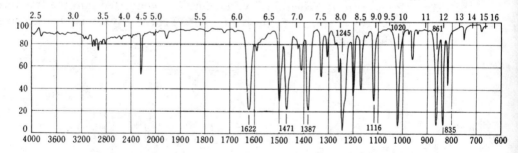

Fig. 69 1.8 mg/600 mg KBr (IRDC 424)

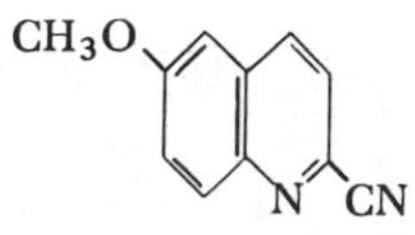

2-cyano-6-methoxyquinoline

PROBLEM 70 / Deduce the structure of $C_3H_7O_2N$.

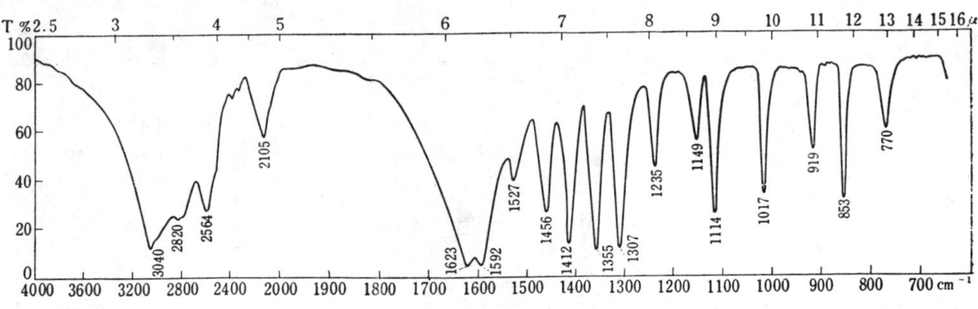

Fig. 70 KBr disk

PROBLEM 71 / Deduce the structure of $C_4H_9NO_2S \cdot HCl$.

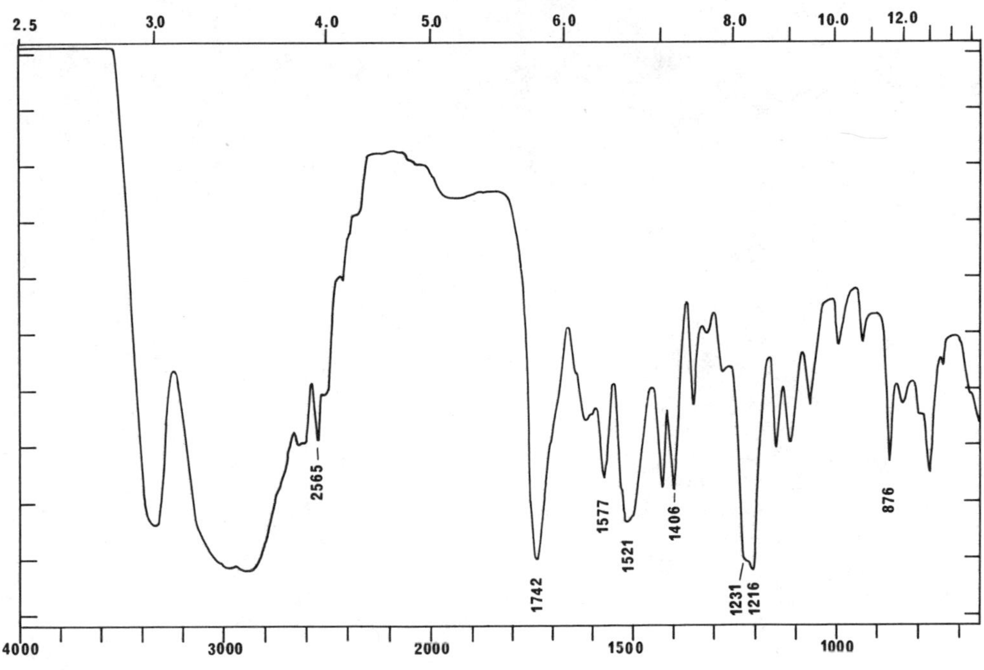

Fig. 71 1.5% in KBr

PROBLEM 72 / Deduce the structure of $C_4H_9O_2N \cdot HCl$ (optically inactive, m.p. 141 ~142°).

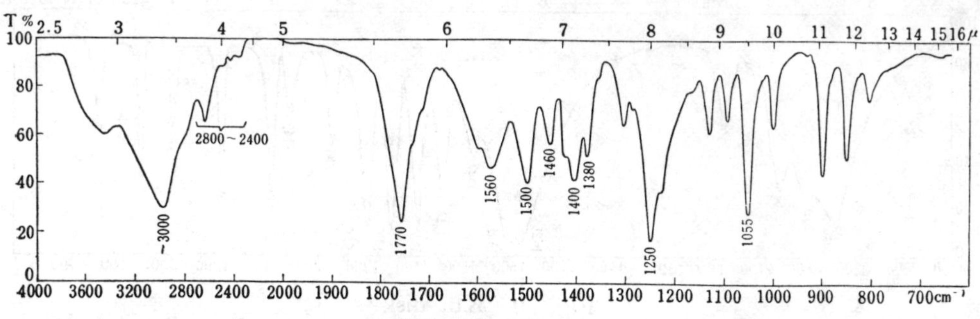

Fig. 72 KBr disk

PROBLEM 73/ Identify the structure of the *t*-butyl-containing compound with molecular formula $C_6H_{13}NO$.

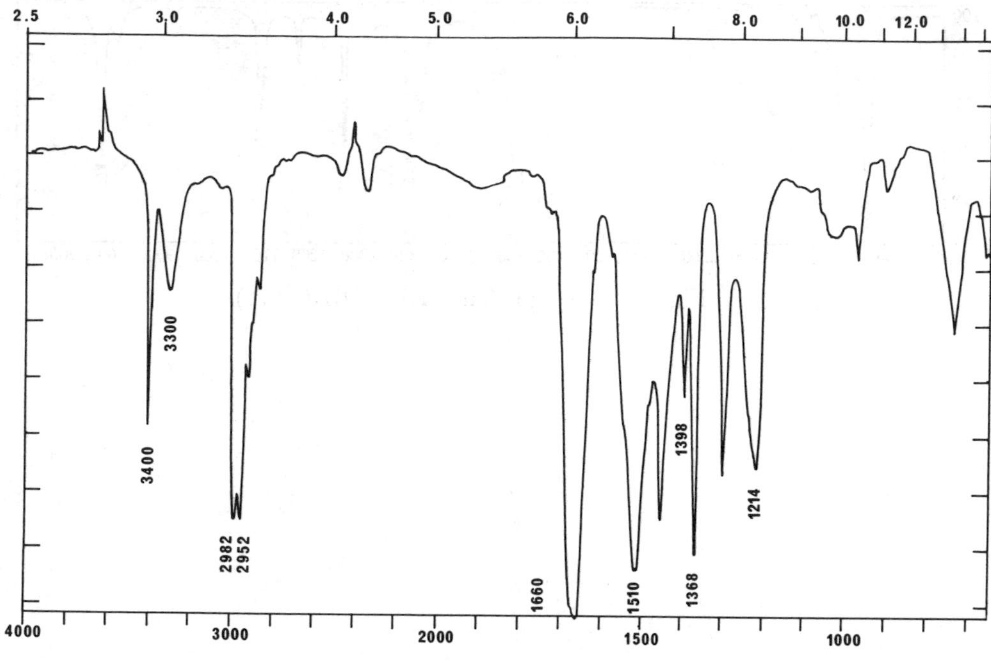

Fig. 73-a $CHCl_3$ solution

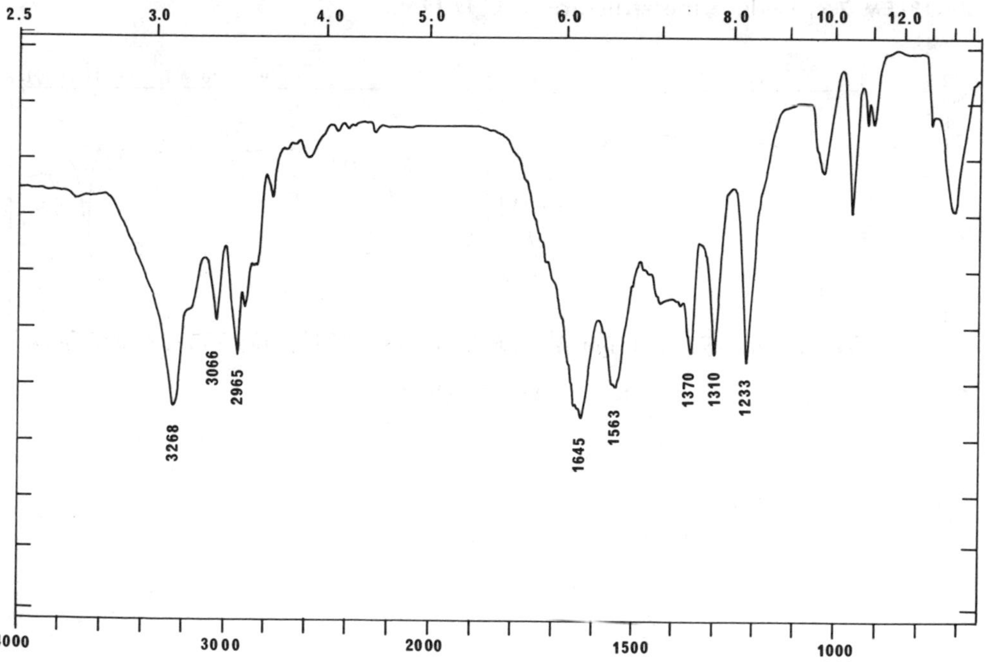

Fig. 73-b 3 mg/200 mg KBr

PROBLEM 74 / Deduce the structure of C_2H_4ClNO, m.p. 118~119°.

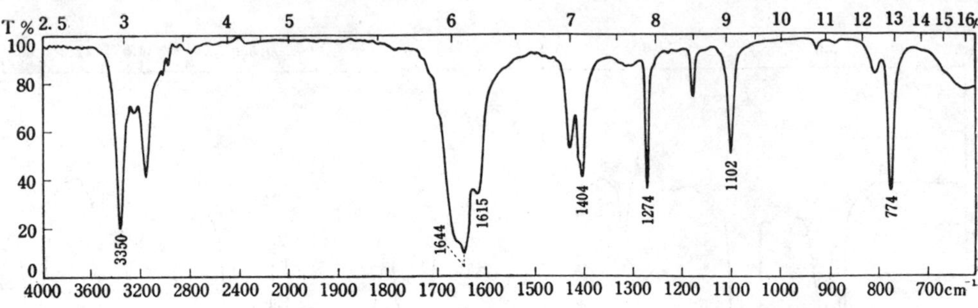

Fig. 74 1.0 mg/600 mg KBr (IRDC 976)

PROBLEM 75 / Deduce the structure of C_8H_9ON.

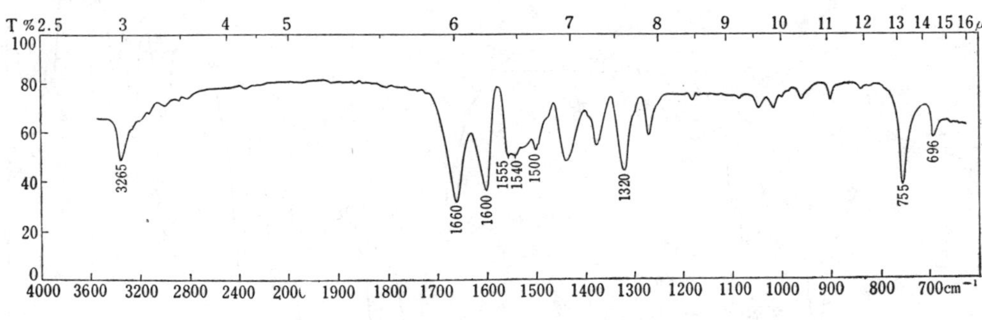

Fig. 75 (KBr)

PROBLEM 76 / Deduce the structure of C_9H_7NO, m.p. 85~86°.

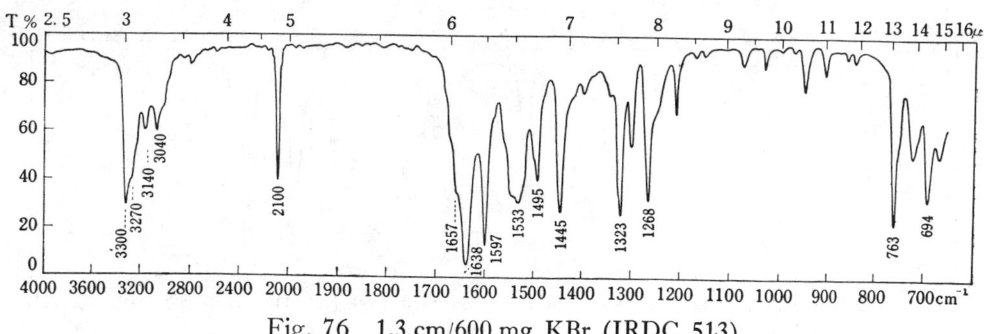

Fig. 76 1.3 cm/600 mg KBr (IRDC 513)

PROBLEM 77 / Deduce the structure of $C_{14}H_{13}ON$. Benzoic acid is obtained upon hydrolysis.

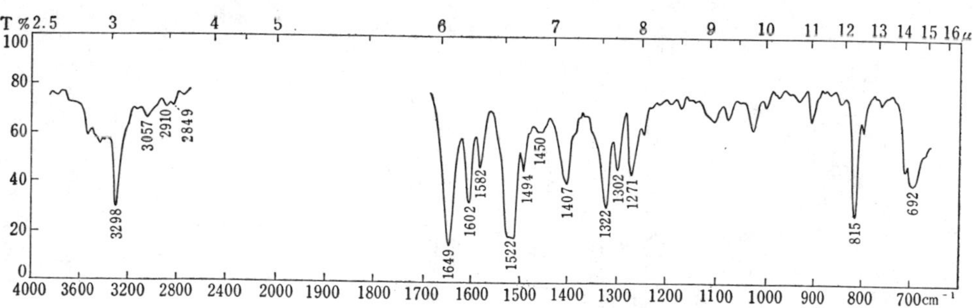

Fig. 77 KBr disk

PROBLEM 78 / Correlate bands with bond vibrations.

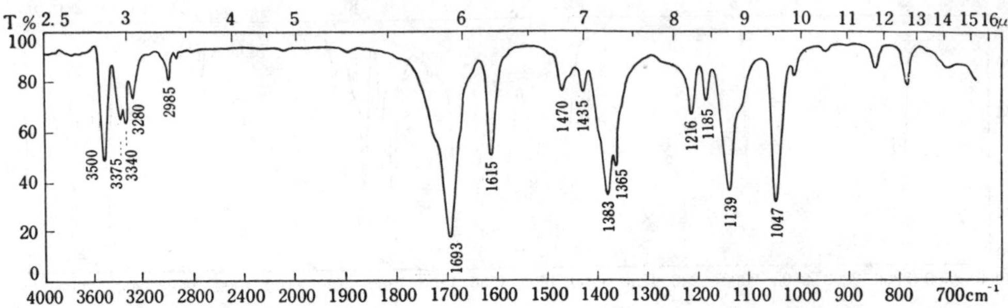

Fig. 78 1,1–Dimethyl–2–propynyl carbamate (0.8 mg/600 mg)

H_3C $C{\equiv}CH$
$>C<$
H_3C $O\text{-}CO\text{-}NH_2$

PROBLEM 79 / Which would be more basic of the two pyrones as deduced from the IR spectra?

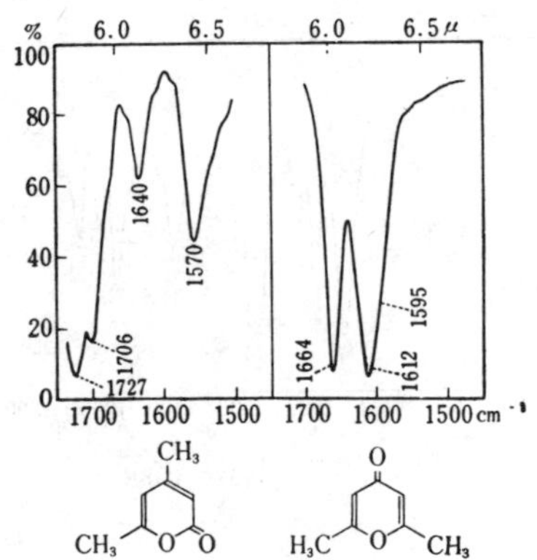

Fig. 79 4, 6–Dimethyl–α–pyrone and 2, 6–dimethyl–γ–pyrone ($CHCl_3$ solution)

PROBLEM 80 / Assign the bands at 1680 and 1625 cm^{-1} to the two carbonyl groups of 2–aminoanthraquinone. In contrast, only a single band at 1673 cm^{-1} is seen in the $\nu C{=}O$ region of 2–hydroxyanthraquinone (in nujol). How could this difference be rationalized?

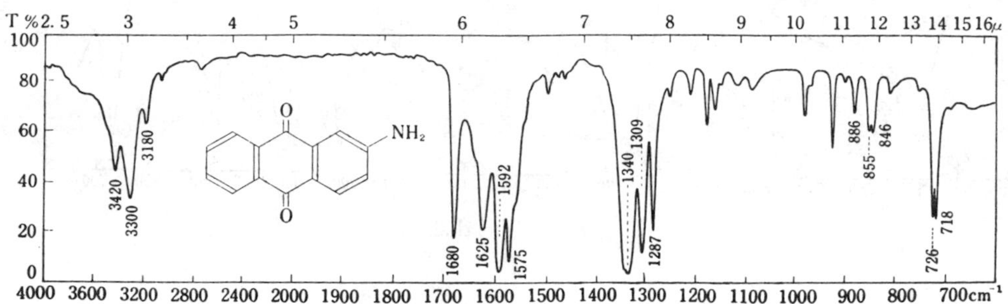

Fig. 80 2–Aminoanthraquinone (KBr)

PROBLEM 81 / Deduce the structure of C_5H_5NO, m.p. 66～67°.

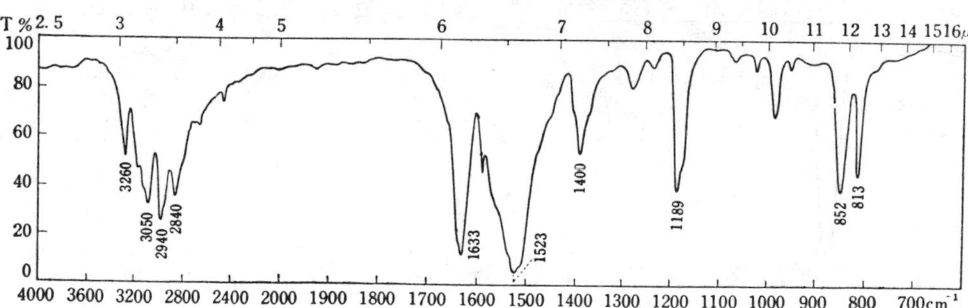

Fig. 81 0.8 mg/600 mg KBr (IRDC 310)

PROBLEM 82 / Deduce the structure of $C_9H_6O_4$.

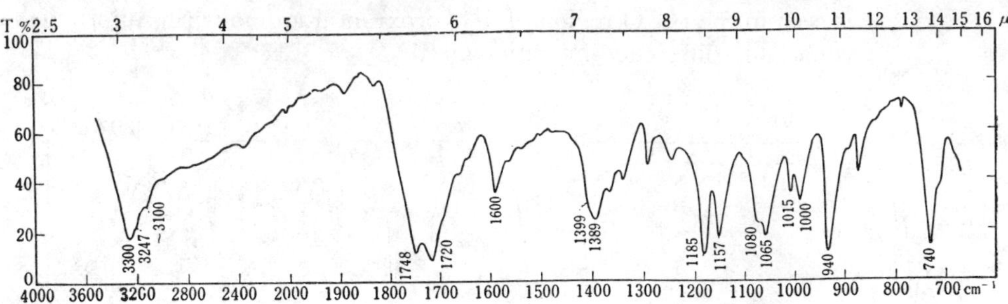

Fig. 82 KBr disk

PROBLEM 83 / Deduce the structure of $C_{18}H_{35}OCl$.

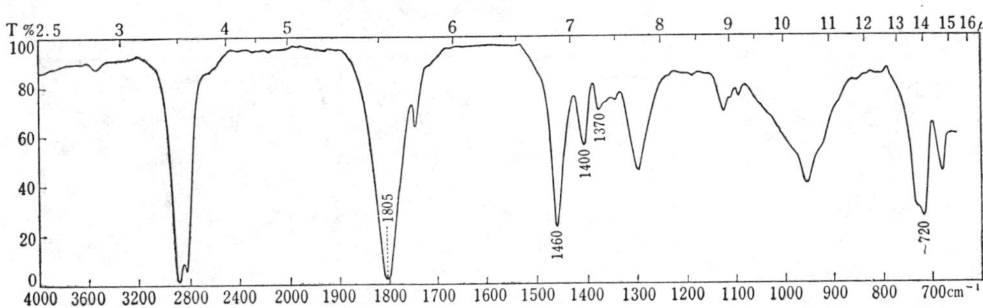

Fig. 83 Liquid film

PROBLEM 84 / Deduce the structure of the hydrate, $C_6H_7NO \cdot H_2O$.

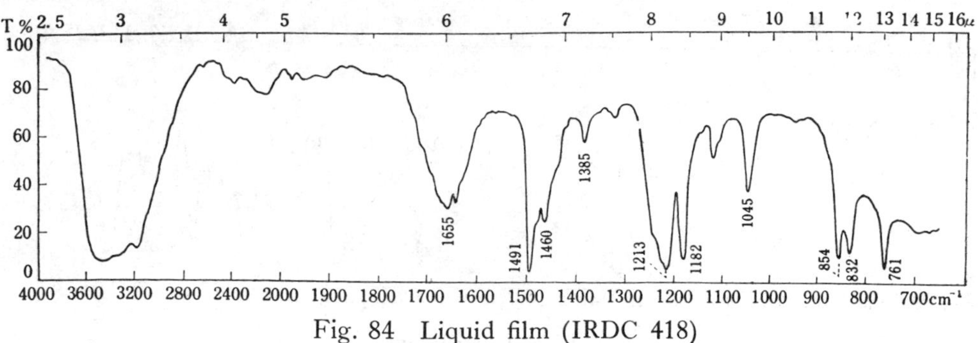

Fig. 84 Liquid film (IRDC 418)

PROBLEM 85 / Deduce the structure of the nitrate, $C_3H_9N_3O \cdot HNO_3$, m.p. 108°.

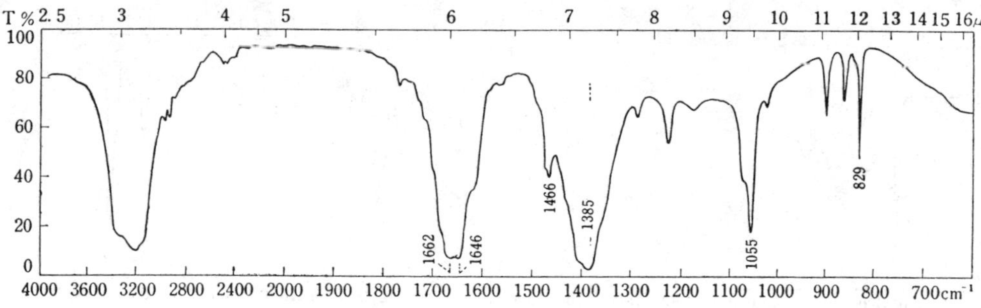

Fig. 85 1.0 mg/600 mg KBr (IRDC 1319)

PROBLEM 86 / Deduce the structure of $C_7H_8O_3$.

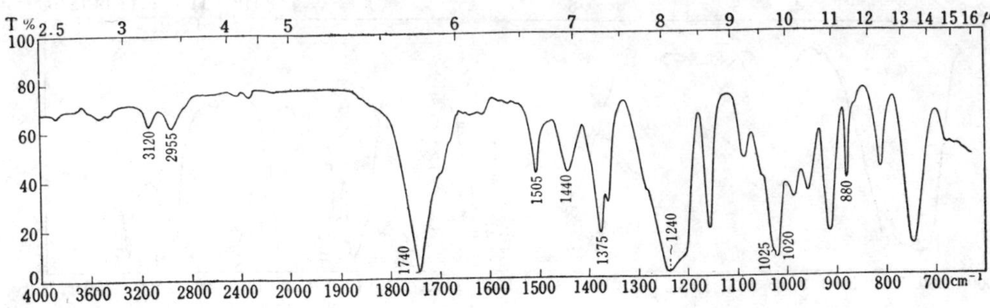

Fig. 86 Liquid film

PROBLEM 87 / Deduce the structure of $C_{11}H_{11}NO_2$, m.p. 134°.

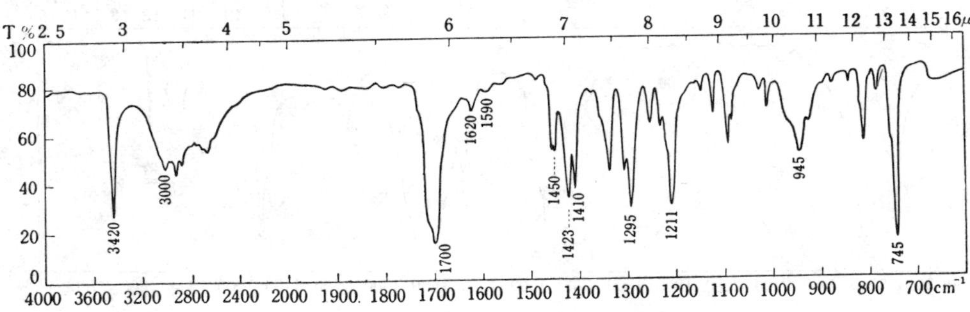

Fig. 87 1.5 mg/600 mg KBr (IRDC 1333)

PROBLEM 88 / Identify the functional groups present in the hygroscopic organic salt with molecular formula $C_{21}H_{33}N_3SO_4$.

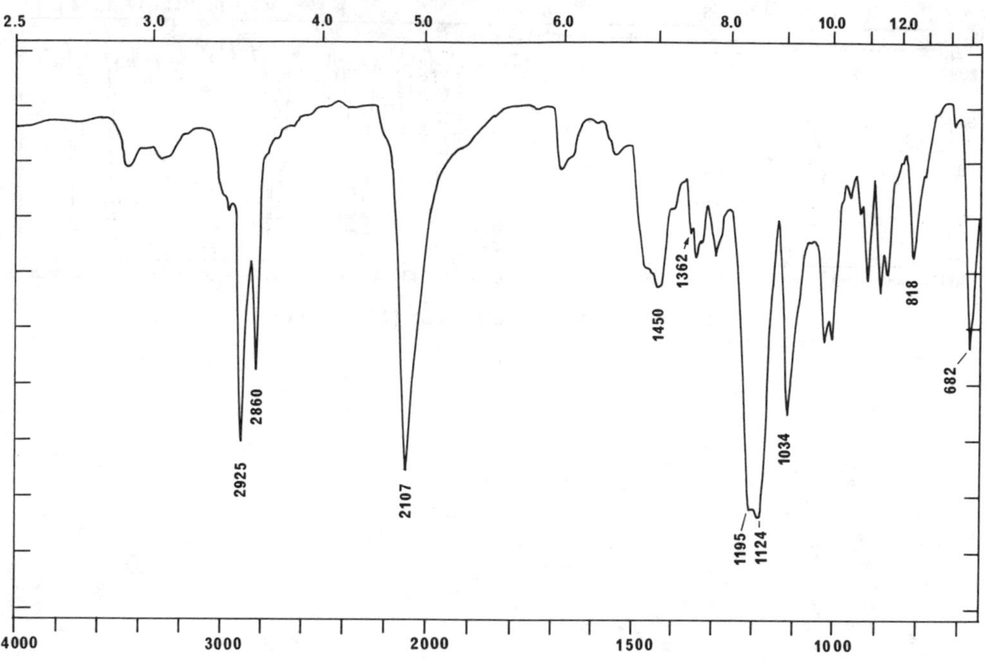

Fig. 88 3 mg/200 mg KBr

PROBLEM 89 / The spectrum is that of a steroid derivative containing three oxygen atoms and one sulfur. What is the nature of the derivative?

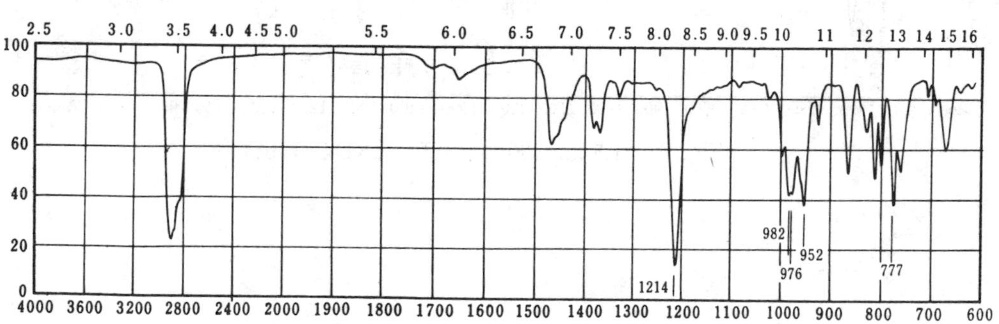

Fig. 89 0.9 mg/300 mg KBr (IRDC 6544)

PROBLEM 90 / What groups can be detected in the spectrum of $C_{17}H_{26}O_2S$?

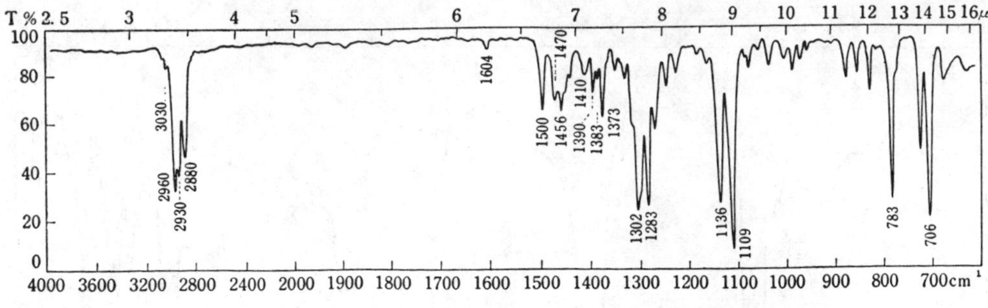

Fig. 90 2.6 mg/600 g KBr (IRDC 1157)

PROBLEM 91 / Deduce the structure of $C_6H_8N_2O_2S$.

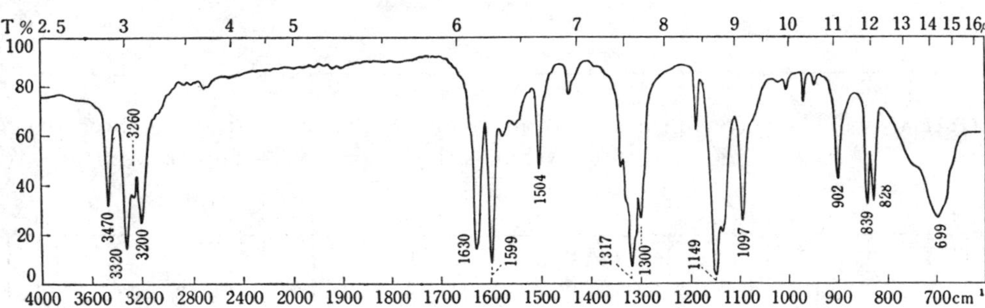

Fig. 91 1.5 mg/600 mg KBr (IRDC 1165)

PROBLEM 92 / Deduce the structure of $C_{14}H_{14}OS$, m.p. 133~134°.

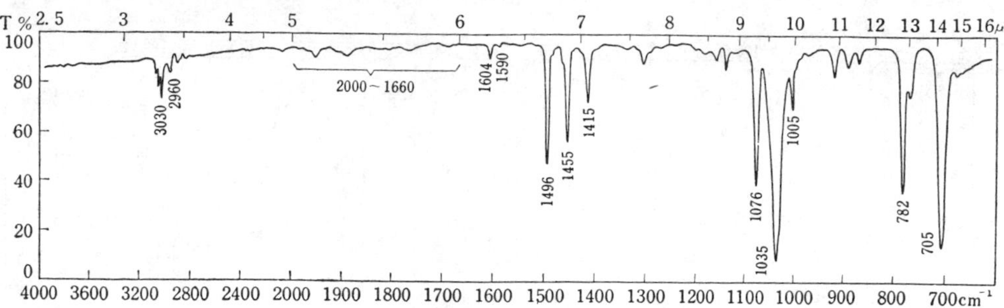

Fig. 92 2.0 mg/600 mg KBr (IRDC 1146)

PROBLEM 93 / Deduce the structure of $C_7H_7ClO_2S$, m.p. 92°.

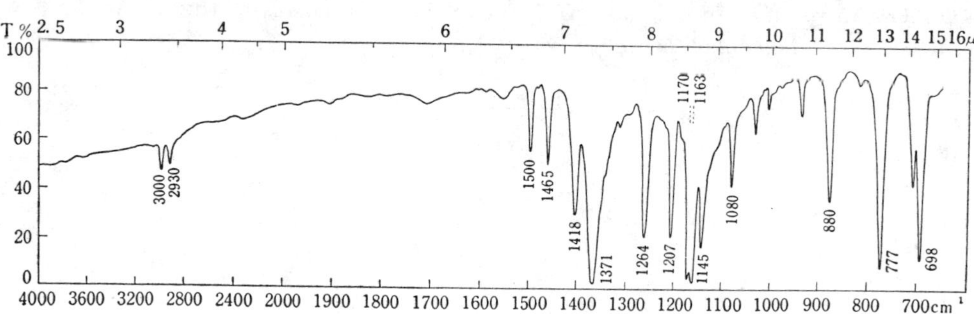

Fig. 93 1 mg/300 mg KBr (IRDC 486)

PROBLEM 94 / Deduce the structure of the derivative of allyl alcohol with formula $C_9H_{15}PO_4$.

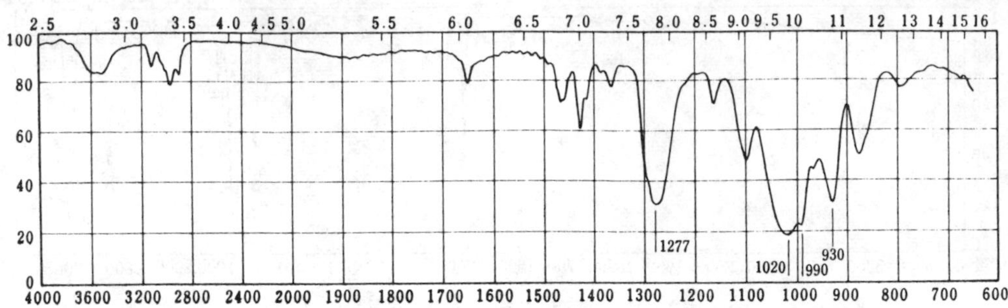

Fig. 94 Liquid film (IRDC 4306)

PROBLEM 95 / What functional groups can be discerned in the compound with molecular formula $C_8H_{17}PO_5$?

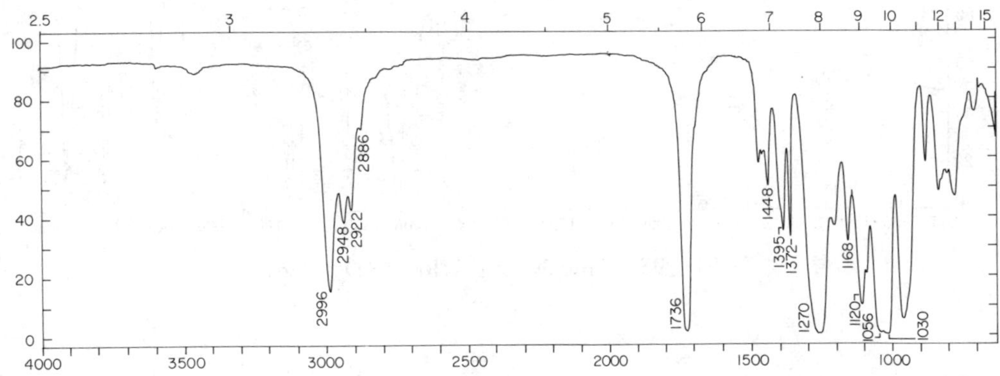

Fig. 95 Liquid film

PROBLEM 96 / Deduce the structure of $C_{13}H_{12}S$, m.p. 55°.

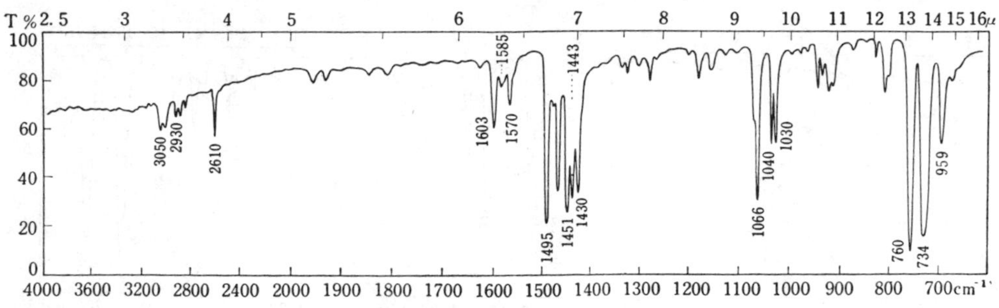

Fig. 96 3.0 mg/600 mg KBr (IRDC 708)

PROBLEM 97 / Deduce the structure of $C_7H_4ClF_3$, b.p. 135～136° (745 mm Hg).

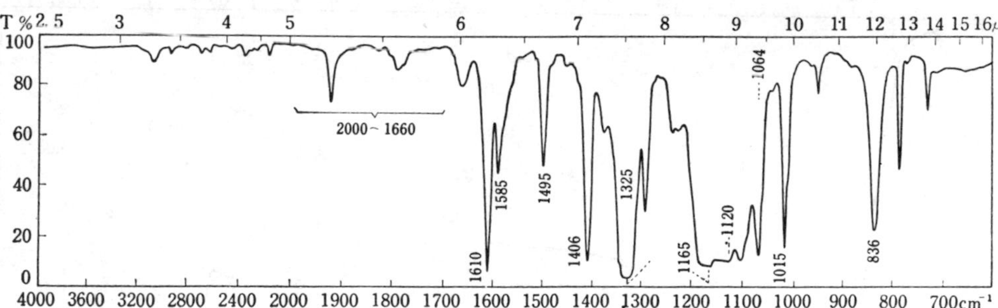

Fig. 97 Liquid film (IRDC 1381)

PROBLEM 98 / Deduce the structure of $C_{13}H_{11}N$, m.p. 56°.

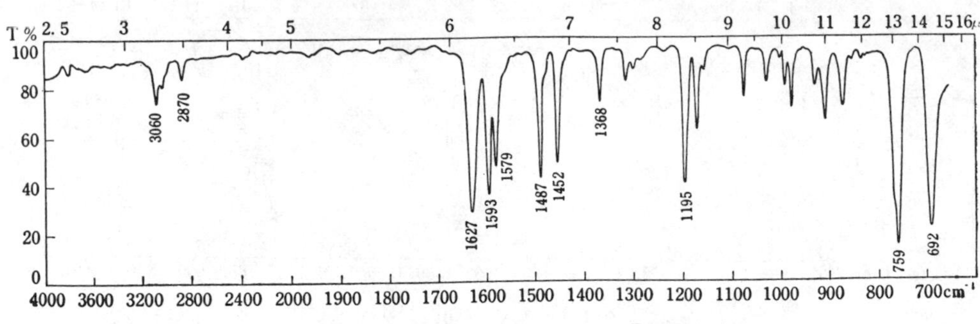

Fig. 98 2.1 mg/600 mg KBr (IRDC 518)

PROBLEM 99 / Identify the structure of the hygroscopic compound with formula $C_4H_7NO_2$.

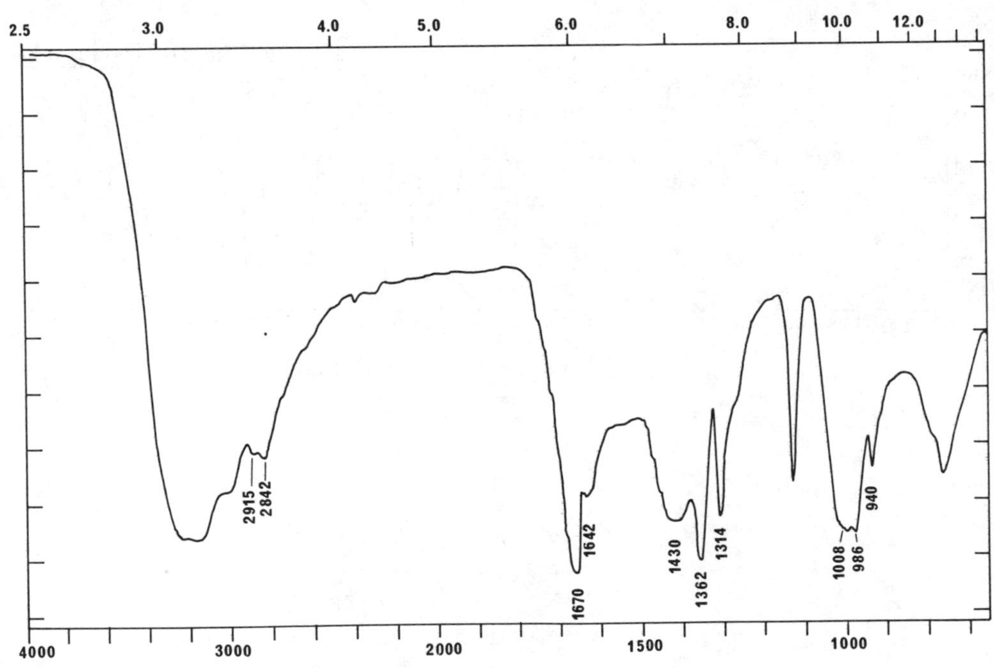

Fig. 99 3 mg/200 mg KBr

PROBLEM 100 / A sample of cyclohexane was flash heated under reduced pressure at 800°C and trapped on a CsI plate cooled to 20°K. The infrared spectrum was measured at 10°K (Fig. 100-a) and then remeasured at 10°K (Fig. 100-b) after warming to 74°K. Account for the differences in the two spectra.

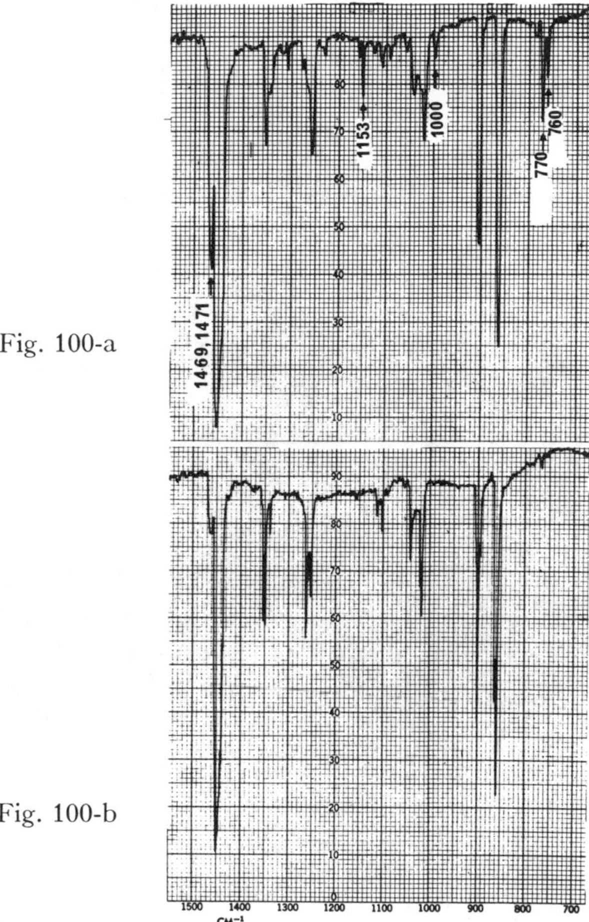

Fig. 100-a

Fig. 100-b

ANSWERS

PROBLEM 1 / Calculate the thickness of the three cells from their interference fringes.

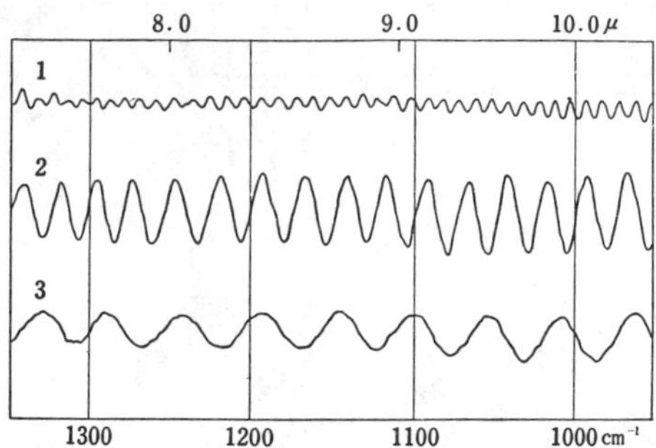

Fig. 1 Interference fringes of 3 cells

ANSWER

The cell thickness can be measured easily from the interference fringe pattern obtained when an empty cell is placed in the IR light path[1,2]. Two arbitrary positions ν_1 and ν_2 (cm^{-1}) on a pattern and the number of interference fringes n between ν_1 and ν_2 are related by the following equation:

$$d=(n/2)\cdot 1/(\nu_1-\nu_2)$$

where d is the cell thickness in cm units.
With infrared spectrophotometers linear in wavelength, this becomes:

$$d=(n/2)\cdot[\lambda_1\cdot\lambda_2/(\lambda_1-\lambda_2)]$$

where λ_1 and λ_2 are the two arbitrary positions expressed in microns. In the latter instruments, the pattern gets broader as the wavelength increases.

Cell no. 1: $n=30$ between 1300—1000 cm^{-1}.
$d=30/2\cdot 1/300=0.050$ (cm)

Cell no. 2: $n=12$ between 1300—1000 cm^{-1}
$d=12/2\cdot 1/300=0.020$ (cm)

Cell no. 3: With thick cells, the calculated value may vary depending on whether the number of peaks or troughs is taken. There are 7 peaks between 1300—1000 cm^{-1}, thus $d=0.012$ (cm). On the other hand, there are only 6 troughs in this range, thus $d=0.010$ (cm). The average value can be taken.

More accurate values are obtained by numbering the peaks and troughs consecutively and plotting these numbers on the ordinate and the respective wave-numbers on the abscissa; a line of slope $4d$ (d in cm) will be obtained. However, the simplified method suffices for qualitative purposes. A uniform interference fringe pattern will not be obtained if the cell window spacing is uneven, or if the cell window surfaces are damaged by solvents, moisture, etc.

[1] D. C. Smith and E. C. Miller, *J. Opt. Soc. Am.* **34**, 130 (1944).
[2] G. B. B. M. Surtherland and H. A. Willis, *Trans. Faraday Soc.* **41**, 181 (1945).

PROBLEM 2 / Deduce the structure of compound C_6H_{14}.

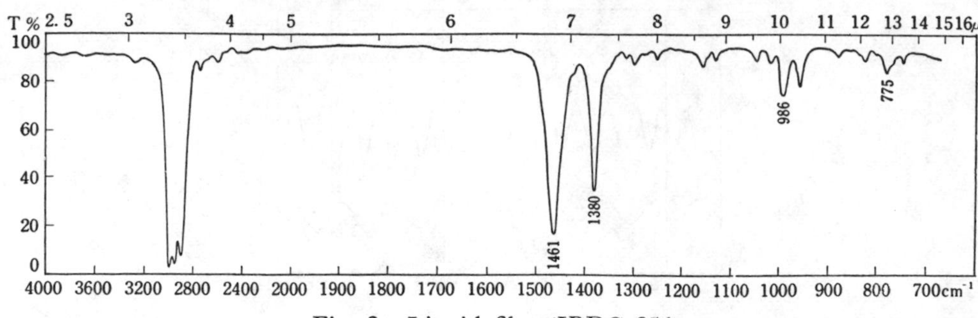

Fig. 2 Liquid film (IRDC 256)

ANSWER

3–Methylpentane

$$\begin{array}{c} \quad CH_3 \\ \quad | \\ CH_3\text{-}CH_2\text{-}CH\text{-}CH_2\text{-}CH_3 \end{array}$$

This is a typical spectrum of a saturated hydrocarbon. No *gem*-dimethyl groups are present because the 1380 cm^{-1} band is a singlet, and the 775 cm^{-1} band suggests an ethyl group. These considerations suggest the shown structure. The spectrum is similar to that of nujol (Chap. 1, Fig. 1.5), which is a mixture of saturated hydrocarbons, except that more fine structure is apparent in the present case due to its lower molecular weight; the IR bands of compounds with high molecular weight tend to be rather broad because of overlap of closely located bands.

1461 cm^{-1}: Methylene and methyl bending (1470 and 1460 cm^{-1}, respectively).
1380 cm^{-1}: Methyl bending (1380 cm^{-1}).
775 cm^{-1}: Rocking of ethyl-CH_2 (780 cm^{-1}).

PROBLEM 3 / Deduce the structure of an optically active monoterpenoid essential oil $C_{10}H_{16}$. No conspicuous UV maximum.

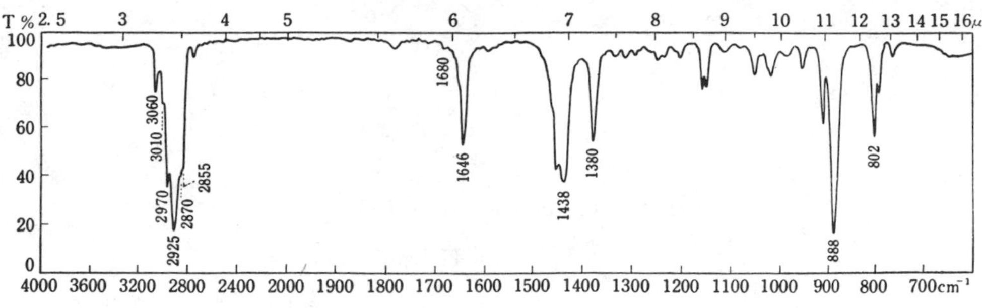

Fig. 3 Liquid film (IRDC 1155)

ANSWER

d–Limonene

Degree of unsaturation as calculated from $C_{10}H_{16}$ is three, two of which are accounted for by terminal methylene (888 cm^{-1}) and tri-substituted double bonds (802 cm^{-1}). The compound is therefore monocyclic. The third degree of unsaturation must be accounted for by a ring, if the terpenoid skeleton, the tri-substituted double bond, lack of conjugation, and optical activity are taken into consideration; the absence of UV absorption precludes a conjugated system. The 1380 cm^{-1} band is a singlet, and hence a *gem*-dimethyl group is absent. The shown structure becomes the most plausible guess.

3060 and 3010 cm^{-1}: Stretching of unsaturated CH (3080 and 3020 cm^{-1}).
2970 and 2870 cm^{-1}: Stretching of methyl (2960 and 2870 cm^{-1}).
2925 and 2855 cm^{-1}: Stretching of methylene (2925 and 2850 cm^{-1}).
1680 cm^{-1}: νC=C of tri-substituted double bond (1670 cm^{-1}).
1646 cm^{-1}: νC=C of terminal methylene (1655 cm^{-1}).
1438 and 1380 cm^{-1}: Methyl and methylene bending.
888 cm^{-1}: Out-of-plane bending of terminal methylene (890 cm^{-1}).
802 cm^{-1}: Out-of-plane bending of tri-substituted double bond (840~800 cm^{-1}).

The in-plane-bending band and overtone band of the terminal methylene are not clear with this film thickness.

PROBLEM 4 / Assign out-of-plane CH bending bands (a)∼(g) to the dienes (I)∼(VII).

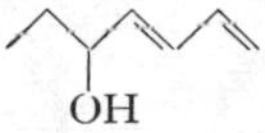

(I) 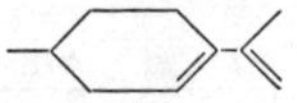 (II) 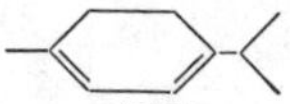 (III) 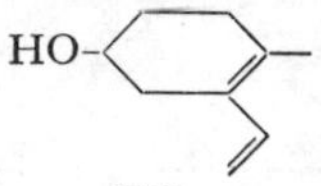 (IV)

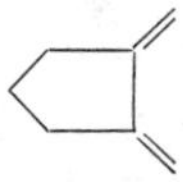

(V) 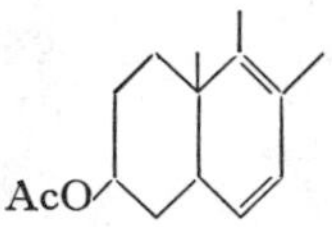 (VI) 3β-Acetoxycholesta-6 : 8-diene 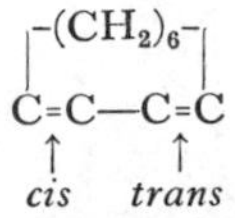 (VII)

-$(CH_2)_6$-
C=C—C=C
cis trans

Out-of-plane δCH bands

(a) 720 cm^{-1}, ε^a 130
(b) 833 cm^{-1}
(c) 980 and 885 cm^{-1}
(d) 976 cm^{-1} s and 836 m, 815 m, 772 m, 720 s, 700 cm^{-1} s.
(e) 890 and 826 cm^{-1}
(f) 1002, 958 (doublet) and 900 cm^{-1}
(g) 880 cm

ANSWER

The out-of-plane CH bending bands of conjugated dienes are not affected much by conjugation and appear in the region of isolated double bonds shown in Table 2, excepting the bands of *cis*-double bonds which are sometimes not clear. Two νC=C bands appear at 1670∼1600 cm^{-1} with an intensity stronger than that of isolated double bonds; however, modes of high symmetry such as the symmetric stretching of butadiene are infrared inactive (but show up in the Raman spectra).

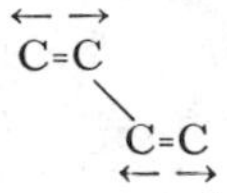

Sym. stretching
IR inactive
(∼1635 cm^{-1})

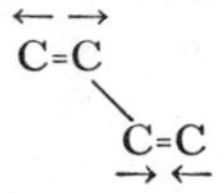

Asym. stretching
IR active
(1600 cm^{-1})

I and (f)[1] (liq.) — 1002 (990), 900 (910); 958 doublet (965)

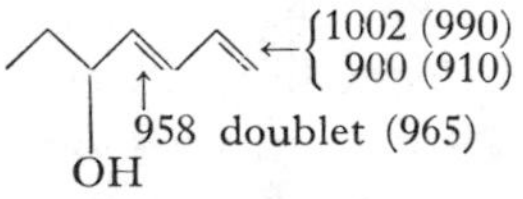

II and (e)[2] (in $CHCl_3$) — 890 (890); 826 (840∼800)

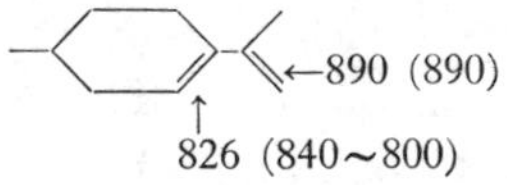

III and (b)[2] (in $CHCl_3$) — 833 (840∼800)

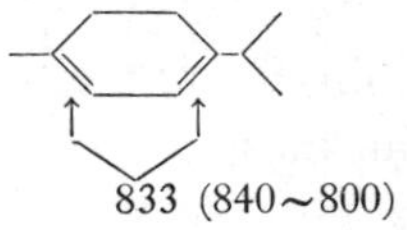

ν C=C band at 1650 and 1610 cm^{-1}.

IV and (c)[3] (in CS_2) — HO- ... δ CH band lacking. 980 (990), 885 (910)

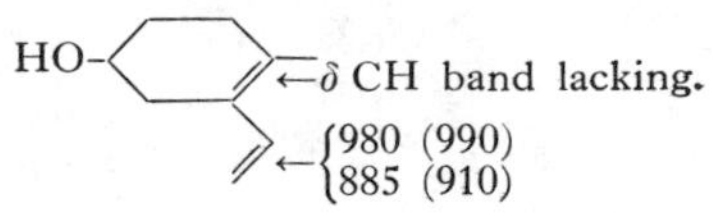

V and (g)[4] (liq.) 880 (890)

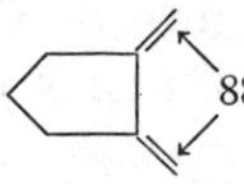

VI and (a)[5] (in CS_2) ←720, ε^a 130 (730~675)

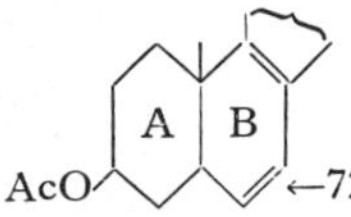

VII and (d)[6] (liq.)

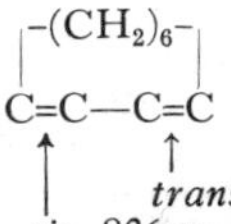

$-(CH_2)_6-$

C=C—C=C

trans, 976 (965)

cis, 836 m, 815 m, 772 m, 720 s, 700 s, (730~675)

976 cm^{-1} *trans* band lacking in *cis-cis*-isomer.

UV of the *cis-cis*-isomer (only end absorption) suggests that the two double bonds are not within the same plane. λ_{max} of *trans*-isomer: 222 mμ, log ε 3.83.

---

[1] K. Alder and H. von Brachel, *Ann.* **608**, 195 (1957); cf. also IR of vitamin D_2, H. H. Inhoffen, K. Brückner, R. Gründel and G. Quinkert, *Ber.* **87**, 1407 (1954).

[2] R. T. O'Conner, L. A. Goldblatt, *Anal. Chem.* **26**, 1726 (1954).

[3] G. Stork, S. S. Wagle and P. C. Mukharji, *J. Am. Chem. Soc.* **75** 3197 (1953).

[4] A. T. Blomquist, J. Wolinsky, Y. C. Meinwold and D. T. Logone, *J. Am. Chem. Soc.* **78**, 6057 (1956); K. Alder and W. Roth, *Ber.* **88**, 407 (1955).

[5] H. B. Henbest, G. D. Meakins, B. Nicholls and R. A. L. Wilson, J. *Chem. Soc.* 997 (1957).

[6] A. T. Blomquist and A. Goldstein, J. *Am. Chem.* Soc. **77**, 998 (1955).

PROBLEM 5 / Deduce the structure of C_7H_{10}.
λ_{max} at 273 mμ (log ε 4.569) and 263 mμ (log ε 4.671)

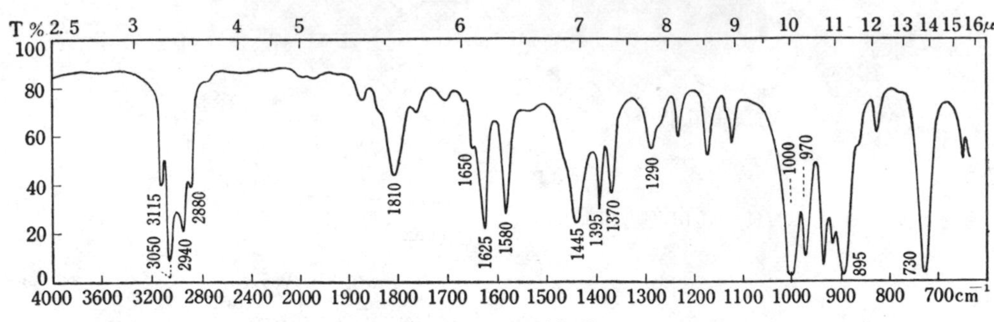

Fig. 5 Liquid film

ANSWER

Hepta-1 : *trans*-3 : *cis*-5-triene[1]

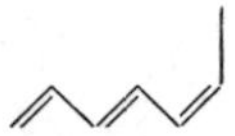

3115 and 3050 cm^{-1}: νCH of alkene (3080 and 3020 cm^{-1}).
2940 and 2880 cm^{-1}: νCH of alkane.
1810 cm^{-1}: Overtone of terminal vinyl 895 cm^{-1}. (1860~1800 cm^{-1}).
1650, 1625, 1580 cm^{-1}: νC=C (1~3 bands at 1650~1600 cm^{-1}).
1445 and 1370 cm^{-1}: Methyl.
1395 and 1290 cm^{-1}: In-plane δCH of terminal vinyl. (1420, 1300 cm^{-1}).
1000, 895 cm^{-1}: Out-of-plane δCH of terminal vinyl (990, 910 cm^{-1}).
970 cm^{-1}: Out-of-plane δCH of *trans* double bond (965 cm^{-1}).
730 cm^{-1}: In-plane δCH of *cis* double bond (730~675 cm^{-1}). In the 1: *trans*-3: *trans* isomer this band is much weaker while the 970 cm^{-1} becomes much stronger.

[1] K. Alder and H. von Brachel, *Ann.* **608**, 195 (1957).

PROBLEM 6 / Correlate bands with groups.

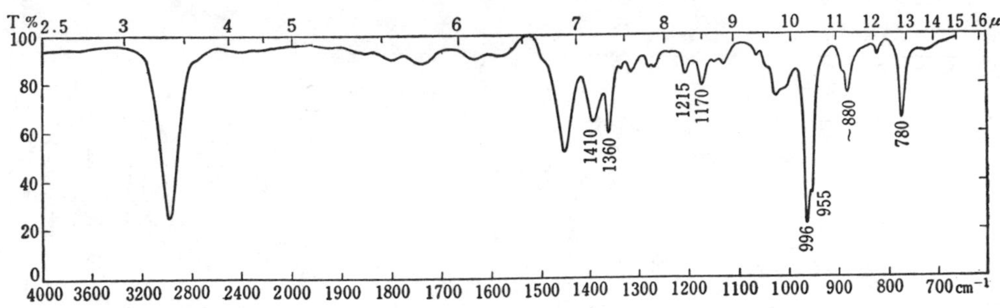

Fig. 6 (in CCl_4 soln.)

15 15′

ANSWER

15–*cis*: 15′–*cis*–β–Carotene[1]

The νC=C bands appear broadly between 1650∼1580 cm⁻¹.

1410, 1360 cm⁻¹: Splitting due to *gem*-dimethyl groups. An additional 1380 cm⁻¹ band appears in carotenes when the moiety (I) in which a methyl is attached to a *cis* double bond is present[2].

$$\begin{matrix} H_3C & & H \\ & C{=}C & \\ {=}C & & C{=} \end{matrix} \qquad (I)$$

1215, 1170 cm⁻¹: *gem*-Dimethyl? (1215 and 1195 cm⁻¹).

996, 955 cm⁻¹: The 965 cm⁻¹ *trans* double bond absorption is usually a singlet in carotenoids. It becomes a doublet when *cis* double bonds are present[2].

880 cm⁻¹: Out-of-plane bending of trisubstituted double bond. The IR of other carotenoids[3] possess this band.

780 cm⁻¹: Central *cis* double bond[2] (730∼670 cm⁻¹). They are located at 780 cm⁻¹ in carotenoids.

---

[1] O. Isler, H. Lindlar, M. Montavon, R. Rüegg and P. Zeller, *Helv.* **39**, 249 (1956).

[2] K. Lunde and L. Zechmeister, *J. Am. Chem. Soc.* **77**, 1647 (1955).

[3] O. Isler, M. Montavon, R. Rüegg and P. Zeller, *Helv.* **39**, 454 (1956); W. Oroshnik, *J. Am. Chem. Soc.* **76**, 5499 (1954); C. D. Robeson, J. D. Cawley, L. Weisler, M. H. Stern, C. C. Eddinger and A. J. Chechak, *ibid.* **77**, 4111 (1955).

PROBLEM 7 / The following three figures are the IR spectra of *o*–, *m*–, and *p*–xylene. Correlate respective figures with the three isomers.

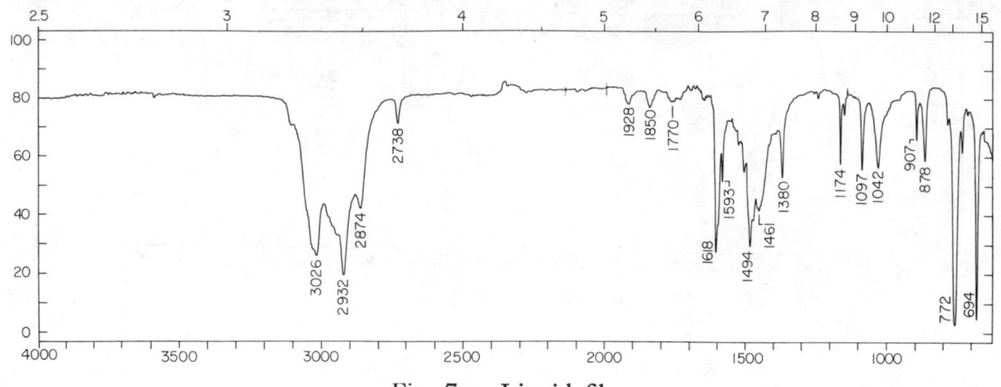

Fig. 7 a Liquid film

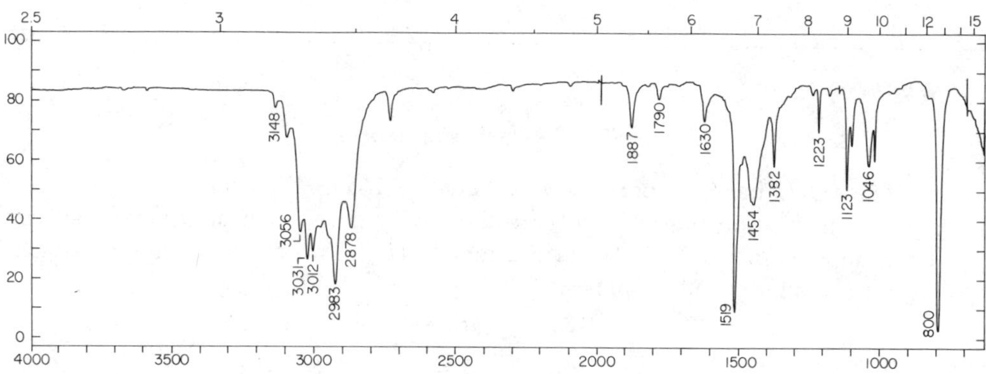

Fig. 7 b Liquid film

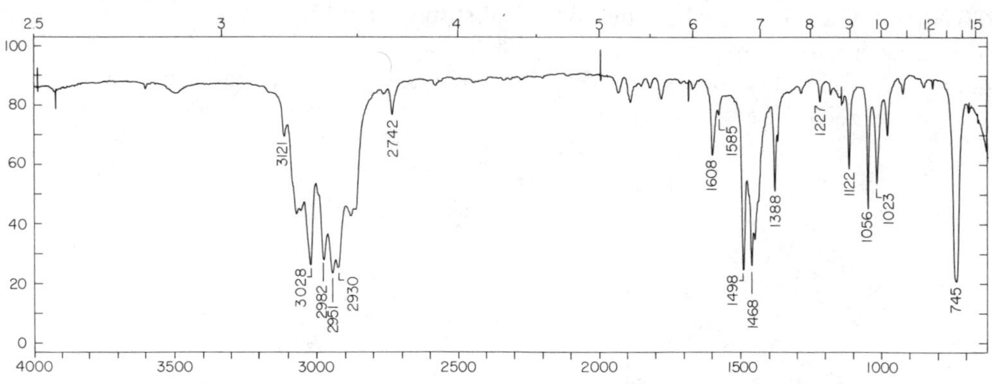

Fig. 7 c Liquid film

ANSWER

The common features of the three IR spectra are:

i) The aromatic and aliphatic νCH bands, respectively, are apparent on the higher and lower frequency side of 3000 cm^{-1}.

ii) The absorption in the 5$\sim$6μ is typical for each substitution pattern (compare with Chap. 2, Fig. 2.1).

iii) The out-of-plane CH bending bands are highly characteristic.

It is to be noted that the position and relative intensities of the set of C=C stretching peaks at 1600, 1580, 1500, and 1450 cm^{-1} vary a great deal, and although there is some dependency on the substitution type, on the whole they afford little structural information.

Fig. 7 a

1618, 1593, 1494 cm^{-1}: Phenyl C=C stretching (1600, 1580, 1500).

1461 cm^{-1}: Phenyl skeletal (1450) and methyl asymmetric bending (1460).

1380 cm^{-1}: Methyl symmetric bending (1380 cm^{-1}).

1174, 1097, 1042 cm^{-1}: In-plane CH bending of phenyl.

772 and 694 cm^{-1}: Out-of-plane CH bending of 1, 3-substituted phenyl (810$\sim$750 and 710$\sim$690, cf. Chap. 1, Fig. 1).

Fig. 7 b

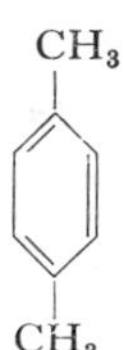

1630, 1519 cm^{-1}: Phenyl skeletal frequency. The 1600 and 1500 cm^{-1} bands are shifted to slightly higher frequencies with para- and 1, 2, 4-tri-substituted phenyls, and to slightly lower frequencies with 1, 2, 3-tri-substituted materials. The 1580 cm^{-1} band, which is usually strong only when groups containing p- or π-electrons are in conjugation, is absent in this particular compound.

1454 cm^{-1}: Phenyl skeletal and methyl bending.

1382 cm^{-1}: Methyl bending.

1223, 1123, 1046, 1023 cm^{-1}: In-plane δCH.

792 cm^{-1}: Two adjacent hydrogens in phenyl ring (840$\sim$810 cm^{-1}).

Fig. 7 c

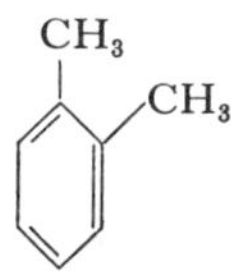

1608, 1585, 1498 cm^{-1}: Phenyl ring.

1468 cm^{-1}: Phenyl and methyl.

1227, 1122, 1056, 1023 cm^{-1}: In-plane δ-CH of 1, 2-substituted phenyl.

745 cm^{-1}: *o*-Substituted phenyl (770$\sim$735 cm^{-1}).

PROBLEM 8 / Deduce the structure of C_7H_7Br, m.p. 28.5°.

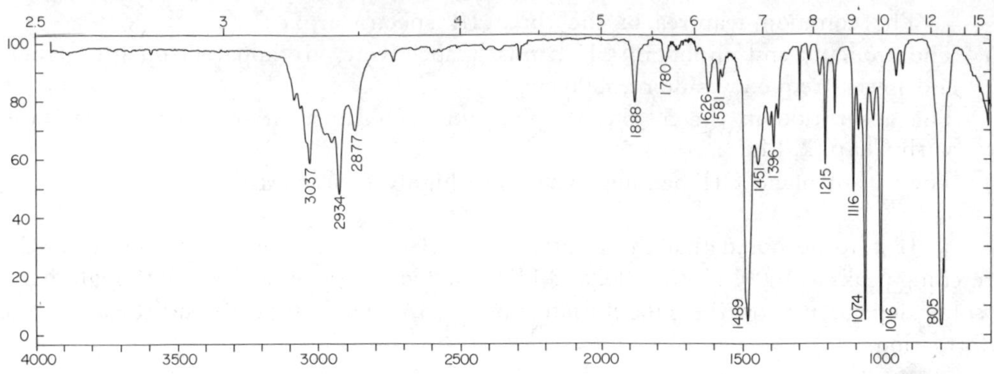

Fig. 8 Liquid film (IRDC 35)

ANSWER

p-Bromotoluene

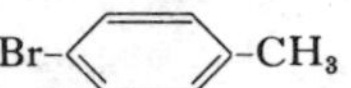

The phenyl ring is *para*-substituted and this leads to the correct structure.

3037 cm^{-1}: Aromatic νCH.
2934 and 2877 cm^{-1}: Methyl νCH.
2000~1660 cm^{-1}: Shape typical for *para*-substitution.
1626, 1581, and 1489 cm^{-1}: Phenyl ring (1600, 1580, 1500 cm^{-1}). The first two bands tend to be higher in *para*-substituted phenyls.
1451 cm^{-1}: Methyl and phenyl ring.
1396 cm^{-1}: Methyl.
1215, 1074, and 1016 cm^{-1}: In-plane CH bending of phenyl.
805 cm^{-1}: Out-of-plane CH bending of 2 adjacent aromatic hydrogens, thus *para*-substituted phenyl.

PROBLEM 9 / Deduce the structure of the solid $C_{14}H_{14}$, m.p. 51.8∼52.0°.

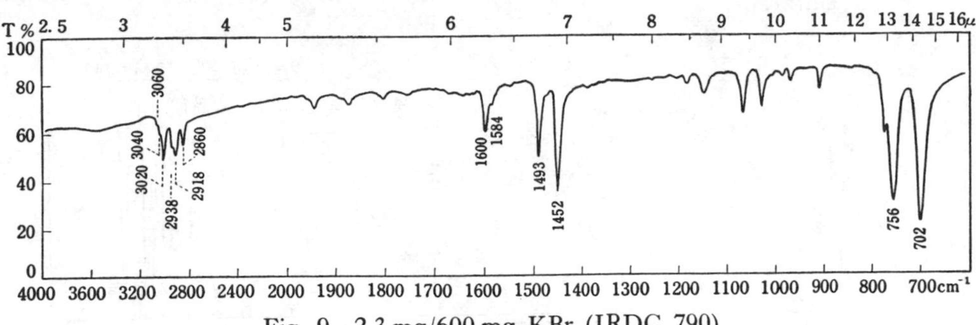

Fig. 9 2.3 mg/600 mg KBr (IRDC 790)

ANSWER

1, 2-Diphenylethane

-CH_2-CH_2-

Degree of unsaturation calculated from molecular formula is eight. Band groups above and below 3000 cm^{-1} are present and this suggests presence of unsaturated and saturated CH groups, respectively. A phenyl ring is present while aliphatic double bonds are absent (ca. 1660 cm^{-1} band lacking); methyl groups are also absent (no 1380 cm^{-1} band).

3060, 3040, 3020 cm^{-1}: Aromatic νCH (several at ca. 3030 cm^{-1}).
2938, 2918, 2860 cm^{-1}: Aliphatic νCH (2925 and 2850 cm^{-1}).
2000∼1660 cm^{-1}: Shape typical for monosubstituted phenyls (p. 27, Fig. 2·1).
1600, 1584, 1493 cm^{-1}: Aromatic νC=C (1600, 1580, 1500 cm^{-1}).
1452 cm^{-1}: Aromatic νC=C and aliphatic δCH_2 (1450 and 1445∼1430 cm^{-1}). The aliphatic methylene bending is lowered because of phenyl group (Table 1∼15).
756, 702 cm^{-1}: Five adjacent H on phenyl (770∼730 and 710∼690 cm^{-1}).

The 1200∼900 cm^{-1} region includes bands due to in-plane aromatic δCH but they are of limited diagnostic value.

PROBLEM 10 / Deduce the structure of the selenium dehydrogenation product $C_{17}H_{16}$ of a natural substance. The UV spectrum is that of an alkylphenanthrene.

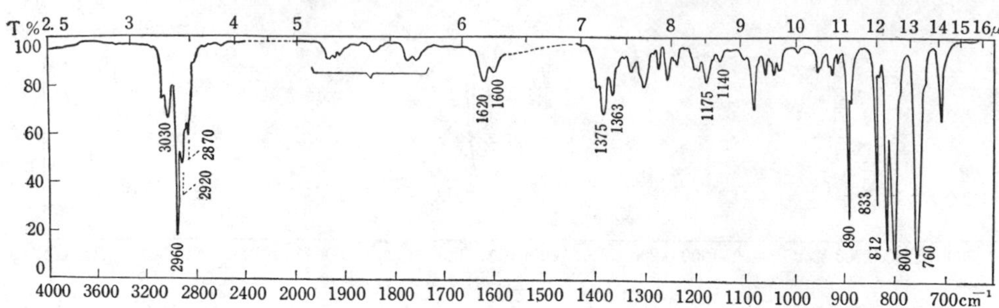

Fig. 10 CS_2 solution in 0.2 mm cell

ANSWER

2-Isopropylphenanthrene (retene)

CH_3
CH
CH_3

The regions of strong solvent absorption* are shown by the dotted line. Splitting of the 1380 cm^{-1} band indicates a *gem*-dimethyl group, which, judging from the molecular formula, can only be an isopropyl; this is supported by the skeletal vibrations at 1175 and 1140 cm^{-1} (1170 and 1140 cm^{-1}). The CH out-of-plane vibrations suggest:

890 cm^{-1}: Isolated H (900∼860 cm^{-1}).
812 and 800 cm^{-1}: Adjacent 2H (860∼800 cm^{-1}, frequently 820∼800 cm^{-1}).
760 cm^{-1}: Adjacent 4H (770∼735 cm^{-1}).

It follows that the product is either 1- or 2-isopropylphenanthrene. The "5∼6 μ bands", although of no diagnostic value in the present case, are apparent. Also note that the aromatic νCH (above 3000 cm^{-1}) are much weaker than the aliphatic νCH (below 3000 cm^{-1}).

---

* Carbon disulfide has a sharp band at 2200 cm^{-1} and a very intense band at ca. 1500 cm^{-1}. Therefore, it cannot be used for measuring the CH_2 and CH_3 bending region, but is one of the best solvents for measurements in the low wave-number range.

PROBLEM 11 / Deduce the structure of C_7H_5NO.

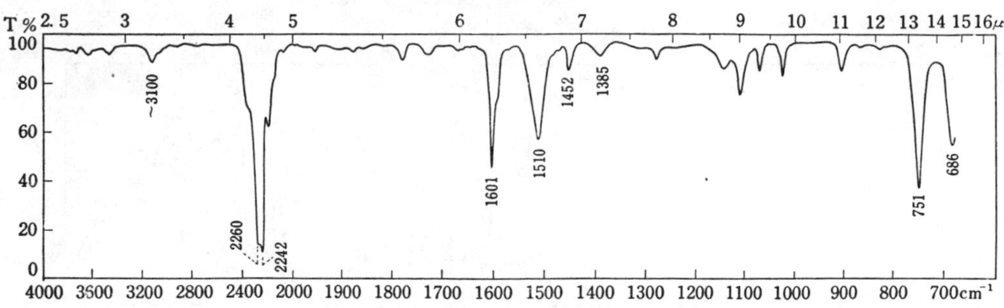

Fig. 11 Liquid film (IRDC 4)

ANSWER

Phenyl isocyanate

-N=C=O

$\simeq$3100 cm^{-1}: Aromatic νCH.

2260, 2242 cm^{-1}: Asymmetric $-N{=}C{=}O$ stretching (2275$\sim$2250 cm^{-1}). Splitting is due to Fermi resonance with overtones of low frequency bands.

1601, 1590 (shoulder), 1601, 1510, 751, 686 cm^{-1}: Monosubstituted phenyl.

1452 or 1385 cm^{-1}: Symmetric $-N{=}C{=}O$ bending ($\simeq$*1400* cm^{-1}). Not practical.

Although these bands have the same frequency as δ_{as} and δ_s of the methyl group, the presence of a methyl group is precluded not only by the molecular formula, but also by the shape of the 1385 cm^{-1} band, which is broader than that of the methyl bending band. (See Problem 7.)

PROBLEM 12 / What is the nature of the two nitrogen atoms in $C_{13}H_{22}N_2$? The compound reacts with water and is converted into a urea derivative.

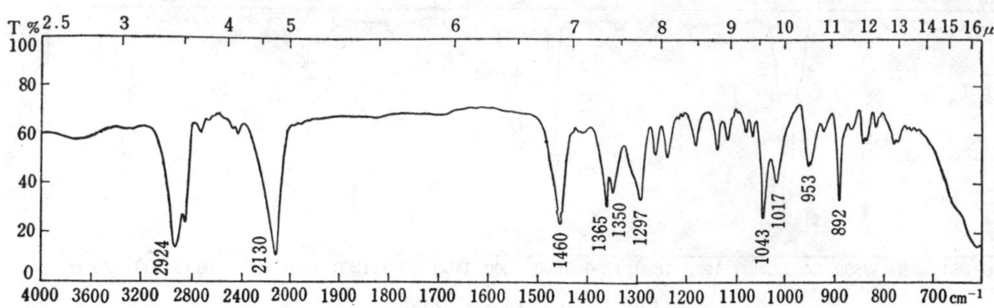

Fig. 12 Liquid film

ANSWER

N, N–dicyclohexylcarbodiimide

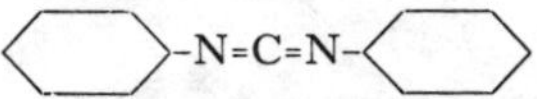

2130 cm^{-1}: Asymmetric stretching of –N=C=N– (2140～2130 cm^{-1}).
1365 cm^{-1}: Symmetric bending of –N=C=N– (≃*1350* cm^{-1}).

PROBLEM 13 / Deduce the structure of C_8H_7N, m.p. 29.5°.

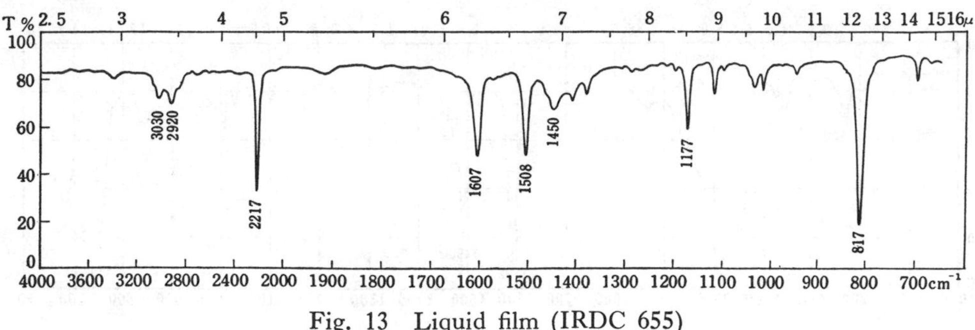

Fig. 13 Liquid film (IRDC 655)

ANSWER

p–Methylbenzonitrile

A *p*-substituted aromatic ring and a nitrile group are at once apparent.

3030 cm^{-1}: Phenyl νCH.
2920 cm^{-1}: Methyl νCH.
2217 cm^{-1}: Stretching of nitrile group (2260~2210 cm^{-1}, lower region when C≡N is conjugated).
1607 and 1508 cm^{-1}: Phenyl ring. Note somewhat high position for *p*-substitution (usually 1600 and 1500 cm^{-1}).
817 cm^{-1}: Adjacent 2H on phenyl (860~800, usually at ≃810 cm^{-1}).

PROBLEM 14 / Deduce the structure of C_7H_5NS.

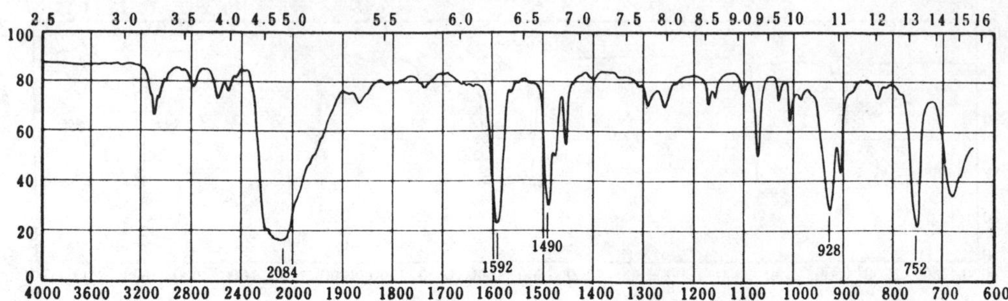

Fig. 14 Liquid film (IRDC 5638)

ANSWER

Phenyl isothiocyanate

N=C=S

3100 cm^{-1}: Absorption above 3000 cm^{-1} indicates only unsaturated ν CH.
2084 cm^{-1}: ν –N=C=S.
1592, 1490 cm^{-1}: Phenyl ring vibration; intensity is increased by conjugated –NCS group.
928 cm^{-1}: Characteristic of isothiocyanate group (930 cm^{-1}, strong).
752 cm^{-1}: δ CH out of plane, monosubstituted phenyl (770~730 cm^{-1}).
685 cm^{-1}: Ring puckering mode, monosubstituted phenyl (710~690 cm^{-1}).

PROBLEM 15 / What are the substituents on the phenyl derivative $C_6H_3FN_3O_2$?

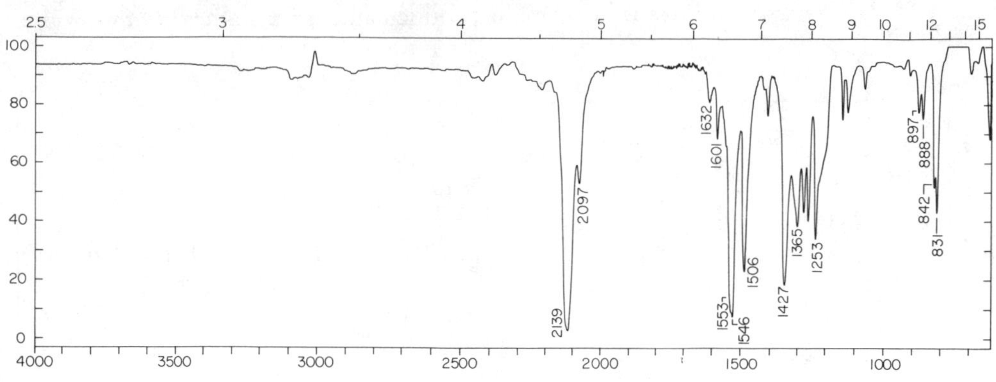

Fig. 15 $CHCl_3$ solution

ANSWER

4-fluoro-3-nitrophenylazide

Absorption near 3000 cm^{-1} is weak and is further obscured by solvent.

2139, 2097 cm^{-1}: aromatic —N=N=N stretch; split due to Fermi resonance is often observed.

1601, 1546, 1506 cm^{-1}: ν C=C aromatic.

1546 cm^{-1}: ν_{as} NO_2.

1427 cm^{-1}: C—F stretch.

1365 cm^{-1}: ν_s NO_2.

The bands below 900 cm^{-1} cannot be used to determine the substitution pattern because of the strongly electronegative substituents.

PROBLEM 16/The following is the infrared spectrum of the crude product (C_9) obtained when $CH_3(CH_2)_3C{\equiv}C{-}MgBr$ is treated with $BrCH_2C{\equiv}CH$. Associate bands with main product and infer nature of by-product.

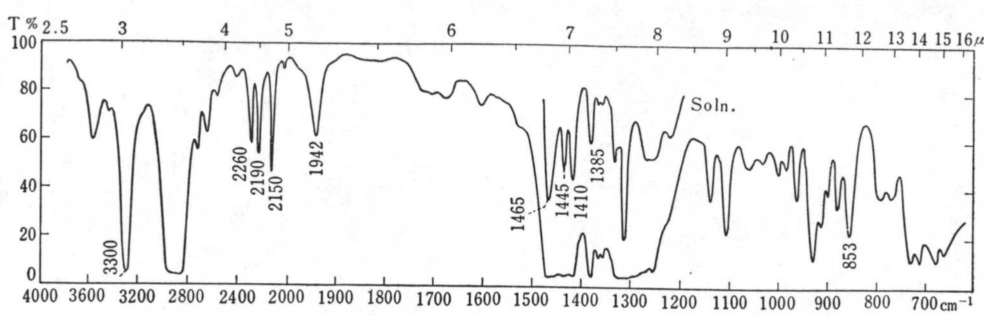

Fig. 16 Liquid film and 0.15 mm cell, CCl_4 solution

ANSWER

Nona-1:4-diyne[1]

$CH_3(CH_2)_3C{\equiv}C{-}CH_2{-}C{\equiv}CH$

3300 cm^{-1}: Acetylenic νCH. Upon deuteration of the terminal acetylene it is replaced by a νC–D band at 2577 cm^{-1}.

2260, 2190, 2150 cm^{-1}: ν C≡C. Number of C≡C stretching bands may exceed that of triple bonds present. They were replaced by a band at 1992 cm^{-1} upon deuteration.

1942 cm^{-1}: Contaminant allene (1950 cm^{-1}). Allenic protons are insufficiently acidic to be substituted by deuterium upon simple exchange reactions, and the bands persisted in the deuterated crude product. Intensity of the 1942 cm^{-1} band indicated that the amount present was roughtly 7%.

1465 cm^{-1}: Methyl and methylenes.

1445 cm^{-1}: Methylene at C–6 (?) (Table 1–15).

1410 cm^{-1}: Methylene at C–3 (?). Flanking by two triple bonds induces a further downward displacement of the 1460 cm^{-1} band.

1385 cm^{-1}: Methyl.

1335 cm^{-1}: This weak shoulder band is CH_2 wag of $-C{\equiv}C{-}CH_2{-}$.

853 cm^{-1}: δ CH of terminal allene (850 cm^{-1}). This allenic 850 cm^{-1} band generally appears to be stronger than the 1950 cm^{-1} band.

The product showed a maximum at 214 mμ (log ε 631) in 95% ethanol but a subsequently prepared pure sample[2] absorbed at 263 mμ (log ε 26) in the UV, and lacked both IR bands at 1942 and 853 cm^{-1}. The diyne is isomerized in two steps by the action of alkali[2]:

[1] W. J. Gensler, A. P. Mahadeuan and J. Casella, Jr., *J. Am. Chem. Soc.* **78**, 163 (1956).
[2] W. J. Gensler and J. Casella, Jr., *ibid.* **80**, 1376 (1958).

$C_4H_9C{\equiv}C\text{-}CH_2\text{-}C{\equiv}CH \longrightarrow C_4H_9C{\equiv}C\text{-}CH{=}C{=}CH_2$

UV (95% EtOH): 220.5 mμ (ε 8200).

IR (liq. film): allene band, 1942, 1710, 853 cm^{-1}

acetylene band, 2230 cm^{-1} (shoulder at 2210 cm^{-1}).

$\longrightarrow C_4H_9C{\equiv}C\text{-}C{\equiv}C\text{-}CH_3$

UV (95% EtOH): 224.5 mμ (ε 366), 232 mμ (ε 294), 237 mμ (ε 335), 252 mμ (ε 194)

IR (liq. film): acetylene band, 2260, 2190, 2150 cm^{-1}.

Also band at 2039 cm^{-1} of unknown origin.

The 237 mμ UV maximum of the final product is in good agreement with the calculated value[3] for conjugated dialkyl-diynes, 239 mμ. Intensity of the 224.5 mμ maximum is larger but this is presumably due to end absorption.

[3] K. Hirayama, "Jikken Kagaku Koza (Experimental Chemistry)" Vol. I–1, p. 44, Maruzen Co., Ltd., (1957). The value 234 mμ is taken as parent and 2.5 mμ is added for each alkyl substituent.

PROBLEM 17 / Deduce the structure of an essential oil $C_{12}H_{10}$.

λ_{max}^{EtOH} 239 mμ (ε 537), 253 mμ (ε 340).

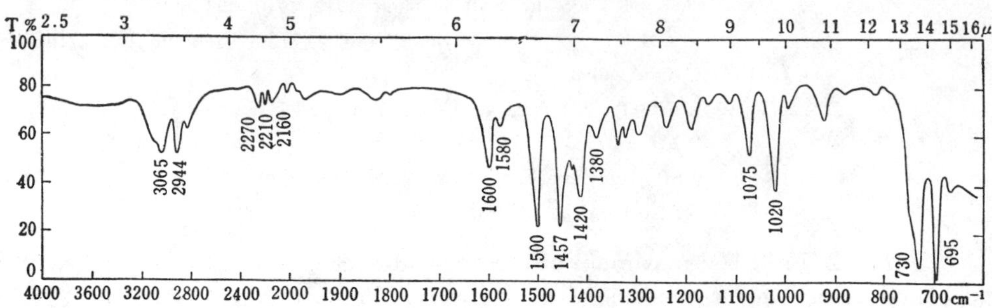

Fig. 17 Liquid film

ANSWER

Capillene[1]

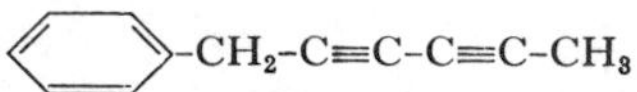

Molecular formula shows that degree of unsaturation is 8. A mono-substituted phenyl group is suggested by the infrared. An acetylenic bond (degree of unsaturation, 2) is present, the λ_{max} agrees with the calculated value of conjugated diynes of the type R–C≡C–C≡C–R′ (234 mμ+2.5 mμ×2)[2], and the absence of a sharp band at 3300 cm^{-1} corresponding to the ν CH of terminal acetylenes also supports the inference. Thus, the shown structure is derived.

3065 cm^{-1}: νCH of phenyl.
2944 cm^{-1}: νCH of methyl and methylene.
2270, 2210, 2160 cm^{-1}: νC≡C.
1600, 1580, 1500 cm^{-1}: Phenyl.
1457 cm^{-1}: Asymmetric bending of methyl.
1420 cm^{-1}: δCH of CH_2, shifted lower from the usual 1470 cm^{-1} position because it is flanked by an aromatic ring and an acetylenic bond (Table 1–15).
1380 cm^{-1}: Symmetric bending of methyl
1075, 1020 cm^{-1}: CH in-plane bending of phenyl.
730 cm^{-1} (770~730 cm^{-1}): Out of plane δ CH of 5 adjacent hydrogen atoms on phenyl ring.
695 cm^{-1}: CH out-of-plane bending of mono-substituted phenyl.

Capillene is the essential oil isolated from Thunberg (*Artemisia capillaris*).

---

[1] R. Harada, Nihon Kagaku Zasshi (*J. Chem. Soc. Japan*), *Pure Chem. Section* **78**, 415 (1957).
[2] K. Hirayama, "Jikken Kagaku Koza (Experimental Chemistry)," Vol. I-1, p. 44, Maruzen Co., Ltd., (1957).

PROBLEM 18 / The figure shows the concentration variation of ethanol in carbon tetrachloride. What facts can be deduced from it?

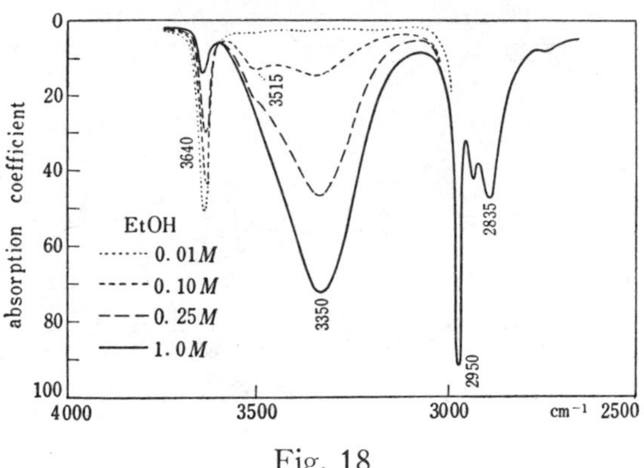

Fig. 18

ANSWER

Absorption of ethanol in the 3300 cm^{-1} region[1]

Since the absorption coefficient is plotted on the ordinate in the figure, the νCH bands at 2950 and 2835 cm^{-1} are not affected by the concentration.

The bands at 3640, 3515, and 3350 cm^{-1} correspond to the absorptions of monomeric, dimeric, and polymeric hydroxyls, respectively.[2] Ethanol exists only in the monomeric form at concentrations less than 0.01 *M*, but strong polymeric absorptions become apparent at 1.0 *M*.

$$\underset{\substack{\uparrow\\ 3515\ \mathrm{cm^{-1}}}}{\overset{\mathrm{Et}\diagdown}{\mathrm{O{-}H}}}\cdots\cdots\left[\cdots\underset{\substack{\uparrow\\ 3350\ \mathrm{cm^{-1}}}}{\overset{\mathrm{Et}\diagdown}{\mathrm{O{-}H}}}\cdots\cdots\right]_n\cdots\cdots\underset{\substack{\uparrow\\ 3640\ \mathrm{cm^{-1}}}}{\overset{\mathrm{Et}\diagdown}{\mathrm{O{-}H}}}$$

The peak at 3640 cm^{-1} due to the terminal " monomeric " hydroxyl is slightly shifted to higher wave-numbers in the true monomer[3]. The change of spectra with temperature is also similar to that shown here. For example, in the case of the 0.25 *M* carbon tetrachloride solution, the intensity of the polymeric 3350 cm^{-1} band decreases and the curve becomes similar to that of a 0.01 *M* solution shown in the figure.

---

[1] U. Liddel, *Ann. N.Y. Acad. Sci.* **69**, 70 (1957).

[2] U. Liddel and E. D. Becker, *Spectrochim. Acta* **10**, 70 (1957); W. C. Coburn, Jr. and E. Grunwald, *J. Am. Chem. Soc.* **80**, 1318 (1958).

[3] The terminal hydroxyl absorbs at 3620 cm^{-1}, whereas the monomer absorbs at 3650 cm^{-1}: F. A. Smith and E. C. Creitz, *J. Res. Natl. Bur. Standards* **46**, 145 (1951).

PROBLEM 19/ Deduce the structure of $C_{10}H_{18}O$, a monoterpene. No conspicuous λ_{max}, and $[\alpha]_D^{25}$ $-18°$.

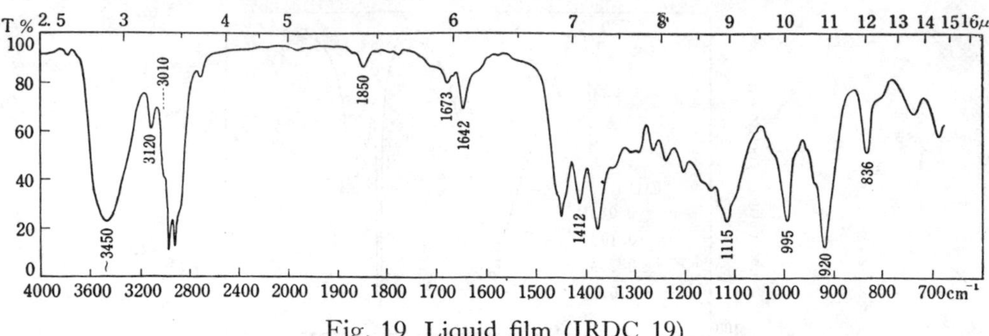

Fig. 19 Liquid film (IRDC 19)

ANSWER

(—)-Linalool

OH

Degree of unsaturation is two, but since mono-substituted (1115 and 995 cm^{-1}) and tri-substituted (836 cm^{-1}) double bonds are present, the terpene is acyclic. The double bonds are not conjugated (no UV). A *tert*-hydroxyl group with α-unsaturation is suggested by the 1115 cm^{-1} band (calculated from Table 5b: $1150-30=1120$ cm^{-1}).

When the monoterpenoid skeleton is taken into consideration the shown structure becomes the most probable.

ca. 3450 and 1115 cm^{-1}: OH.
3120 and 3120, 3010 cm^{-1}: νCH of $=CH_2$ and $=CH-$, respectively (3080 and 3020 cm^{-1}).
1850, 1642, 1412, 995 and 920 cm^{-1}: Terminal vinyl.
1673 and 836 cm^{-1}: Tri-substituted double bond.

PROBLEM 20/ What is the structure of the product $C_{13}H_{22}O$ obtained when dihydro-α-ionone is allowed to react with hydrogen chloride in *sec*-butanol at −13°C for ten days? When ethanol was employed as solvent the ethyl ether of the product was obtained instead.

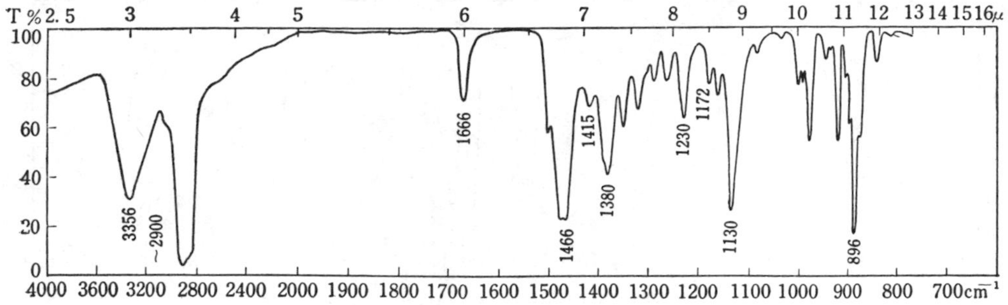

Fig. 20 Nujol

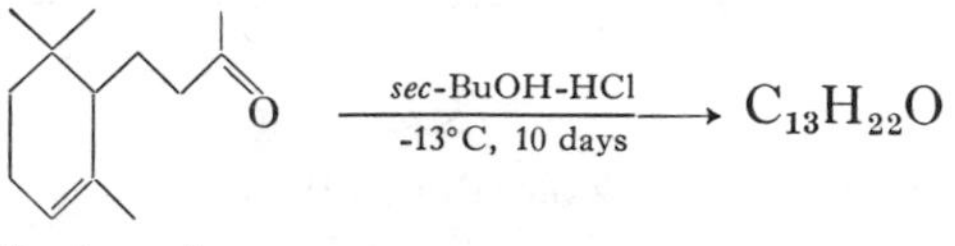

Dihydro-α-ionone

ANSWER

2 : 6 : 6–Trimethyl–9–methylenebicyclo [3 : 3 : 1]–nonan–2–ol[1]

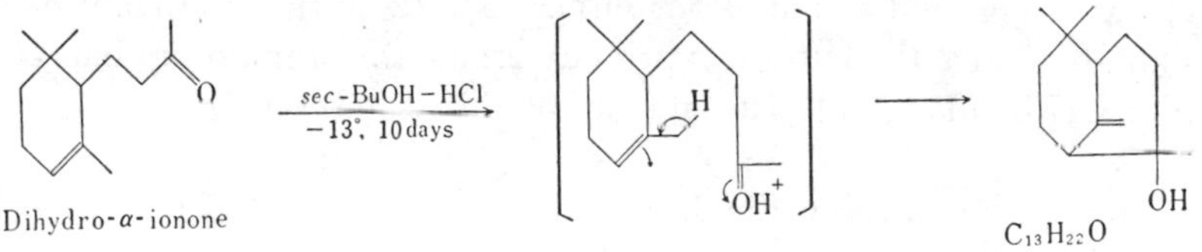

Dihydro-α-ionone

$C_{13}H_{22}O$

The carbonyl group has been replaced by a hydroxyl group (tertiary from position of C–O stretching band at 1130 cm^{-1}). Formation of a terminal methylene group is also indicated.

3350 cm^{-1}: Associated OH.
1665, 1415 and 895 cm^{-1}: $=CH_2$ (1655, 1415 and 890 cm^{-1}).
1230 and 1170 cm^{-1}: $-C(Me)_2-$ (1215 and 1195 cm^{-1}).
1130 cm^{-1}: *tert*–OH; calculated value according to Table 5b is 1135 cm^{-1}.
Nujol absorptions are present in the 2900, 1465 and 1380 cm^{-1} regions.

---

[1] M. Stoll, B. Willhalm, and G. Büchi, *Helv.* **38**, 1573 (1955).

PROBLEM 21 / Deduce the structure of C_7H_9NO. An ammonium band well separated from the C–H stretching bands appears in the IR spectrum of its hydrochloride.

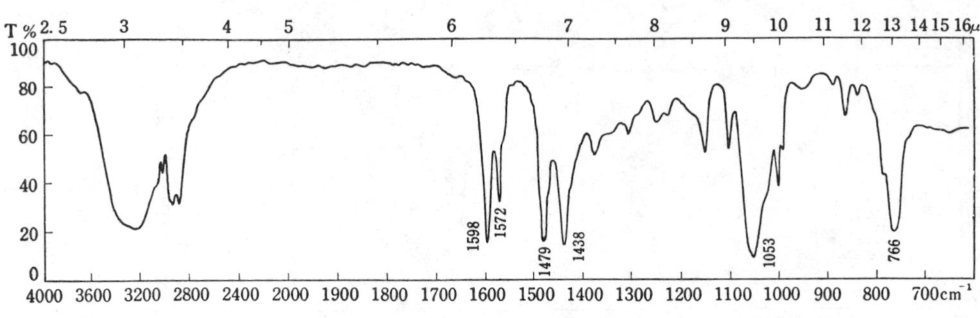

Fig. 21 Liquid film (IRDC 1111)

ANSWER

2–(2–Pyridyl) ethanol

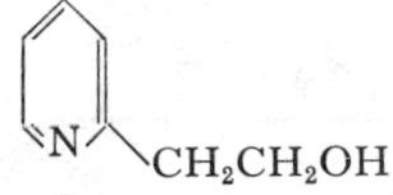

Degree of unsaturation is four. A hydroxyl group, probably primary because of 1053 cm^{-1} band (Table 5b), and an aromatic group with four adjacent hydrogens on it (766 cm^{-1}) are suggested. This leads either to 2–(2–pyridyl) ethanol or to 2–amino benzyl alcohol, but since the latter is a primary amine the ammonium band would overlap with the ν CH bands (Table 7d) and can be discarded.

$\simeq$3200 and 1053 cm^{-1}: Primary hydroxyl group.

1598, 1572, 1479 and 766 cm^{-1}: Pyridine ring vibrations are quite similar to those of the phenyl ring.

766 cm^{-1}: Four adjacent hydrogen atoms on aromatic ring (770$\sim$735 cm^{-1}). The hetero atom is simply regarded as an additional substituent when the out-of-plane CH bending bands (Table 3) are used for identifying the substitution pattern of heterocyclic aromatics.

PROBLEM 22/Deduce the structure of C_3H_6O.

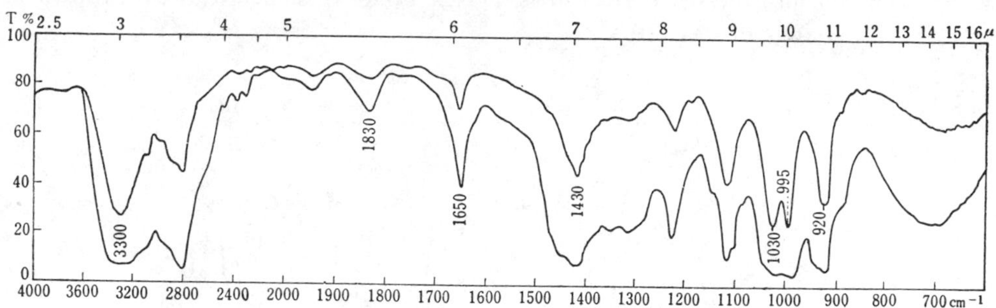

Fig. 22 Liquid film

ANSWER

Allyl alcohol

CH_2=CH-CH_2OH

3300 cm^{-1}: Associated ν OH (3400∼3200 cm^{-1}). A slight shoulder that originates from the unsaturated ν CH is observed around 3105 cm^{-1} on the lower frequency side of the main ν CH.

2860, 2810 cm^{-1}: Saturated methylene.

1830 cm^{-1}: Overtone of terminal methylene 920 cm^{-1} band (1860∼1800 cm^{-1}).

1650 cm^{-1}: ν C=C (1645 cm^{-1}).

1430 cm^{-1}: δ CH of saturated methylene, shifted from the usual position by adjacent double bond (1445∼1430 cm^{-1}, Table 1–15). In-plane δ CH (1420 cm^{-1}) of =CH_2 is overlapping.

1030 cm^{-1}: Rough calculation of primary alcoholic ν C–O according to Table 5 b is: 1050–30=1020.

995 and 920 cm^{-1}: CH out-of-plane bending of terminal methylene (990 and 910 cm^{-1}).

PROBLEM 23 / The following infrared spectrum is that of a simple derivative of allyl alcohol. What derivative is it? Compare this spectrum with the curve shown in Fig. 20.

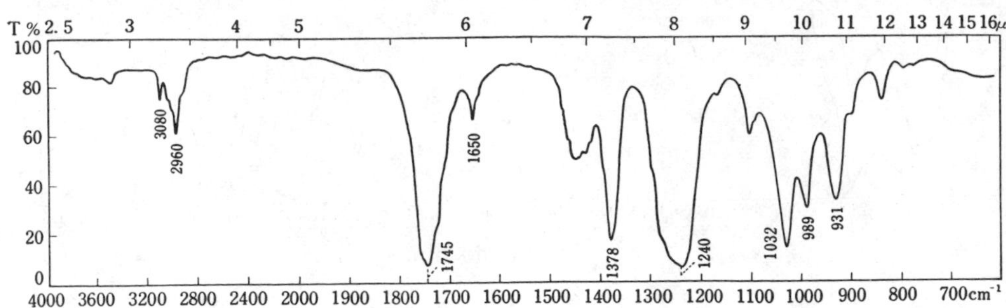

Fig. 23 Liquid film

ANSWER

Allyl acetate

$CH_2=CH-CH_2-O-CO-CH_3$

The hydroxyl absorptions are replaced by acetoxyl absorptions. The 1380 cm^{-1} methyl bending bands of acetates are quite pronounced and assist in their identification (Table 1–11).

3080 cm^{-1}: Unsaturated ν CH.
2960 cm^{-1}: Saturated ν CH.
1745 cm^{-1}: ν C=O (1735 cm^{-1}).
1650 cm^{-1}: ν C=C.
1378 cm^{-1}: Symmetric bending of acetoxyl methyl.
1240 and 1032 cm^{-1}: C–O–C stretching of acetoxyl group (two between 1300 and 1050 cm^{-1}).
989 and 931 cm^{-1}: Out-of-plane bending of terminal vinyl.

PROBLEM 24 / Deduce the structure of C_4H_5N.

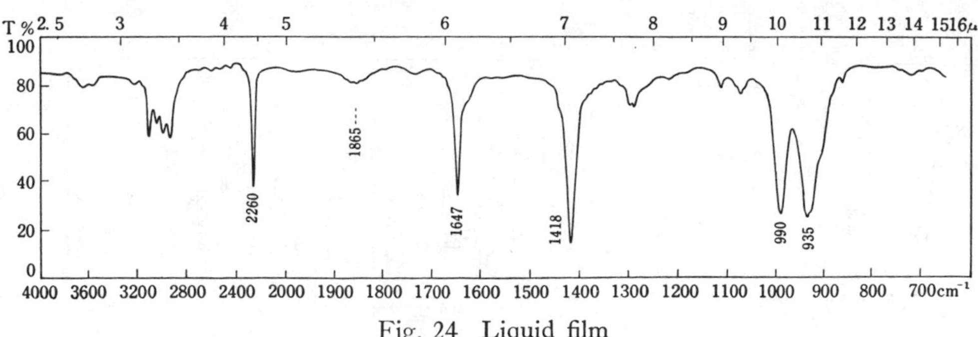

Fig. 24 Liquid film

ANSWER

Allyl cyanide

CH_2=CH-CH_2-CN

Compare spectrum with previous two charts.

2260 cm^{-1}: Stretching vibration of nitrile group (2260∼2210 cm^{-1}).
1865, 1647, 990 and 935 cm^{-1}: Terminal vinyl group.
1418 cm^{-1}: Overlapping of vinyl in-plane bending and methylene scissoring (perturbed, by flanking of two unsaturated groups).

PROBLEM 25 / What is the nature of the two oxygen atoms contained in $C_{18}H_{24}O_2$? Figs. 25–a and –b are the IR spectra of the same compound.

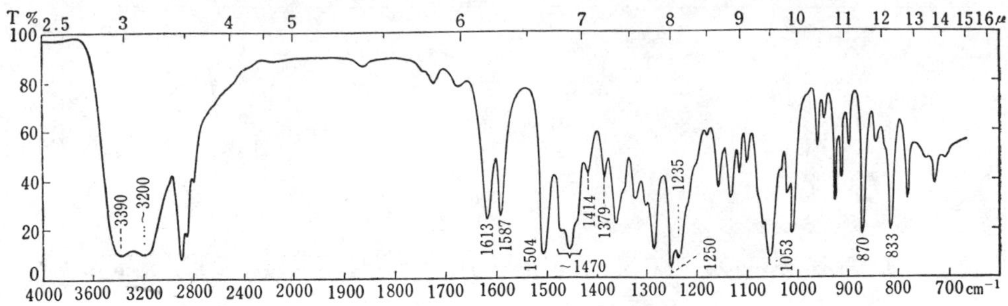

Fig. 25-a 1 mg/300 mg KBr

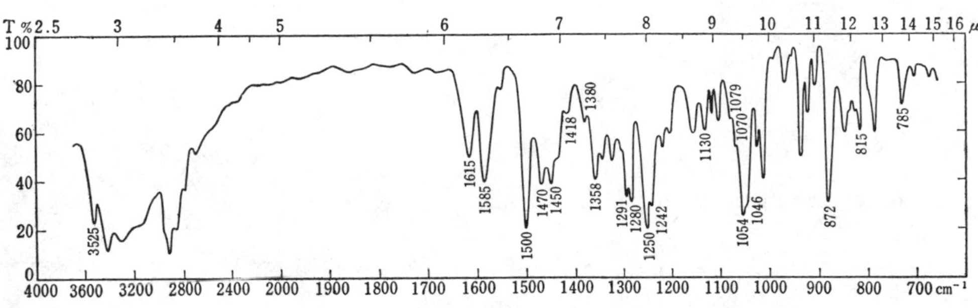

Fig. 25-b Nujol

ANSWER

Polymorphs of estra–1 : 3 : 5(10)–triene–3, 17 β–diol[1]

OH

HO

This compound is obtained in 5 polymorphic forms according to the method of crystallization. " Form A " shown in Fig. 25–a is obtained when recrystallized from ethyl acetate at room temperature. " Form C " shown in Fig. 23–b can be obtained either by rubbing a fused sample, or by heating Forms A, B, or D from 100° up to 179° (common melting point). Assignments of main bands are shown in the following Table. Although Form C has been measured in nujol, the nujol bands (2919, 2861, 1458, 1378, 720 cm^{-1}) are weak; such spectra with weak nujol bands are obtained when the sample is soluble in nujol.

[1] E. Smakula, A. Gori and H. H. Woltiz, *Spectrochim. Acta* **9**, 346 (1958).

Form A	Form C	
	3525	Present only in Form C. νOH of "free" hydroxyl, the high frequency has been attributed to steric hindrance within the crystal.
3390	3390	H-bonded νOH of alcoholic or phenolic O-H[1].
~3200	~3200	νOH of mixed H-bonds between alcoholic and phenolic O-H.
1613, 1587, 1504, 1435 (sh.)	1615, 1585, 1500, 1435 (sh.)	Phenyl ring.
1414	1418	Has been atributed to bending of 17-OH shifted lower by 17-CH_3[1]. Thus a small shoulder at 2805 cm^{-1} on the main νCH band is present only in Forms A and C, and this has been assigned to the 17-CH_3 interacting with the 17-OH. However, the band could be the bending of the 6-CH_2 adjacent to the phenyl group.
1470, 1450	1470, 1450	δ CH of methyl and methylene.
1379	1380	Angular methyl.
1250	1250	νC-O of phenolic OH.
1053	1054	νC-O of alcoholic OH.
870	872	Isolated ring H(900~860).
833	815	Two adjacent rings H(860~800).

PROBLEM 26/ The figure shows the absorption of *cis*- and *trans*-cyclohexane-1, 2-diols in the 3600 cm^{-1} region (0.005 M CCl_4 solution, LiF prism). Which is the *cis* and which is the *trans* isomer? Also explain the difference in the $\Delta\nu$ values, 38 and 32 cm^{-1}.

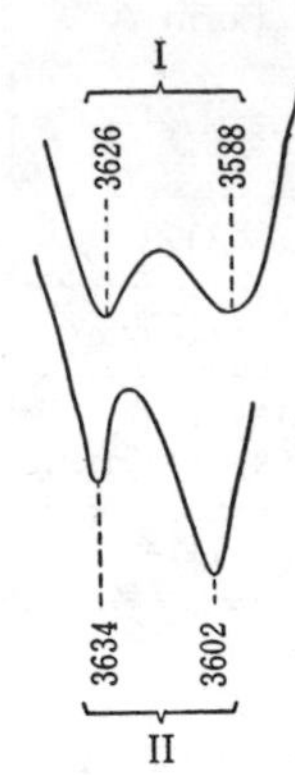

ANSWER

Curve (I) is the IR of the *cis*, and curve (II) that of the *trans* isomer[1,2].

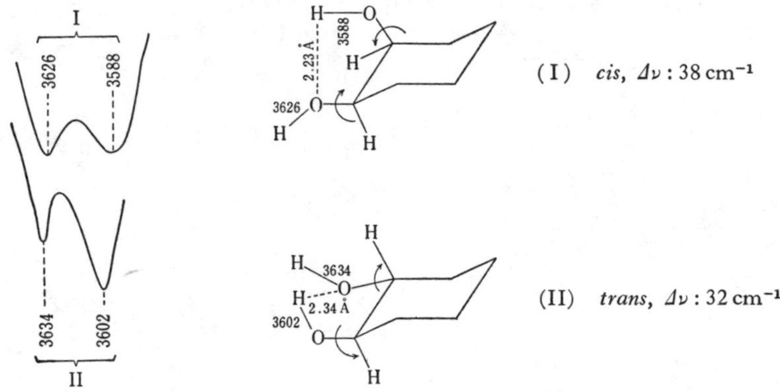

Kuhn[1] has shown that the wave-number difference $\Delta\nu$ between the ν OH of free and intramolecular hydrogen-bonded hydroxyl groups is approximately inversely proportional to the OH···O distance (Table 1).

Free and Bonded OH of Aliphatic Polyols[1]

Compound	Free OH	Bonded OH	$\Delta\nu$	-OH···O distance (Å)
1) *trans*-cyclopentane-1, 2-diol	3620	none		3.3
2) *trans*-cyclohexane-1, 2-diol (*ee*)	3634	3602	32	2.34
3) *cis*-cyclopentane-1, 2-diol	3633	3572	61	1.84
4) *cis*-cyclohexane-1, 3-diol (*aa*)	3619	3544	75	1.64
5) HO-$(CH_2)_4$-OH	3634	3478	156	1.1

[1] L. P. Kuhn, *J. Am. Chem. Soc.* **74**, 2492 (1952); **76**, 4323 (1954); **80**, 5950 (1958).
[2] A. R. H. Cole and P. R. Jeffries, *J. Chem. Soc.* 4391 (1956).

The –OH···O distances shown in the Table were measured by employing standard values for the bond angles and bond distances, and assuming that the hydroxyl group approaches as near as possible to its hydrogen-bonding partner. The results show that hydrogen-bonds are formed when the OH···O distance is less than 3.3 Å, and that the shift caused by hydrogen-bondings ($\Delta\nu$) is larger when the distance is shorter. Taking the given 5 compounds as references, one can obtain a standard curve by plotting $\Delta\nu$ against the OH···O distance, and from this it is possible to deduce the OH···O distance of a diol from its found $\Delta\nu$ value. The curve can be expressed by the following empirical relation[3]:

$$\Delta\nu\ (\text{cm}^{-1}) = \frac{250 \times 10^{-8}}{\text{OH}\cdots\text{O distance}} - 74$$

The OH···O distances, which furnish valuable information for conformational analyses, can be deduced by inserting the found values.

The OH···O distance of the *trans*-1, 2–diol (2) that was used as one of the reference samples is shown from models to be 2.34 Å. The distance of the *cis* isomer is likewise 2.34 Å, but the experimental $\Delta\nu$ value of 38 cm^{-1} is larger than that of the *trans* isomer, and insertion of this into the equation gives 2.38 Å. The shorter OH···O distance of the *cis* isomer can be accounted for by the easy formation of the hydrogen-bonded 5-membered ring. Thus, the movement in the direction shown by arrows in structure (I) will eventually lead to the inverted chair form and requires but little energy, whereas that shown in structure (II) is in the direction of a more puckered chair. This ease of formation of 5–membered rings from *ea* bonds (*cis*–1, 2 bonds) as compared to that from *ee* bonds (*trans*–1, 2 bonds) is also reflected in the greater susceptibility of *cis*–diols to HIO_4 oxidation and boric acid complex formation. The $\Delta\nu$ of ethylene glycol is 32 cm^{-1} (3644 and 3612 cm^{-1}), and the –OH···O distance of 2.34 Å is the same as that of *trans*-cyclohexane–1, 2–diol (2). This is because the ethylene glycol molecule is prevented from assuming the eclipsed conformation by nonbonded interactions. On the other hand, in compound (5), the two hydroxyl groups can approach each other very closely because of the absence of non-bonded interactions (1.1 Å is near the value of ordinary O–H distances, 0.96 Å). The $\Delta\nu$ of $(t\text{–Bu})_2$COH–COH$(t\text{–Bu})_2$ is 170 cm^{-1} (3630 and 3460 cm^{-1}) indicating the –OH···O distance to be even shorter. This has been explained by assuming the central C–C bond to be bent so as to bring the two hydroxyl groups within an anomalously short distance because of interactions between the two bulky *t*–Bu groups.

As regards compound (3), Angyal[4] has stated that the 5-membered ring *cis*–cyclopentane–1, 2–diol is strained so that the two hydroxyl groups are not in the same plane, and accordingly it should not be selected as a standard compound for derivation of –OH···O distances. They mention that the values of the more rigid *cis*–norcamphanediol,[5] –OH···O distance 1.84 Å and $\Delta\nu$ 102 cm^{-1}, should be adopted.

OH
OH

[3] The relation holds for OH···O distances between 1.6～3.3 Å, and the values in dilute CCl_4 solutions (ca. 0.005 mole) are employed for $\Delta\nu$.

[4] S. J. Angyal and R. J. Young, *J. Am. Chem. Soc.* **81**, 5467 (1959) footnote 26.

[5] H. Kwart and W. G. Vogburgh, *J. Am. Chem. Soc.* **76**, 5400 (1954).

PROBLEM 27 / The 1069 cm^{-1} band intensity of a 30.05 mg/g CS_2 solution of cyclohexanol (M.W. 100.16) was identical with the 1062 cm^{-1} band intensity of a 31.04 mg/g CS_2 solution of *trans*-4-*t*-butylcyclohexanol (M.W. 156.26) (20°). From this, calculate the equilibrium constant between the two conformational isomers, II_e and II_a, of cyclohexanol.

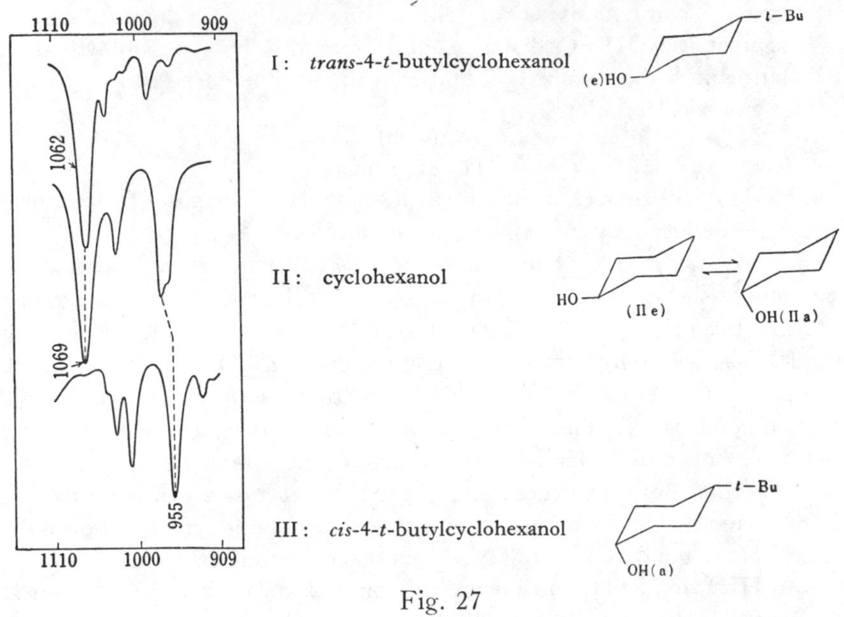

Fig. 27

ANSWER

It has been shown[1] that due to the bulky *t*-butyl group, *trans*-4-*t*-butylcyclohexanol exists entirely in conformation I (equatorial OH), whereas the *cis*-epimer is fixed in conformation III (axial OH). The CS_2 solution IR spectra of the *trans* and *cis* isomers show, respectively, conspicuous bands at 1062 cm^{-1} and 955 cm^{-1}, while the IR spectrum of cyclohexanol, which is present in both conformations, shows both bands. The concentration of 31.04 mg/g for the *trans* isomer having a moleculer weight of 156.26 corresponds to a concentration of 19.90 mg/g for cyclohexanol (M.W. 100.16). If cyclohexanol had existed entirely as II_e, a concentration of 19.90 mg/g would have sufficed. The fact that 30.05 mg/g was actually required shows that II_a is contained in 1 g of solution to the extent of (30.05−19.90=10.15) mg. The equilibrium constant at 20° becomes[2]:

$$K = II_e/II_a = 19.90/10.15 = 2.0$$

and

$$\Delta F^\circ = -RT \ln K = -2.303\ RT \log 2.0 = -0.4\ \text{kcal/mol}$$

In the above calculation, it has been assumed that the molecular absorption coefficients of the *trans* 1062 cm^{-1} and the cyclohexanol 1069 cm^{-1} bands are identical.

This method should be applied with some caution since cases have been observed where conformationally rigid systems (e.g. *exo*- and *endo*-nonborneol) show more than one apparent C—O stretching band[3], and the proportion of the equatorial conformer of cyclohexanol by the infrared method is substantially lower than that given by other methods[4].

For a related example dealing with conformational analysis by infrared spectroscopy, see Problem 100.

---

[1] S. Winstein and N. J. Holness, *J. Am. Chem. Soc.* **77**, 5562 (1955).
[2] R. A. Pickering and C. C. Price, *J. Am. Chem. Soc.* **80**, 4931 (1958).
[3] L. K. Dyall and R. G. Moore, *Aust. J. Chem.*, **21**, 2569 (1968).
[4] E. L. Eliel, *Angew Chem. Int. Ed.*, **4**, 761 (1965).

PROBLEM 28 / Deduce the structure of $C_{10}H_{10}O$. A three proton methyl singlet is present in the nuclear magnetic resonance spectrum.

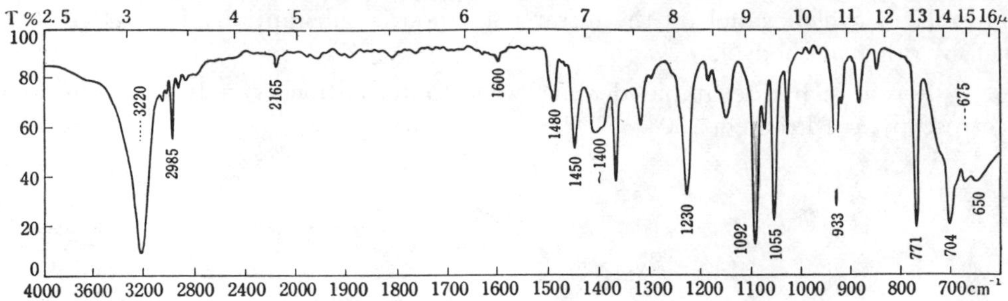

Fig. 28 2.8 mg/600 mg KBr (IRDC 1345)

ANSWER

2–Phenyl–3–butyn–2–ol

```
            OH
            |
  (C6H5) -C-C≡CH
            |
           CH3
```

Degree of unsaturation is 6. Hydroxyl, terminal acetylene, and mono-substituted phenyl groups are apparent, and in conjunction with the presence of a methyl group, the shown structure is easily derived.

3220 cm^{-1}: Stretching of acetylenic CH (3300 cm^{-1}, sharp) overlapping with band of associated ν OH (3400 ~ 3200 cm^{-1}); the latter is broad and would not show so sharply.

2985 cm^{-1}: νCH of methyl.

2165 cm^{-1}: νC≡C (2140 ~ 2100 cm^{-1}). Although the triple bond is attached to a terminal position, surprisingly the band is rather weak. Number of bands in this region does not coincide with that of triple bond, and di-alkyl acetylenes occasionally show one or two additional bands.

2000 ~ 1660 cm^{-1}: Shape characteristic for aromatic substitution pattern is vaguely noticed.

1600, 1480, 1450 cm^{-1}: Phenyl ring. An additional band at 1570 cm^{-1} is evident in the spectrum of phenylacetylene in which the triple bond is conjugated to the phenyl.

≃1400 cm^{-1}: In-plane-bending of associated OH (1500 ~ 1300 cm^{-1}).

1092 cm^{-1}: Stretching of tertiary hydroxyl group (1100 cm^{-1}) displaced to lower frequencies by adjacent unsaturation.

771, 704 cm^{-1}: Mono-substituted phenyl.

675 or 650 cm^{-1}: Bending of ≡CH (700 ~ 600 cm^{-1}). Its overtone may appear around 1250 cm^{-1} but diagnostic value is limited. The 1230 cm^{-1} band here is too strong for an overtone and is the δCH_{arom} (in-plane). In phenylacetylene, the two bands occur at 669 and 1238 cm^{-1} (broad, medium), respectively.

PROBLEM 29 / To what type of compound does the following spectrum belong?

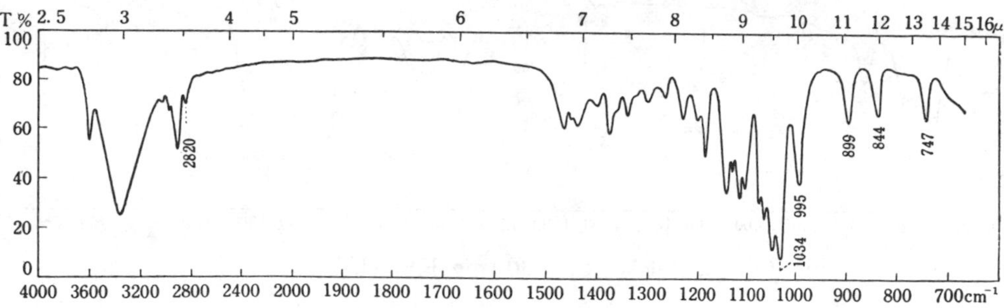

Fig. 29 1.2 mg/600 mg KBr (IRDC 53)

ANSWER

Methyl α–D–(+)–glucoside

The only prominent bands are those of the hydroxyl group. Moreover, the 844 cm^{-1} absorption suggests it to be an α–sugar. Results for the D–glucopyranose can be extended to the glucosides.

2820 cm^{-1}: Symmetric stretching of methoxyl CH_3 (2830 ~ 2815 cm^{-1}).
1150 ~ 1030 cm^{-1}: Various C–O–C stretching.
899 cm^{-1}: Type 1 band of sugars (Table 5e, 917 ± 13 cm^{-1}).
844 cm^{-1}: Type 2 band (844 ± 8 for α–sugars and 891 ± 7 cm^{-1} for β–sugars). Most useful band for distinguishing the two type of sugars.
747 cm^{-1}: Type 3 band (766 cm ± 10 cm^{-1}). This band corresponds to the 813 cm^{-1} band of tetrahydropyran (ring-breathing vibration), and the expected downward displacement is observed when the tetrahydropyran hydrogens are substituted by heavier groups. Thus, it is at 770 cm^{-1} in α–glucose, and at 740 cm^{-1} in α–glucose penta-acetate[1].

The type 2 band is also apparent in the spectrum of cellulose, which carries β–1 : 4 linkages. A comparison with the methyl glucoside spectrum shows that the IR bands in the cellulose spectrum are relatively diffuse because of the high molecular weight.

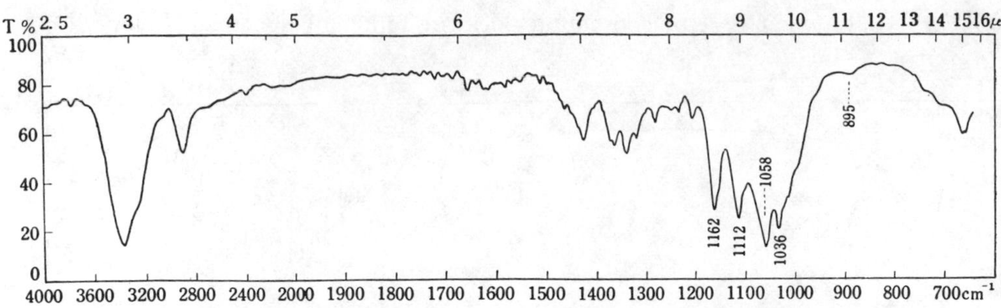

Fig. 29 a Cellulose, 1 mg/300 mg KBr (IRDC 144)

At —180° the spectra of methyl glycosides show additional bands in the regions 3600∼3100 cm^{-1} and 750∼650 cm^{-1}, due to coupled vibrations of ν OH and δ OH frequencies. These low temperature bands, in conjunction with measurements on D_2O treated samples, can yield information on H-bonding and hydroxyl configurations.[2] For a review of carbohydrate spectra, see Ref. 3.

[1] S. A. Barker, E. J. Bourne, R. Stephens and D. H. Whiffen, *J. Chem. Soc.* 3468 (1954); *cf.* also, K. Nakanishi, N. Takahashi and F. Egami, *Bull. Chem. Soc. Japan* **29**, 434 (1956).

[2] A. J. Michell, *Austr. J. Chem.* **28**, 335 (1975).

[3] R. S. Tipson, "Infra-red Spectroscopy of Carbohydrates—A Review of the Literature." N.B.S. Monograph 110, U.S. Government Printing Office, Washington, D.C., 1969.

PROBLEM 30 / Deduce the structure of $C_6H_{12}O$. The nmr spectrum shows the presence of one methyl group.

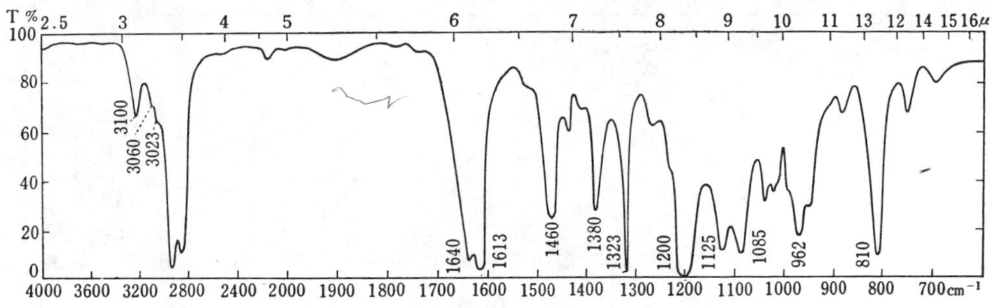

Fig. 30 Liquid film

ANSWER

n–Butyl vinyl ether[1]

n-C_4H_9-O-CH=CH_2

3100, 3060, 3023 cm^{-1}: ν CH of alkene.

1640 and 1613 cm^{-1}: ν C=C, linking of oxygen or nitrogen enhances intensity and decreases frequency. The two ν C=C bands in vinyl ethers are due to rotational isomerism.[2]

1460 and 1380 cm^{-1}: δ CH of alkane.

1200 and 1085 cm^{-1}: Vinyl ether (1275～1200 and 1075～1020).

962 and 810 cm^{-1}: The δ CH out-of-plane vibrations of vinyl ethers are shifted outside the usual region (990 and 910 cm^{-1}). All of the 7 other alkyl or aryl vinyl ethers that were measured absorbed in the same region.[1]

---

[1] Y. Mikawa, *Bull. Chem. Soc. Japan* **29**, 110 (1956).
[2] A. D. H. Claque and A. Danti, *Spectrochim. Acta* **24A**, 439 (1968).

PROBLEM 31 / Deduce the structure of $C_{11}H_{10}O$.

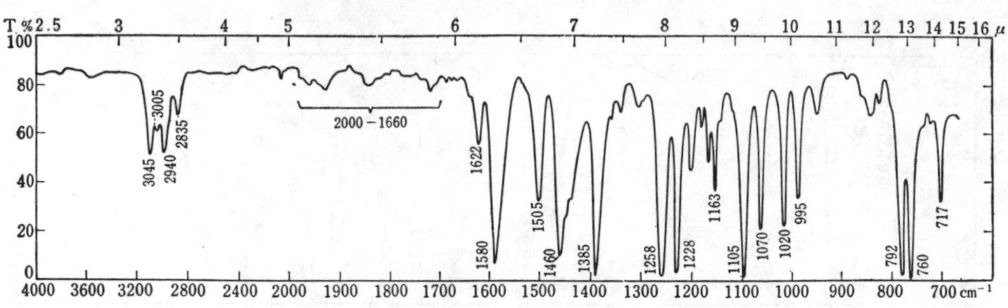

Fig. 31 Liquid film

ANSWER

α–Naphthol methyl ether

OCH_3

3045, 3005 cm^{-1}: Aromatic ν CH.
2940 cm^{-1}: 2950 cm^{-1} triplet of aromatic $-OCH_3$ (Table 6).
2835 cm^{-1}: ν CH of aromatic $-OCH_3$
2000∼1660 cm^{1}: Absorption of *o*– and 1, 2, 3–tri-substituted phenyl.
1622, 1580, and 1505 cm^{-1}: Phenyl ring.
1460 cm^{-1}: Phenyl ring and methyl.
1385 cm^{-1}: Methyl.
1258 cm^{-1}: Asymmetric stretching of =C–O–C (1250 cm^{1}).
1228, 1163, 1105, 1070, and 995 cm^{-1}: In-plane bending of aromatic C–H.
1020 cm^{-1}: Symmetric stretching of =C–O–C.
792 cm^{-1}: Adjacent 3H (810∼750 cm^{-1}).
760 cm^{-1}: Adjacent 4H (770∼735 cm^{-1}).
717 cm^{-1}: Band that is present in mono–, 1 : 3–, 1 : 3 : 5–, and 1 : 2 : 3–substituted phenyls (710∼690 cm^{-1}).

PROBLEM 32 / Three compounds with molecular formulas C_7H_8O, C_7H_8S, and $C_8H_{11}N$ have the spectra shown below. Deduce the structures.

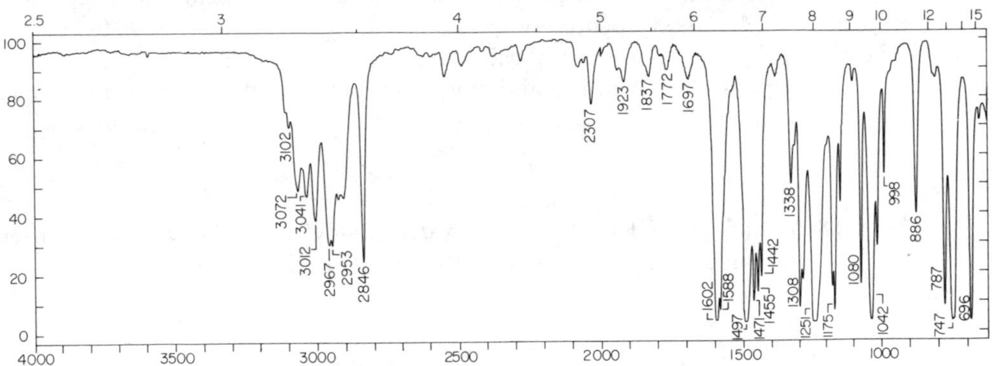

Fig. 32-a Liquid film

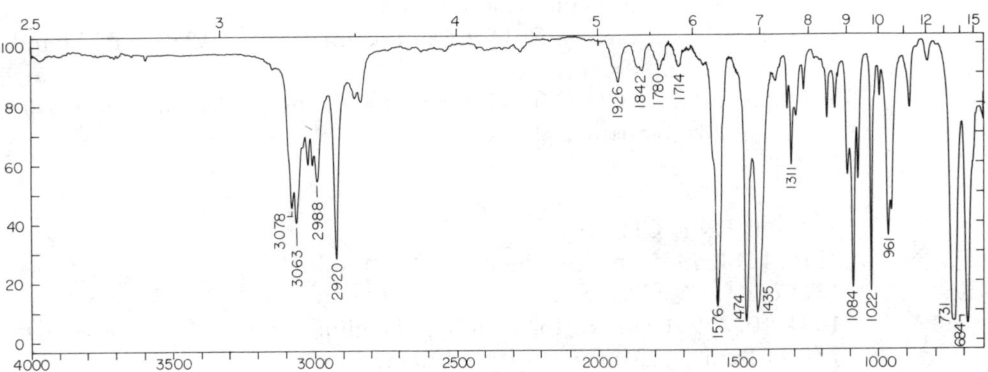

Fig. 32-b Liquid film

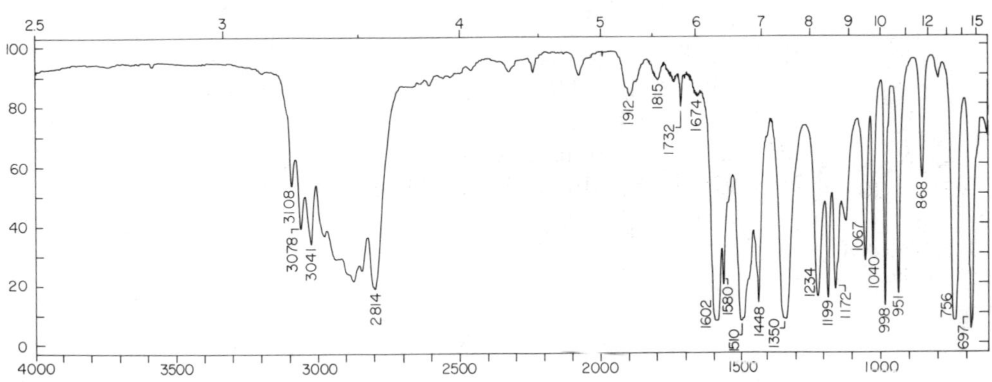

Fig. 32-c Liquid film

ANSWER

The spectra show three common features:

i) On the higher and lower frequency side of 3000 cm^{-1} appear the aromatic and aliphatic CH bands, respectively.

ii) The absorption at 5~6 μ is typical of monosubstituted phenyl.

iii) The out-of-plane CH bending and ring puckering bands are typical of monosubstituted phenyl.

Fig. 32-a

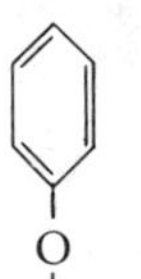

2846 cm^{-1}: ν_s CH_3–O–; the low frequency and relatively high intensity distinguish it from other bands in this region.

1602, 1588, 1497 cm^{-1}: Phenyl C=C stretch.
1455 cm^{-1}: δ_s CH_3–O– .
1251 cm^{-1}: ν_{as} aromatic C–O–C, (1250 cm^{-1} band).
1175, 1080, 1022 cm^{-1}: Set of strong sharp bands in this region is mainly due to δ CH in-plane bending modes.
1042 cm^{-1}: ν_s aromatic C–O–C; not as intense as "1250 cm^{-1}" band.
787 cm^{-1}: ?
747, 696 cm^{-1}: δ CH (770~730 cm^{-1}) and ring puckering (710~690 cm^{-1}) of monosubstituted phenyl.

O
CH3
Anisole

Fig. 32-b

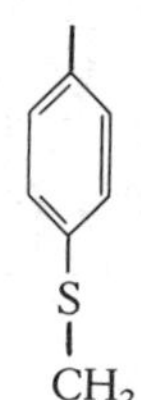

2920 cm^{-1}: ν_s CH_3–S–.
1576, 1474, 1435 cm^{-1}: Phenyl C=C stretch.
1311 cm^{-1}: δ_s CH_3–S–.
1084, 1022, 961 cm^{-1}: In-plane CH bending modes.
731, 684 cm^{-1}: Monosubstituted phenyl.

S
CH3
Thioanisole

Fig. 32-c

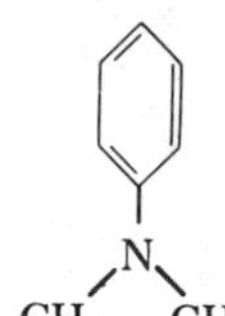

2814 cm^{-1}: ν_s CH_3–N.
1602, 1580, 1510, 1448 cm^{-1}: Phenyl C=C stretch.
1448 cm^{-1}: δ_s CH_3–N.
1350 cm^{-1}: ν_{as} aromatic C–N–C.
756, 697 cm^{-1}: Monosubstituted phenyl.

N,N-Dimethylaniline

PROBLEM 33 / Deduce the structure of $C_9H_{12}O_2$.

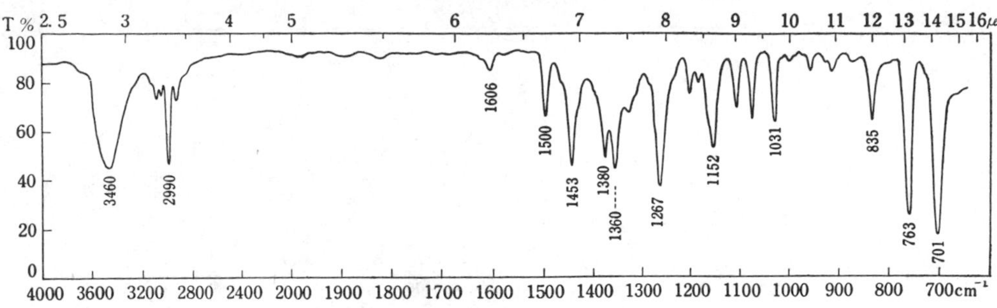

Fig. 33 Liquid film (IRDC 207)

ANSWER

2–Hydroperoxycumene

$$C_6H_5\text{-}\underset{CH_3}{\overset{CH_3}{\underset{|}{\overset{|}{C}}}}\text{-O-OH}$$

The presence of hydroxyl, mono-substituted phenyl, and *gem*-dimethyl groups (splitting of 1380 cm^{-1} band) is apparent. It is actually a hydroperoxide. Although a band at 890～830 cm^{-1} has been assigned to the O–O vibration of aliphatic peroxides, hydroperoxides, and peracids, the bands are weak and the substances cannot be characterized clearly by the IR method.[1,2] The *t*–hydroperoxides and *t*–peroxides give more distinct bands at 880～840 cm^{-1}, but since *t*–alcohols also give rise to bands in the same region—however, at slightly higher frequencies—it appears that this band should be attributed to the $\text{C–}\underset{C}{\overset{C}{\underset{|}{\overset{|}{C}}}}\text{–O}$ group.[3,4]

3460 cm^{-1}: ν OH.
1606, 1500, 763, and 701 cm^{-1}: Mono-substituted phenyl.
1453, 1380, and 1360 cm^{-1}: *gem*-Dimethyl.
1152 cm^{-1}: ν C–O.
835 cm^{-1}: $\text{C–}\underset{C}{\overset{C}{\underset{|}{\overset{|}{C}}}}\text{–O}$ group. Phenyl dimethyl carbinol has a band at the higher frequency of 860 cm^{-1}[4].

---

[1] G. J. Minkoff, *Proc. Roy. Soc.* **A244**, 176 (1954).
[2] H. R. Williams and H. S. Mosher, *Anal. Chem.* 27, 517 (1955).
[3] G. J. Minkoff, *Disc. Far. Soc.* **9**, 320 (1950).
[4] A. R. Philpotts and W. Thain, *Anal. Chem.* 24, 638 (1952).

PROBLEM 34 / Deduce the structure of $C_{12}H_{24}O$.

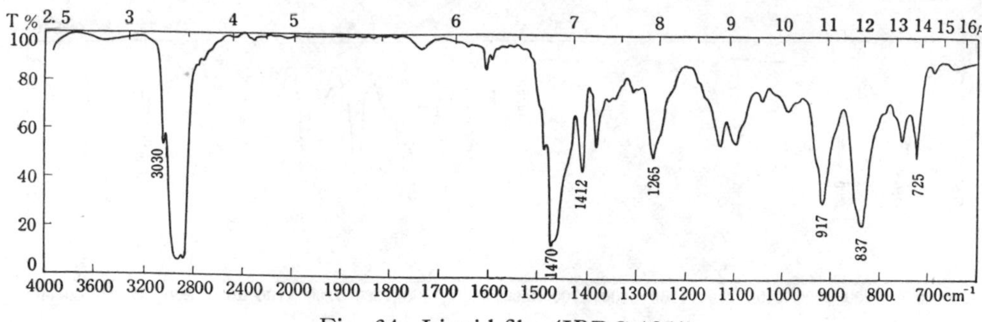

Fig. 34 Liquid film (IRDC 1251)

ANSWER

1, 2–Epoxydodecane

$$\underset{\diagdown O \diagup}{CH_2\text{-}CH}\text{-}(CH_2)_9\text{-}CH_3$$

Degree of unsaturation is one. Although weak absorptions are seen at 1600 cm^{-1}, the position is too low for an isolated double bond. The pronounced bands at 917 and 837 cm^{-1} (ca. 11 and 12μ), in conjunction with the bands at 1265 cm^{-1} (ca. 8μ) and 3030 cm^{-1} suggest a terminal epoxide ring.

3030 cm^{-1}: Epoxy CH_2.
1265, 917, and 837 cm^{-1}: Epoxy " 8μ ", " 11μ ", and " 12μ " bands.
1412 cm^{-1}: Methylene adjacent to epoxy ring.
725 cm^{-1}: *n*-Aliphatic chain, methylene rocking.

PROBLEM 35 / Deduce the structure of $C_6H_{15}N$ (Fig. 35 a). Fig. 35 b shows the IR spectrum of its simple derivative. Rationalize the peak changes.

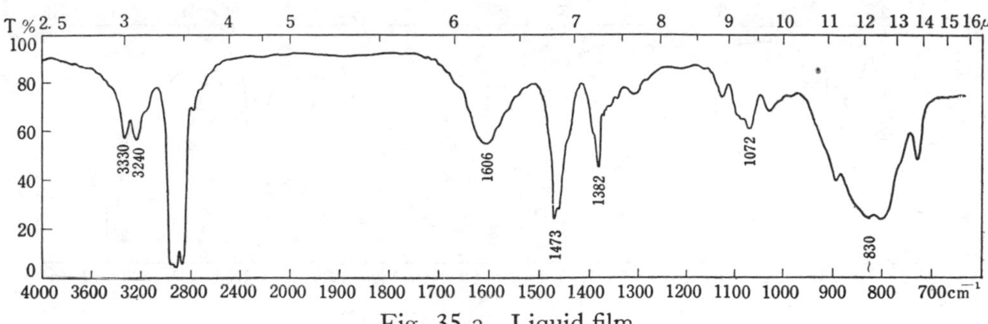

Fig. 35-a Liquid film

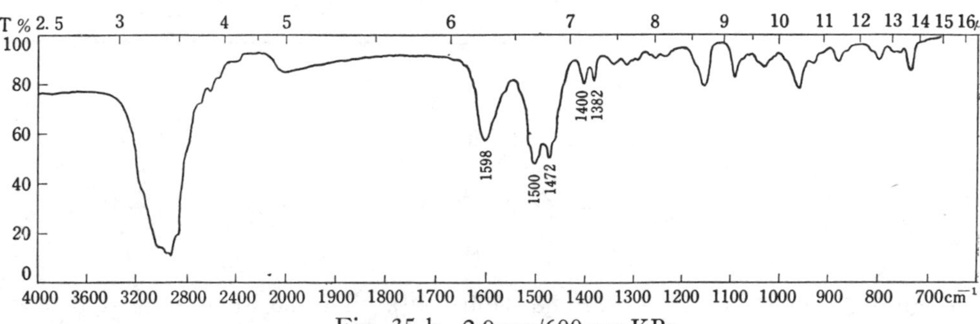

Fig. 35-b 2.0 mg/600 mg KBr

ANSWER

n-Hexylamine (Fig. 35 a).

$CH_3(CH_2)_5NH_2$

3330 and 3240 cm^{-1}: Asymmetric and symmetric stretching of NH_2. Both bands are shifted by ca. 170 cm^{-1} from the position of thc free amino group (3500 and 3400 cm^{-1}). The calculated position of the $\nu_s NH_2$ by inserting the frequency of the higher $\nu_{as} NH_2$ band gives 3252 cm^{-1} (Table 7 a), which is in good agreement with the observed 3240 cm^{-1}. This will suggest that the shift to lower wave-numbers is caused either by participation of both N–H groups in hydrogen-bonding to the same extent or by some phenomenon other than hydrogen-bonding.

1606 cm^{-1}: δNH_2 (1640 ~ 1560 cm^{-1}).
1473 and 1382 cm^{-1}: Methyl and methylene.
1072 cm^{-1}: ν C–N (1230 ~ 1030 cm^{-1}).
≃830 cm^{-1}: Out-of-plane bending of NH_2. Broad and typical for primary amines; however, sometimes lacking (900 ~ 650 cm^{-1}).

Fig. 32 b is the hydrochloride of *n*-hexylamine.
≃3000 cm^{-1}: Ammonium band of primary amine salt.
1598 and 1500 cm^{-1}: Asymmetric and symmetric bending of NH_3^+ (1600 ~ 1575 and 1500 cm^{-1}).
1472 and 1382 cm^{-1}: Methyl and methylene.
1400 cm^{-1}: Bending of $-CH_2-$ adjacent to N^+ (1440 ~ 1400 cm^{-1}).

PROBLEM 36 / Deduce the structure of C_7H_9N.

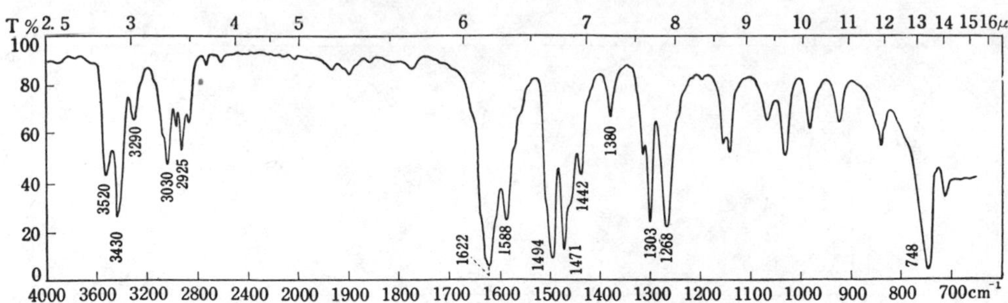

Fig. 36 Liquid film (IRDC 562)

ANSWER

o-Toluidine

NH_2

CH_3

3520 and 3430 cm^{-1}: ν NH_2 (3500 and 3400 cm^{-1}). The positions suggest that the amino groups giving rise to these bands are not involved in intermolecuar hydrogen-bonding because of steric hindrance. The calculated value of the lower symmetric stretching by insertion of 3520 cm^{-1} into the equation, $\nu_s = 345.53 + 0.876\,\nu_{as}$, is 3429 cm^{-1}.

3290 cm^{-1}: Bonded NH. The shape of these three NH stretching bands are similar in aniline and *m*- and *p*-toluidine, excepting that the positions of the two high-frequency bands are lower in *o*-toluidine.

1622 cm^{-1}: NH_2 bending (1640∼1560 cm^{-1}).

1588, 1494, 1471, and 748 cm^{-1}: *o*-Substituted phenyl.

1442 and 1380 cm^{-1}: Methyl.

1268 cm^{-1}: C–N stretching (1360∼1250 cm^{-1}).

PROBLEM 37 / The IR spectrum of $C_{10}H_{16}N_2 \cdot HCl$ is shown. The amine also forms a dihydrochloride. When alkali is added to the dihydrochloride solution the UV spectrum changes in two steps; namely, the UV spectra of the diamine, and its mono- and dihydrochlorides are different. What is the structure of the diamine?

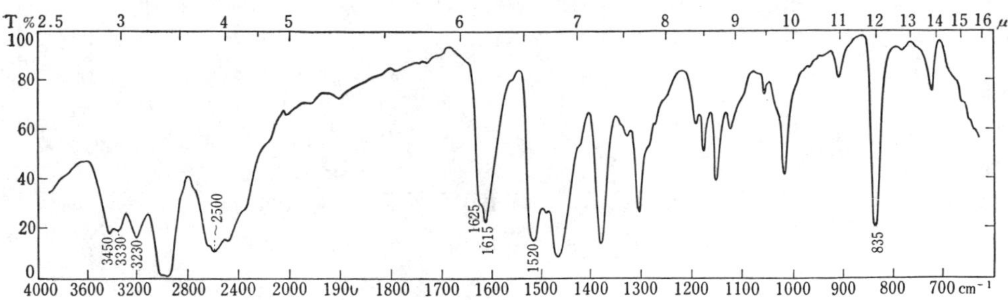

Fig. 37 Nujol

ANSWER

N,N–Diethyl–*p*–phenylenediamine monohydrochloride

$$H_2N\text{-}C_6H_4\text{-}N^{+}H(CH_2CH_3)_2 \quad Cl^{-} \qquad \text{(a)}$$

The peaks at 2900, 1460, and 1380 cm^{-1} are due to nujol and will not be considered.

The ammonium band at ca. 2500 cm^{-1} shows that the more basic amino group is tertiary. The peaks at 3450, 3330, 3230, and 1625 cm^{-1} show that the remaining amino group is primary. Presence of a *para*-substituted phenyl is indicated by bands at 1615, 1520, and 835 cm^{-1}.

In view of the fact that aliphatic amines are stronger bases than aromatic amines, and that in the present case it is the *tert*-amino that is the stronger, the compound cannot contain an aliphatic *prim*-amino group and an aromatic *tert*-amino group.

Thus, the three remaining combinations have to be considered:

i) Both groups are aromatic, as in (a).
ii) Both groups are aliphatic, as in (b).
iii) The primary group is aromatic, while the tertiary group is aliphatic, as in (c).

$$H_2N\text{-}CH_2\text{-}C_6H_4\text{-}CH_2\text{-}N^{+}H(CH_3)_2 \qquad \text{(b)}$$

$$H_2N\text{-}C_6H_4\text{-}CH_2\text{-}N^{+}H(CH_2CH_3)(CH_3) \qquad \text{(c)}$$

The 1305 cm^{-1} peak is attributable to the C_{arom}–N stretching, and this excludes (ii). If the compound were of type (iii), the UV would only change in the step involving the aromatic amino group. This is against facts and structure of type (i) is deduced.

The ca. 1400 cm^{-1} band of the $-N^{+}-CH_2-$ group is hidden by nujol absorptions.

PROBLEM 38 / Correlate peaks with structure.

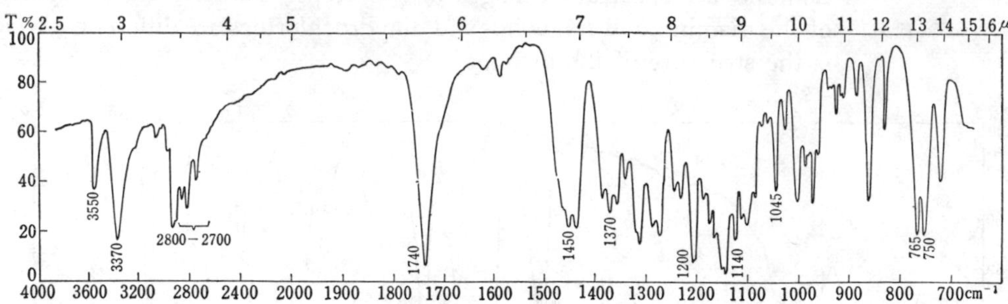

Fig. 38 Yohimbine (KBr)

ANSWER

Yohimbine[1]

3550 cm^{-1}: ν NH of indole; association is sterically hindered.
3370 cm^{-1}: ν OH, associated.
2800$\sim$2700 cm^{-1}: *trans*-Quinolizidine C/D rings (Table 7 e).
1740, 1200, 1140 cm^{-1}: $-COOCH_3$.
$\simeq$1450 and 1370 cm^{-1}: Methylene and methyl.
1045 cm^{-1}: ν C–O of *sec*-alcohol.
765 and 750 cm^{-1}: Four adjacent ring hydrogens.

PROBLEM 39 / Correlate IR peaks with structure of yohimbine hydrochloride (cf. previous problem).

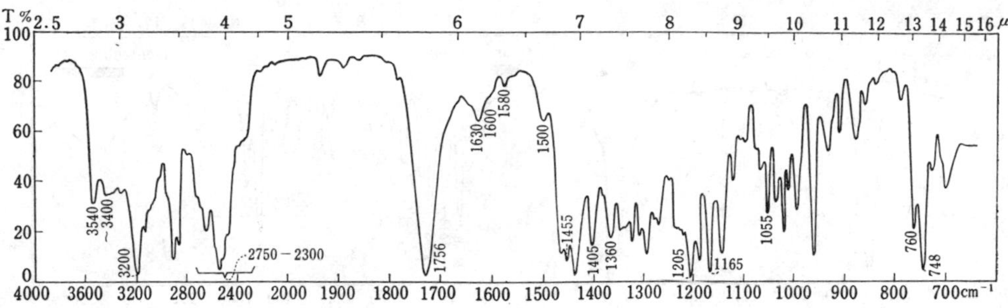

Fig. 39 Yohimbine hydrochloride (KBr)

ANSWER

Yohimbine hydrochloride.[1]

3540 and 3400 cm^{-1}; ν NH, free and associated.

3200 cm^{-1}: ν OH.

2750～2300 cm^{-1}: Typical *tert*-ammonium band; it is clearly separated from the main νCH bands. The 2800～2700 cm^{-1} *trans*-quinolizidine band group (Fig. 36) disappears when the lone pair is bound by protons, e.g., hydrochloride, or by oxygen, e.g., N–oxide.

1756, 1360, 1205, and 1165 cm^{-1}: $-COOCH_3$.

1405 cm^{-1}: $-CH_2-$ adjacent to N^+- (Table 1～16).

1630, 1600, 1580, 1500: Indole ring.

1055 cm^{-1}: ν C–O of *sec*-alcohol.

760 and 748 cm^{-1}: Four adjacent ring hydrogens.

[1] F. Yamasaki, *Nippon Kagaku Zasshi* **82** 72 (1961).

PROBLEM 40 / Deduce the structure of the monoterpene, $C_{10}H_{16}O$. Three methyl singlets are apparent in the nuclear magnetic resonance spectrum.

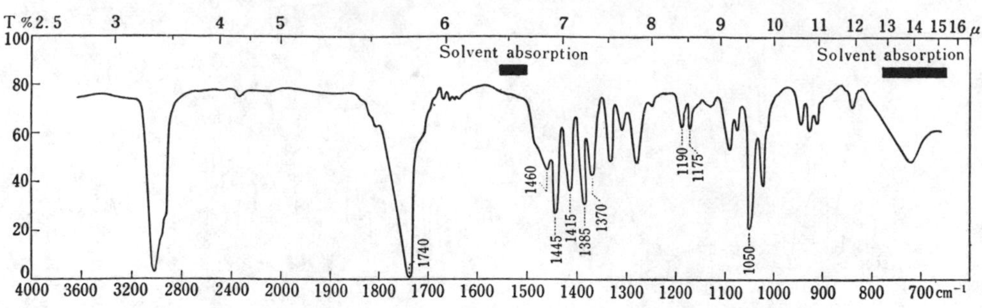

Fig. 40 CCl_4 solution, 0.5 mm cell.

ANSWER

Camphor

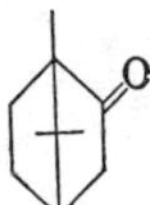

The degree of unsaturation is three, one of which is attributable to a five-membered ring ketone. Of the three methyl groups attached to quaternary carbon atoms (NMR spectrum), two constitute a *gem*-dimethyl group (split 1380 cm^{-1} band, and 1190 and 1175 cm^{-1} bands). An active methylene group is also present.

1740 cm^{-1}: Five-membered-ring ketone (1745 cm^{-1}).
1460 and 1445 cm^{-1}: CH_2 and CH_3.
1415 cm^{-1}: Active methylene.
1385 and 1370 cm^{-1}: *gem*-Dimethyl.
1190 and 1175 cm^{-1}: Two methyl groups attached to quaternary carbon atom (1215 and 1195 cm^{-1}).
1050 cm^{-1}: Aliphatic ketone (1100 cm^{-1}).

PROBLEM 41 / Correlate steroids with figures.

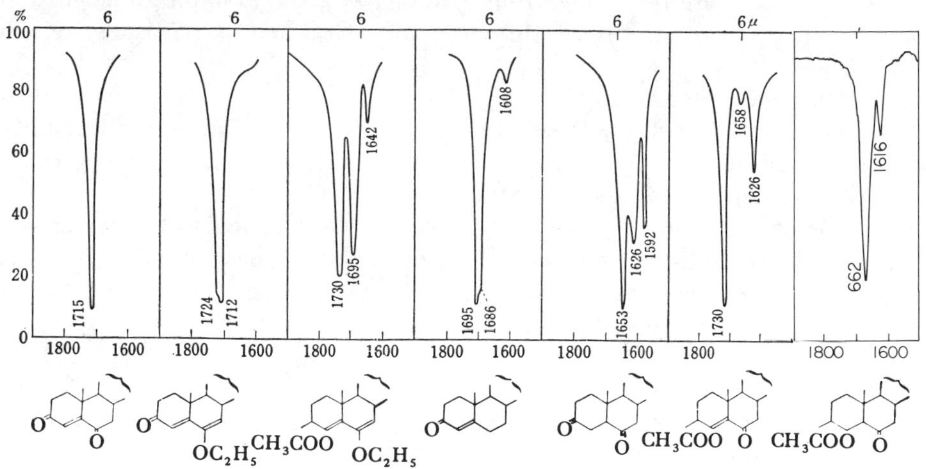

Fig. 41 Cholestane derivatives with unsaturated A, B rings ($CHCl_3$ solutions).

ANSWER

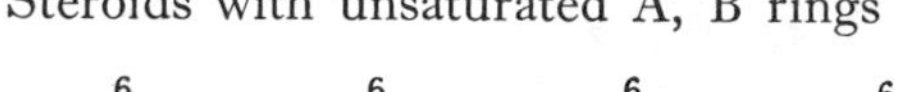

Steroids with unsaturated A, B rings

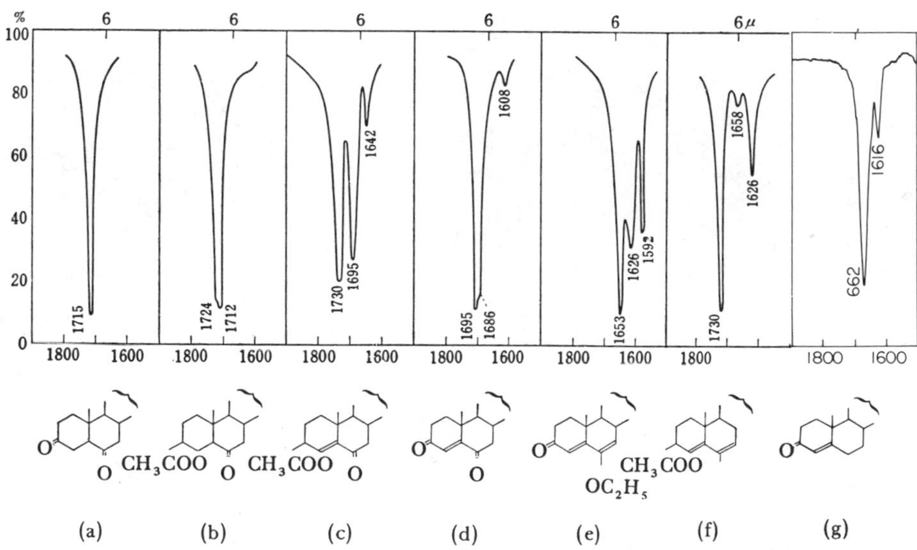

(a) (b) (c) (d) (e) (f) (g)

a) Isolated carbonyl groups do not interact, and the two bands overlap at the standard position.

b) 1724 cm^{-1}: Acetate.
   1712 cm^{-1}: Ketone.

c) 1730 cm^{-1}: Acetate.
   1695 cm^{-1}: Ketone.
   The 1695 cm^{-1} is high for α, β-unsaturated ketones (1675 cm^{-1}); this could be due to *cis*-locked structure.

1642 cm^{-1}: ν C=C. Although the ν C=C band of *cis*-locked α, β-unsaturated ketones are generally strong (cf. Fig. 40), the example shows that this is not necessarily the case. Moreover, the ν C=O and ν C=C bands are only 53 cm^{-1} apart.

d) 1695 and 1686 cm^{-1}: Ketone.

1608 cm^{-1}: The double bond character of the C_4–C_5 bond is small because it is flanked by two carbonyl groups. For the same reason the degree of polarization is low and therefore the band is weak.

e) 1653 cm^{-1}: Since the ethoxyl group is not conjugated with the carbonyl group, the ν C=O band position is not much lower than usual (α,β-γ,δ-unsaturated ketones, 1665 cm^{-1}).

1626 and 1592 cm^{-1}: ν C=C at 4,5 and 6,7.

f) 1730 cm^{-1}: Acetate.

1658 cm^{-1}: ν C=C at 4, 5.

1626 cm^{-1}: ν C=C at C_6–C_7; enhanced intensity due to linking of oxygen function (i.e., enol ether).

g) 1662 cm^{-1}: Ketone.

1616 cm^{-1}: νC=C. Intensity is typical of transoid α,β-unsaturated enones; cisoid forms usually have greater intensity relative to C=O.[1]

[1] K. Noack, *Spectrochim. Acta*, **18**, 1625 (1962).

PROBLEM 42 / What is the structure of the thujopsene derivative[1]? The molecular formula is $C_{14}H_{22}O$, and it lacks olefinic protons, m.p. 105.0~105.5°.

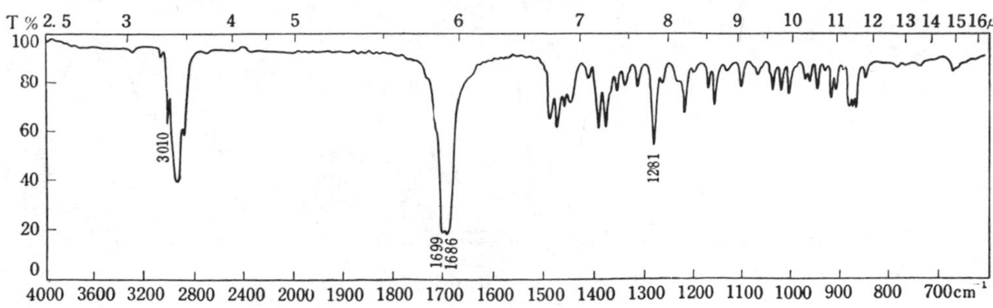

Fig. 42 1.8 mg/600 mg KBr (IRDC 856)

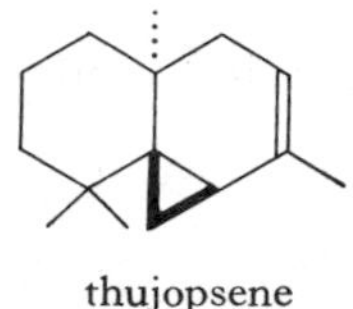

thujopsene

ANSWER

Thujopsene anhydrodihydroketone[1].

Since no olefinic proton is present, the 3010 cm^{-1} band would suggest a cyclopropane ring; the ν C=O is low and thus conjugation with the 3-membered ring is indicated.

3010 cm^{-1}: Cyclopropane ν CH (3050 cm^{-1}).

1699 and 1689 cm^{-1}: ν C=O. The cyclopropane acts as an electron source and conjugation with carbonyls lowers the ν C=O by ca. 20 cm^{-1}. Cyclopentanones carrying α-cyclopropanes absorb at 1720 cm^{-1}.

[1] H. Erdtman and T. Norin, *Chem. and Ind.* 622 (1960).

[2] T. Nozoe, H. Takeshita, S. Ito, T. Ozeki, and S. Seto, *Chem. Pharm. Bull.* **8**, 936 (1960).

PROBLEM 43 / Deduce the structure of $C_6H_{10}O$.

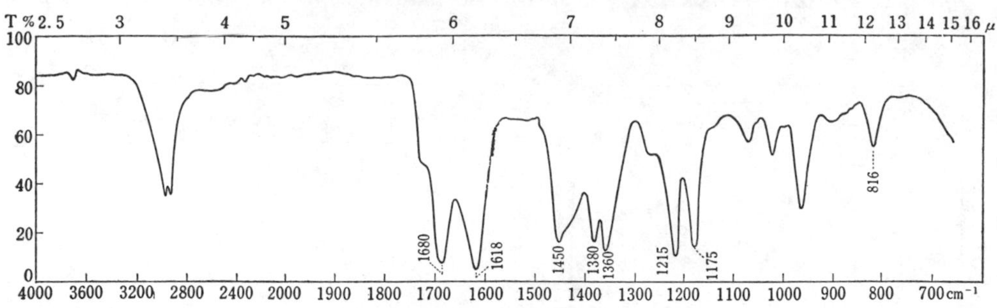

Fig. 43 Liquid film

ANSWER

Mesityl oxide

$$\begin{matrix} CH_3 \\ CH_3 \end{matrix}\!\!>C=C\!<\!\begin{matrix} H \\ C(=O)CH_3 \end{matrix}$$

1680 cm^{-1}: ν C=O (α, β-unsaturated ketones, 1675 cm^{-1}).
1618 cm^{-1}: ν C=C (α, β-unsaturated ketones, 1650~1600 cm^{-1}).
1450 and 1380 cm^{-1}: Methyl.
1360 cm^{-1}: CH_3 of methyl ketone.
1215 and 1175 cm^{-1}: Ketone and methyl (rocking.)
816 cm^{-1}: Tri-substituted double bond.

The ν C=O bands of α, β-unsaturated ketones are usually stronger than the ν C=C bands. However, if the ketone does not assume the s-*trans* form and is locked in the s-*cis* form the relative intensities will be reversed.[1] Furthermore, it has been noted[2] that the ν C=O and ν C=C bands of the s-*cis* form are shifted, respectively, to lower and higher frequencies so that the two bands become rather widely separated (band separation 75 cm^{-1}). Although the relations are not always true, they are empirically quite useful in characterizing the s-*cis* form.[3] The very strong ν C=C band of mesityl oxide implies that it assumes the s-*cis* form because this is sterically less hindered; methyl-methyl interactions are present in the planar s-*trans* form. A more rigorous treatment of the conformational equilibria of *trans*-Δ^3-penten-2-one and Δ^3-buten-2-one is given in reference [4].

C=C>C=O

(1)

νC=O stronger than νC=C. Differrence between νC=O and ν C=C is less than 75 cm^{-1}.

C=C–C(=O)–

(2)

Intensities of νC=O and νC=C are comparable. Difference between two bands is larger than 75 cm^{-1}.

[1] O. Wintersteiner and M. Moore, *J. Am. Chem. Soc.* **78**, 6193 (1956).
[2] E. A. Braude and C. J. Simmons, *J. Chem. Soc.* 3766 (1955).
[3] e.g., lactucin: D. H. R. Barton and C. R. Narayaman, *ibid.* 963 (1958).
[4] K. Noack and R. N. Jones, *Can. J. Chem.* **39**, 2225 (1961).

PROBLEM 44 / Deduce the structure of C_8H_8O.

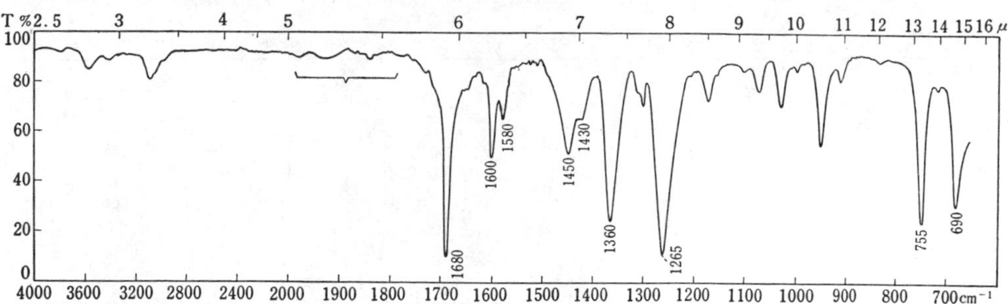

Fig. 44 Liquid film

ANSWER

Acetophenone.

-CO-CH_3

2000～1660 cm^{-1}: The characteristic pattern for mono-substituted phenyls is seen in the region not overlaid by the strong carbonyl band.

1680 cm^{-1}: ν C=O of aromatic ketone.

1600, 1580, 1450, 755, 690 cm^{-1}: Mono-substituted phenyl.

1430, 1360 cm^{-1}: Methyl. The position and intensity of the latter is typical for methyl ketones (Table 1～12).

1265 cm^{-1}: Aromatic ketone.

PROBLEM 45 / Deduce the structure of $C_9H_{10}O_3$, m.p. 50°. The compound forms a crystalline derivative with 2,4-dinitrophenylhydrazine.

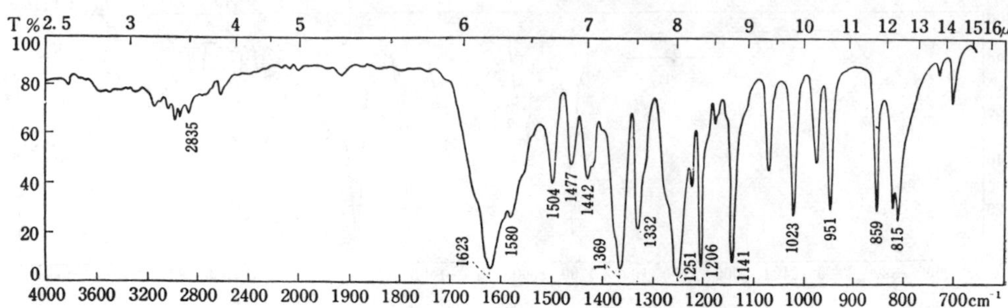

Fig. 45 1.6 mg/600 mg KBr (IRDC 307)

ANSWER

2–Hydroxy–4–methoxyacetophenone

H_3C O

C H

O

OCH_3

An aromatic ring is present. The very low frequency of the ν C=O indicates chelation with an OH group. Subtraction of these groups from the molecular formula leaves a methoxyl, the presence of which is supported by the 2835 cm^{-1} peak. Substitution type of phenyl is 1, 2, 4.

2835 cm^{-1}: OCH_3; the ν OH, which is around 2700 cm^{-1} in strongly chelated systems, are frequently diffuse and overlaid by other bands.
1623 cm^{-1}: ν C=O.
1580, 1504, 859 and 815 cm^{-1}: 1, 2, 4-Trisubstituted phenyl.
1477, 1442, and 1369 cm^{-1}: Methyls.
1251 cm^{-1}: ν=C–O– (or asymmetric stretching of =C–O–C–).
1023 cm^{-1}: ν C–O (or symmetric stretching of =C–O–C–).

PROBLEM 46 / Deduce the structure of $C_8H_8O_2$.

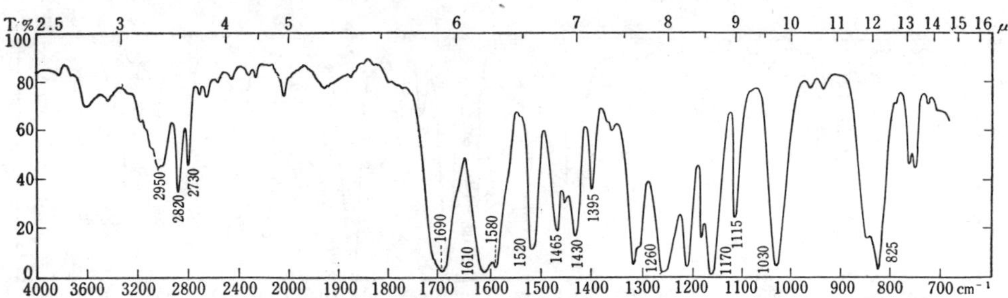

Fig. 46 Liquid film

ANSWER

p-Anisaldehyde.

2950 cm^{-1}: Stretching of aromatic and aliphatic C–H.
2820 and 2730 cm^{-1}: Aldehyde C–H; the CH_3O– 2850 cm^{-1} is overlaid by 2820 cm^{-1} band.
1690 cm^{-1}: Aromatic –CHO (1700 cm^{-1}).
1610, 1580, 1520, 1430, 1170, 1115, and 825 cm^{-1}: *p*-Substituted phenyl.
1465 and 1395 cm^{-1}: Methyl.
1260 and 1030 cm^{-1}: Asymmetric and symmetric stretching of =C–O–C (or, to a first approximation, stretching of =C–O– and –O–C–, respectively).

PROBLEM 47 / Deduce the structure of $C_8H_6O_3$, m.p. 37°C.

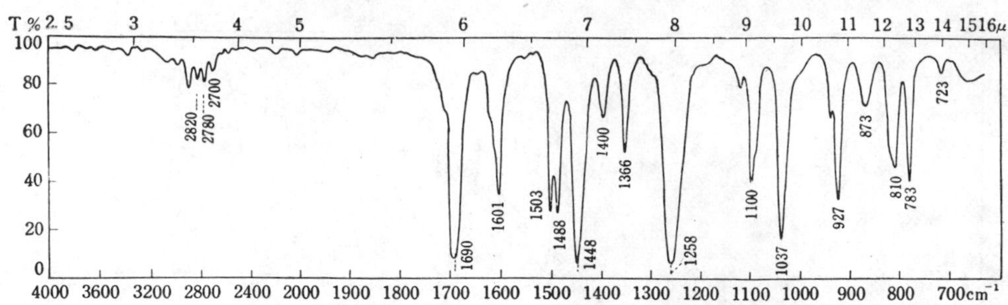

Fig. 47 Molten solid between plates (IRDC 102)

ANSWER

3, 4-Methylenedioxybenzaldehyde.

O CHO

O

The following groups are suggested.

Methylenedioxy

- 2780 cm^{-1}: CH_2 symmetric stretching (2780 cm^{-1}).
- 1488 cm^{-1}: CH_2 bending (1480 cm^{-1}).
- 1400 and 1360 cm^{-1}: These bands are sometimes absent.
- 1258 and 1037 cm^{-1}: Asymmetric and symmetric stretching, respectively, of =C–O–C (1250 and 1040 cm^{-1}).
- 1100 cm^{-1}: Probably in-plane bending of phenyl CH enhanced by the polar substituent.
- 927 cm^{-1}: Most characteristic (925 cm^{-1}).
- 723 cm^{-1}: Sometimes absent (1130 cm^{-1}).

Phenyl ring

- 1601, 1503, and 1448 cm^{-1}.
- The out-of-plane CH bending bands suggest 3 adjacent hydrogen atoms (810 ~ 750 cm^{-1}), thus a 1, 2, 3-tri-substituted phenyl, but this is not definite in view of the presence of substituents exerting strong *M* effects.

Aromatic aldehyde

- 2820 and 2700 cm^{-1}.
- 1690 cm^{-1}: Lower than normal value because of mesomerism involving the methylenedioxy group (solution data should be compared!).

Thus, the first guess is 2,3-methylenedioxybenzaldehyde, while the other possibility is the 3, 4-isomer.

PROBLEM 48 / Deduce the structure of $C_7H_5NO_3$, m.p. 106°.

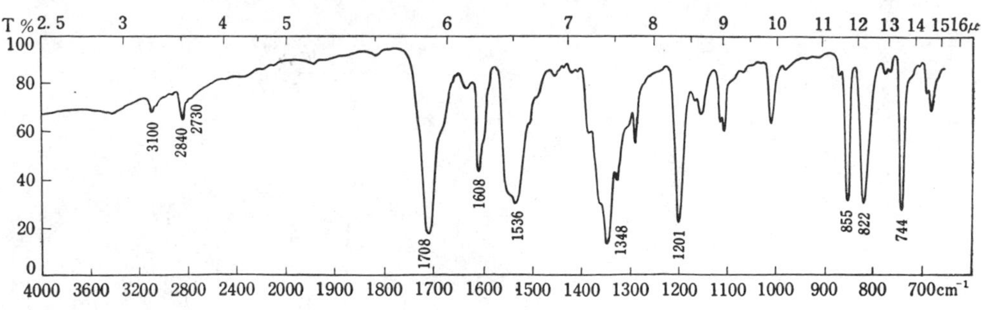

Fig. 48 1.0 mg/300 mg KBr

ANSWER

p-Nitrobenzaldehyde.

O_2N-⟨ ⟩-CHO

The presence of aldehyde, nitro, and phenyl groups is apparent, but the substitution type is not clear.

3100 cm^{-1}: Aromatic CH.
2840 and 2730 cm^{-1}: Aldehydic CH.
1708 cm^{-1}: Aldehyde; 1721 cm^{-1} in $CHCl_3$ (1700 cm^{-1}). High shift is due to $-I$ and $-M$ effects of nitro group.
1608 cm^{-1}: Phenyl.
1536 and 1348 cm^{-1}: Asymmetric (1550∼1515 cm^{-1}) and symmetric stretching (1385∼1345 cm^{-1}) of nitro. Band position of the nitro group has been studied extensively and correlated with electric and steric effects[1].
855 cm^{-1}: C–N stretching.
822 cm^{-1}: Adjacent 2H (860∼800 cm^{-1}).
744 cm^{-1}: C–N–O bending. Nitrobenzenes often have bands at ca. 850 (and 750) cm^{-1}.

---

[1] Review: A. R. Katritzky and P. Simmons, *Quart. Rev.* **13**, 353 (1959); *Rec. trav. Chim.* **79**, 361 (1960). N. Yoda, " Infrared Spectra," Vol. 13, p. 43 (1961).

PROBLEM 49 / The following spectrum is that of a simple derivative of cholesterol. What is the derivative?

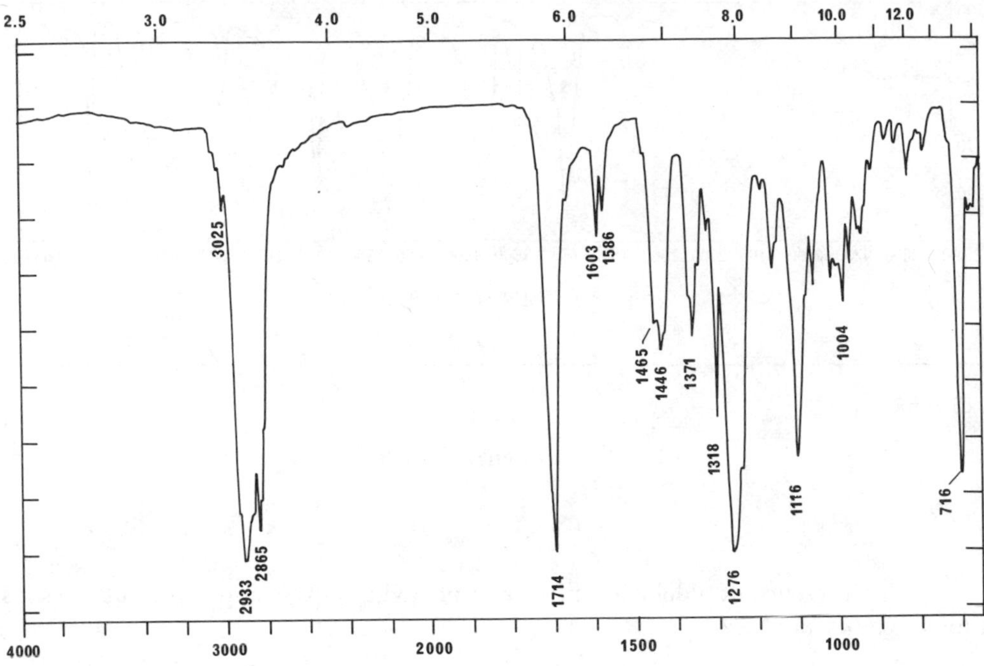

Fig. 49 3 mg/200 mg KBr

ANSWER

Cholesteryl benzoate

$3025\ cm^{-1}$: ν C—H aromatic.
$1714\ cm^{-1}$: ν C=O of aromatic ester.
1603, $1586\ cm^{-1}$: Phenyl C=C.
$1276\ cm^{-1}$: ν_{as} C—O, typically stronger than ν_s C—O.
$1116\ cm^{-1}$: ν_s C—O.
$716\ cm^{-1}$: Monosubstituted benzene ring. Benzoates typically show only one band at $715\ cm^{-1}$; they thus deviate from the behavior of other monosubstituted phenyls which show two bands at $770 \sim 730\ cm^{-1}$ and $710 \sim 690\ cm^{-1}$.

PROBLEM 50 / Deduce the structure of the compound with molecular formula $C_9H_9NO_3$.

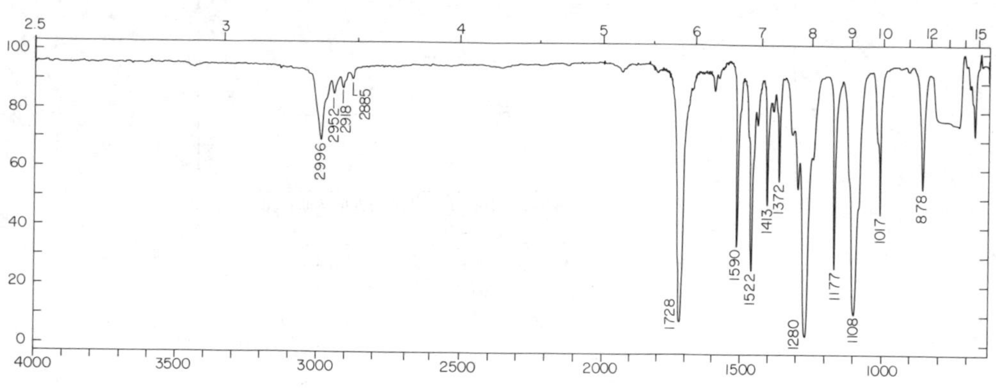

Fig. 50 5% solution in CCl_4

ANSWER

Ethyl *p*-nitrosobenzoate

N=O

$COOCH_2CH_3$

The molecular formula shows a degree of unsaturation of six and suggests an aromatic structure. The presence of the aromatic ester is apparent from the 1728 cm^{-1} band but the —N=O group must be inferred indirectly (from the degree of unsaturation and lack of OH, NH, C≡N, and other group stretching frequencies), since the region of its characteristic frequency (~1500 cm^{-1}) is obscured by strong phenyl C=C absorption.

1728 cm^{-1}: ν C=O of aryl ester.
1605, 1590, 1522, 1470 cm^{-1}: Phenyl C=C stretch; these strong bands obscure the region where —N=O stretch is expected.
1280 cm^{-1}: ν_{as} C—O—C of aryl ester.
1108 cm^{-1}: ν_s C—O—C of aryl ester.
878 cm^{-1}: δ CH of *p*-disubstituted phenyl; appears at higher frequency than usual (860~810 cm^{-1}) due to the presence of electron-withdrawing groups.

PROBLEM 51 / Correlate structures with IR spectra.

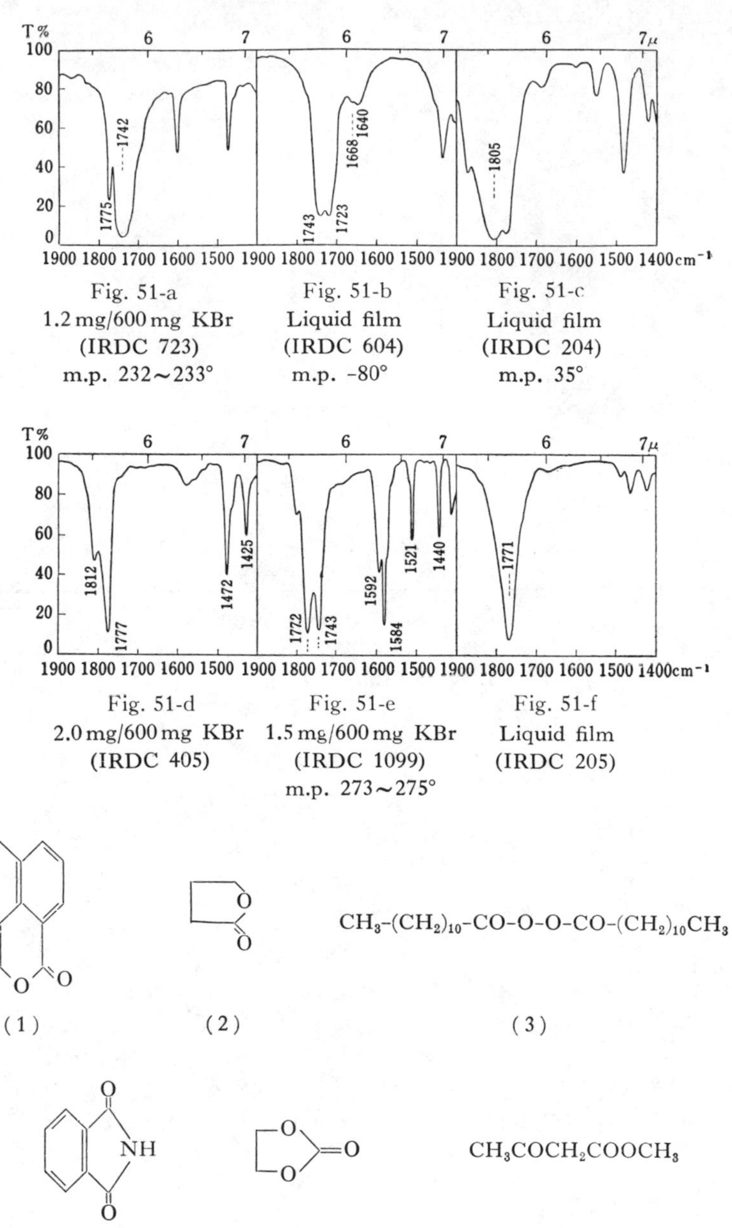

Fig. 51-a
1.2 mg/600 mg KBr
(IRDC 723)
m.p. 232~233°

Fig. 51-b
Liquid film
(IRDC 604)
m.p. -80°

Fig. 51-c
Liquid film
(IRDC 204)
m.p. 35°

Fig. 51-d
2.0 mg/600 mg KBr
(IRDC 405)

Fig. 51-e
1.5 mg/600 mg KBr
(IRDC 1099)
m.p. 273~275°

Fig. 51-f
Liquid film
(IRDC 205)

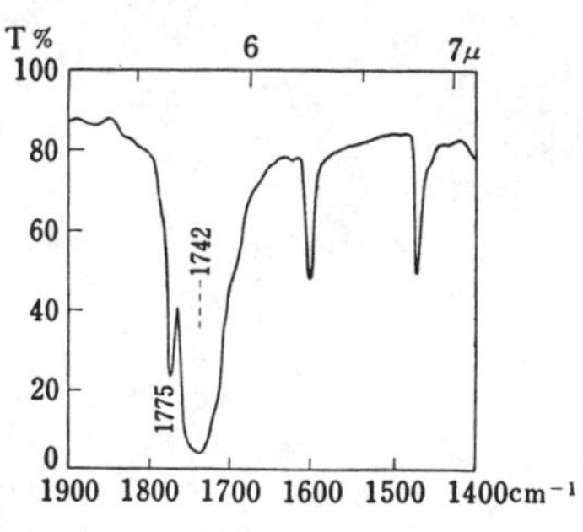

Fig. a

O
NH
O

(4) Phthalimide

1775 and 1742 cm^{-1}: Asymmetric and symmetric ν C=O. The lower band is usually broader and stronger.

1600 cm^{-1}: Phenyl.

NH at 3150 cm^{-1}, out-of-plane CH at 721 cm^{-1}.

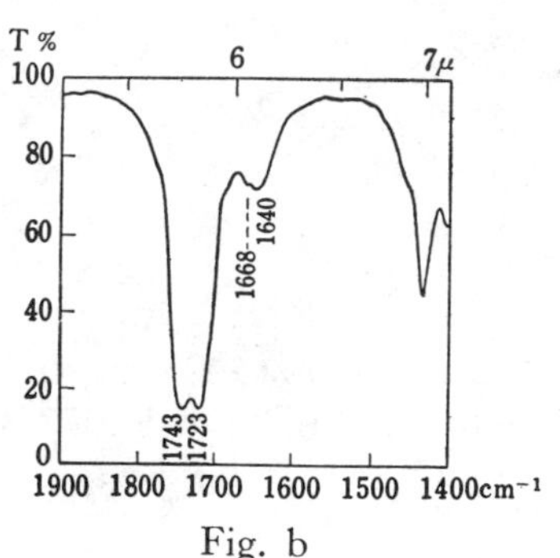

Fig. b

$CH_3COCH_2COOCH_3$

(b) Methyl acetoacetate

1743 and 1723 cm^{-1}: Ester and ketone, respectively. In ethyl acetoacetate the two ν C=O bands are at 1738 and 1717 cm^{-1} (KBr); these data show that interaction of the two carbonyls to shift the ν C=O to higher frequencies occurs only to a slight extent.

1668 and 1640 cm^{-1}: ν C=O and ν C=C of the small amount of the enol form, which is in equilibrium. Strong bands at 1251 and 1042 cm^{-1} are associated with the $-COOCH_3$ group, and another strong band at 1153 cm^{-1} probably with the CH_3-COC– group. The symmetric bending of methyl is at 1360 cm^{-1}, and bending of the active methylene is at 1420 cm^{-1}.

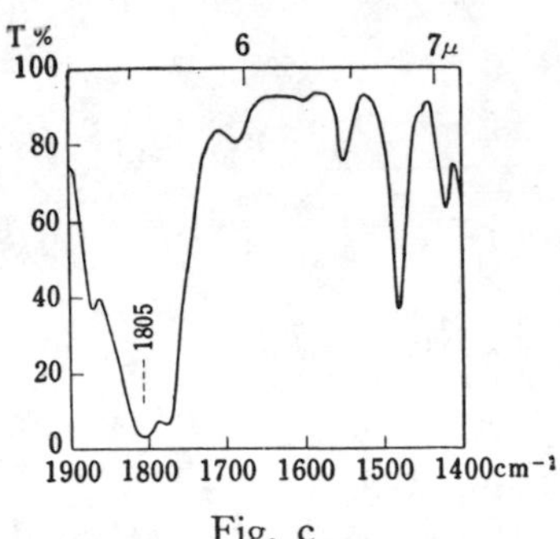

Fig. c

(5) Ethylene carbonate

1805 cm^{-1}: ν C=O.
The ν C–O peaks are at 1165 and 1075 cm^{-1}; both strong.

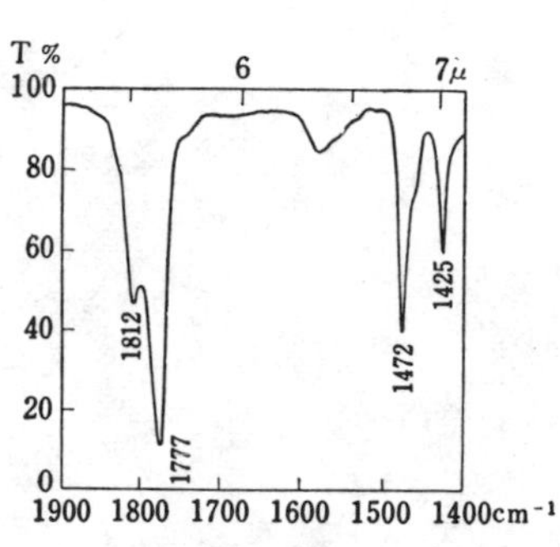

Fig. d

$CH_3(CH_2)_{10}CO$-O-O-$CO(CH_2)_{10}CH_3$
Lauroyl peroxide

1812 and 1777 cm^{-1}: Perturbed stretchings of carbonyls. A band at 1425 cm^{-1} is due to the active methylene, and three distinct bands at 1133, 1099, and 1077 cm^{-1} are most probably due to ν C–O vibrations. The ν O–O band is not clear, as with other peroxides.

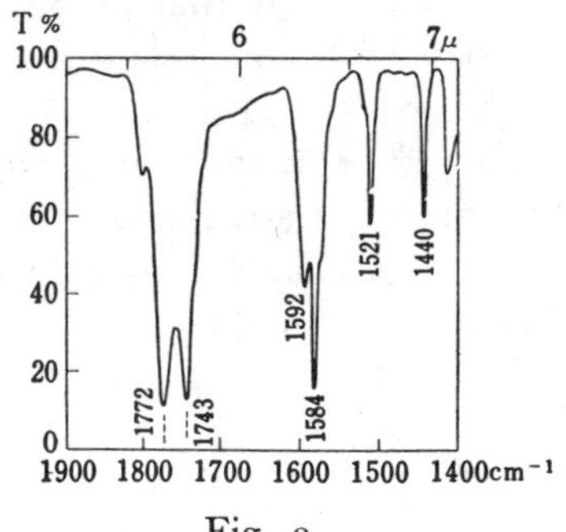

Fig. e

Naphthalene-1,8-dicarboxylic anhydride

1772 and 1743 cm^{-1}: Perturbed stretchings of carbonyls.
1592, 1584, 1521, 1440 (and 781) cm^{-1}: Aromatic ring.

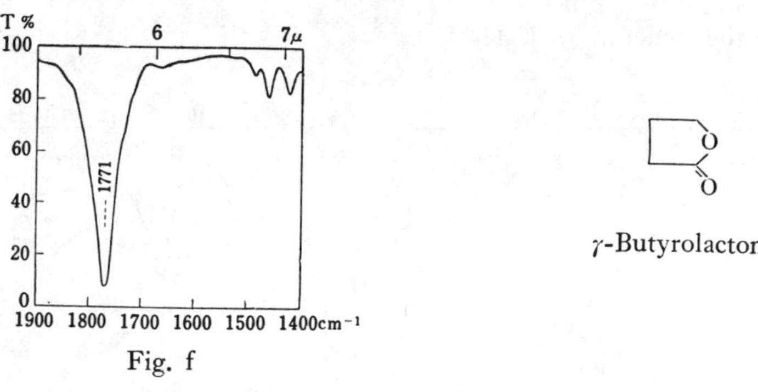

γ-Butyrolactone

Fig. f

1771 cm^{-1}: ν C=O.
The two other ester bands are at 1169 and 1037 cm^{-1}: both strong.

PROBLEM 52 / Deduce the structure of $C_3H_6O_2$.

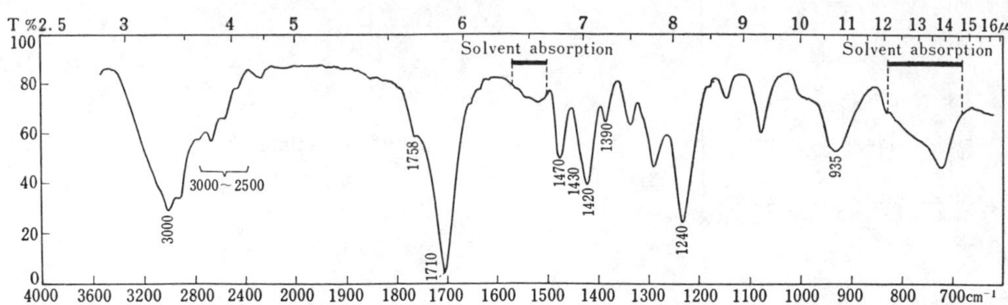

Fig. 52 CCl_4 1% solution, 0.5 mm cell

ANSWER

Propionic acid.

CH_3CH_2COOH

3000~2500 cm^{-1}: Group of small bands typical for carboxylic acid dimer. Only ammonium groups show similar absorption. The band at 3000 cm^{-1} is due to the dimeric ν OH, the other small bands being combination tones of the fundamental vibrations at 1420~1250 cm^{-1}.[1]

1758 and 1710 cm^{-1}: ν C=O of dimer and monomer; the relative intensities therefore vary with concentration. In dioxane they are replaced by a single band at 1735 cm^{-1} assigned to the solvated carboxyl. Thus, the solvating power of ethers such as dioxane and tetrahydrofuran are strong and the dimeric hydrogen-bonding is cleaved.

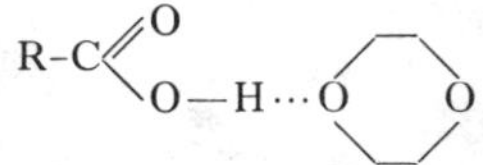

1470 and 1390 cm^{-1}: Methyl.

1420 cm^{-1}: Dimeric OH in-plane bending coupled with C–O stretching. The methylene adjacent to the carboxyl occurs as the slight shoulder at ca. 1430 cm^{-1}.

1240 cm^{-1}: ν C–O coupled with δ OH (in-plane). Formally it corresponds to the 1240 cm^{-1} band that is characteristic for all =C–O– moieties, e.g., enol ethers, esters, phenols.

935 cm^{-1}: δ OH (out-of-plane) of dimer. It corresponds to the 750~650 cm^{-1} band of associated alcohols (δ OH out-of-plane), the displacement to higher frequencies being caused by the stronger hydrogen-bonding. The 1420 cm^{-1} band is also related to the OH in-plane bending of associated alcohols at 1400 cm^{-1}.

The 3000~2500, 1758, 1710, 1420, 1240, and 935 cm^{-1} bands are characteristic for the carboxyl group measured in solvents. The regions of strong solvent absorption are indicated by the solid lines in the Figure, and the ill-defined absorptions in the regions are due to the fact that the reference cell used in this measurement was thinner than the sample cell.

---

[1] S. Bratoz, D. Hadži, and N. Sheppard, *Spectrochim. Acta* **8**, 249 (1956).

PROBLEM 53 / Figs. 53 a∼d are the IR spectra of propionic acid measured as liquid film, gas, carbon tetrachloride solution, and dioxane solution. Correlate IR with state of measurement.

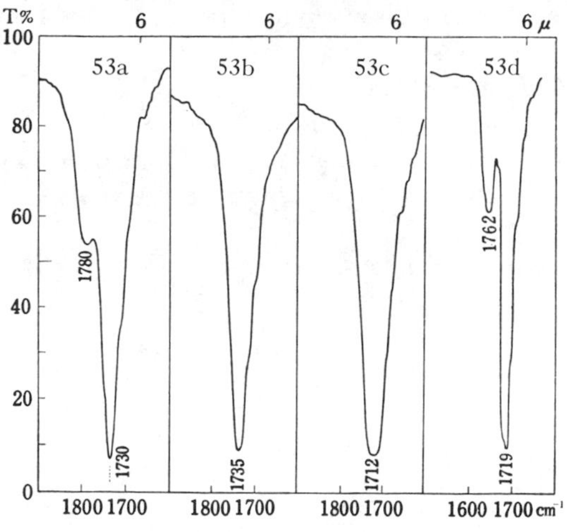

Fig. 53 C=O bands of propionic acid

ANSWER

Propionic acid measured under various conditions.

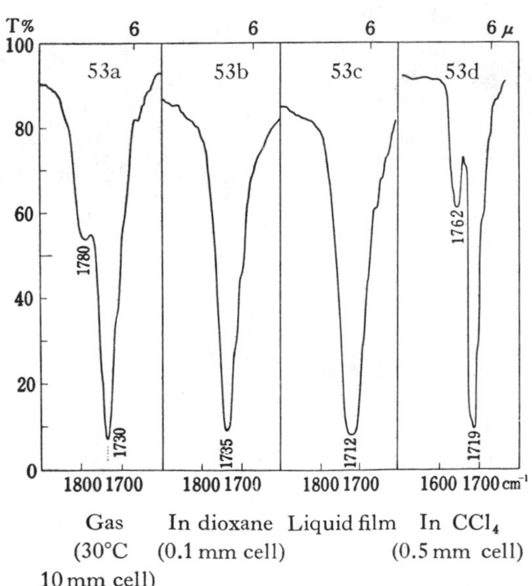

The C=O band of the carboxyl group changes remarkably with conditions of measurement. This is due to the strong tendency for association. Other similar groups behave likewise.

53 a: Gas (30°C, 10 mm cell)
1780 and 1730 cm^{-1} bands are attributed to the monomer and dimer, respectively. The relative intensity of the former is enhanced at higher temperatures.

53 b: Dioxane solution (0.1 mm cell)
The single band at 1735 cm^{-1} is attributed to the carboxyl group having its hydroxyl oxygen solvated with the solvent.

53 c: Liquid film, 1712 cm^{-1} is due to dimer.

53 d: Carbon tetrachloride (0.5 mm cell)
1758 and 1710 cm^{-1} are due to monomer and dimer, respectively. Relative intensities of the two bands are dependent on concentration.

The situation can be summarized as follows:

Gas	–COOH monomer, 1780	$-C\langle{}^{O\cdots H-O}_{O-H\cdots O}\rangle O-$ dimer, 1730
Liquid film		$-C\langle{}^{O\cdots H-O}_{O-H\cdots O}\rangle C-$ dimer, 1710
Nonpolar solvent	–COOH monomer, 1760	$-C\langle{}^{O\cdots H-O}_{O-H\cdots O}\rangle C-$ dimer, 1710
Ether solution	$-C\langle{}^{O}_{O-H\cdots O\langle}$	1735
Alcohol solution	$-C\langle{}^{O\cdots H-OR}_{O-H}$	1720
Basic solution	$-C\langle{}^{O}_{O}\}^{\ominus}$	1610∼1550 and 1400 cm^{-1}

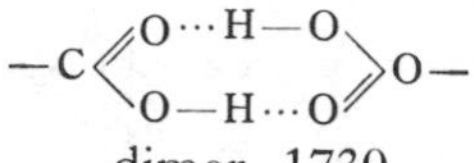

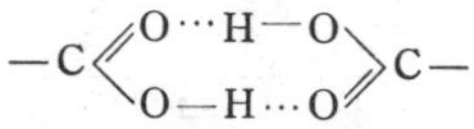

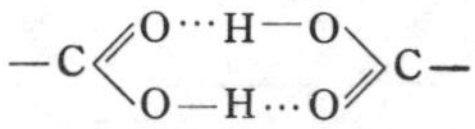

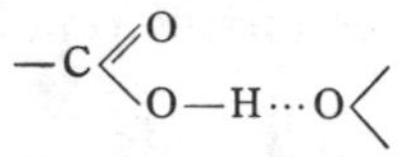

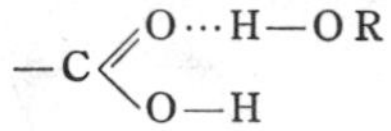

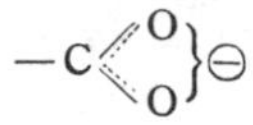

PROBLEM 54 / Deduce the structure of $C_{16}H_{34}O_2$.

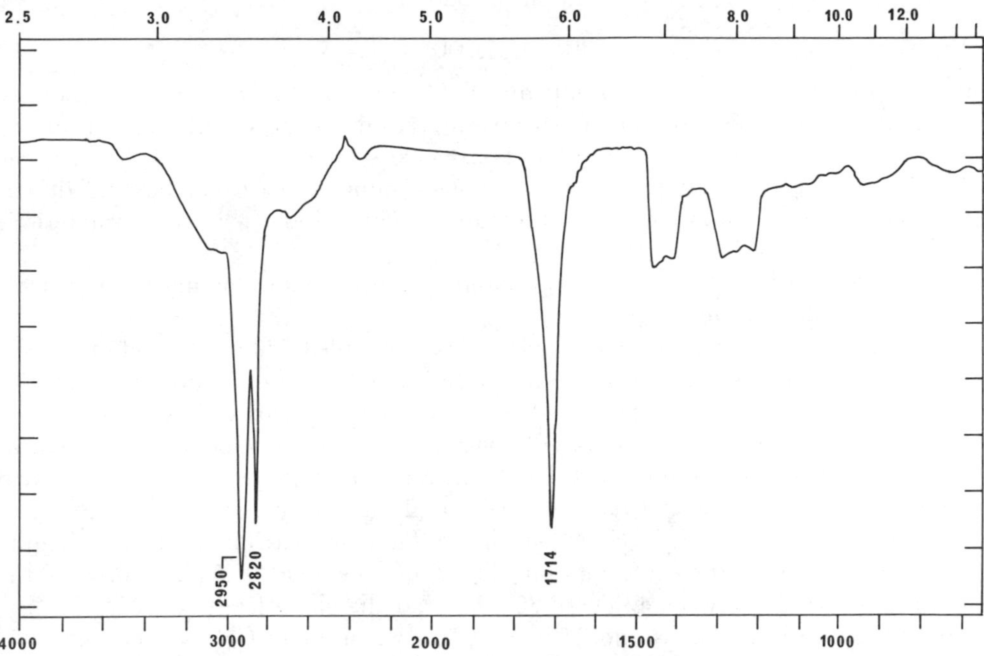

Fig. 54-a 5% solution in $CHCl_3$, 0.1 mm cell

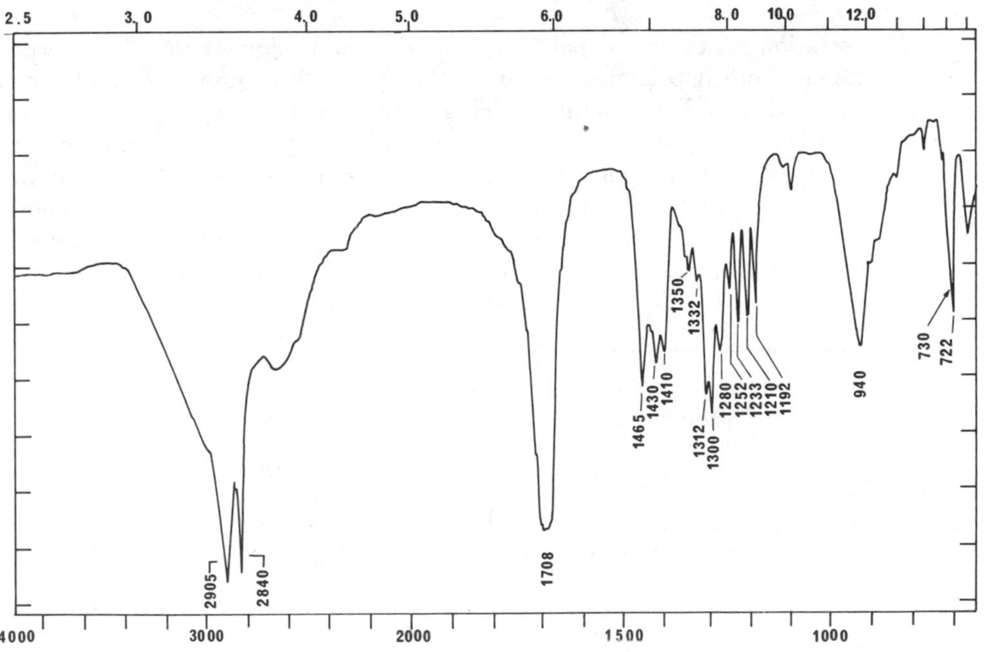

Fig. 54-b 3mg/200 mg KBr

ANSWER

Palmitic acid

$CH_3(CH_2)_{14}COOH$

Solution Spectrum: The 3000∼2500 and 1714 cm^{-1} bands indicate a carboxylic acid dimer. The remainder of the spectrum gives little additional information.

KBr Spectrum: In addition to the bands at 3300∼2500 cm^{-1} and 1708 cm^{-1}, the bands at 1431, 1300, and 938 cm^{-1} obviously indicate a carboxylic acid dimer. The evenly spaced absorptions in the range 1350∼1180 cm^{-1} are characteristic of long *n*-alkyl chains.

1410 cm^{-1}: CH_2 adjacent to COOH. This band is highly reliable and has been securely assigned by deuteration experiments.[6]

730 and 722 cm^{-1}: Methylene chain rocking. The so-called 720 cm^{-1} band of *n*-alkyl chains is sometimes split in the solid phase but is a singlet in liquid or solution spectra.[1] The splitting is attributed to interactions of nearby molecules.

1350∼1180 cm-1: The solid spectra of long-chain *n*-alkyl compounds (e.g., paraffins, fatty acids, acid amides, alkyl halides, α,ω-polymethylene dicarboxylic acid esters) exhibit a series of evenly spaced bands in the region 1350∼1180 cm^{-1} that are characteristic of the chain length.[2, 3] They are attributed to the wagging of methylenes in *trans* conformation. The following relation holds between the number of bands between 1350∼1180 cm^{-1} and the chain length:

no. of methylene groups = (no. of bands) × 2 when no. of CH_2 is even, or,

" " = [(no. of bands) × 2] — 1 when no. of CH_2 is odd.

In the present case 7 bands with intervals of approximately 20 cm^{-1} are present; one is hidden by the 1300 cm^{-1} COOH band, and in order to avoid ambiguity the usage of Ba salts has been recommended in determining the chain length of fatty acids.[3]

The solution spectrum of palmitic acid does not show these characteristic small band groups, and it becomes similar to the spectrum of propionic acid shown in Fig. 52. This is because the methylene chains assume the most stable *trans* conformation in the crystalline state, whereas under fluid conditions they assume numerous noncharacteristic conformations.[4] Lower homologs of *n*-paraffins (e.g. *n*-butane), on the other hand, exhibit more bands in the fluid state.[5] This is because the energy barriers between the various conformations of the lower *n*-paraffins are relatively large and they exist as a mixture of several well-defined conformational isomers.

---

[1] The 720 cm^{-1} bands of all *n*-paraffins having 17∼36 carbon atoms are singlets at temperatures higher than their solid transition points. At lower temperatures, the odd-numbered *n*-paraffins with less than 26 carbons and all *n*-paraffins with more than 26 carbons show split 720 cm^{-1} bands, whereas all even-numbered *n*-paraffins shorter than C_{27} show singlet 720 cm^{-1} bands: J. M. Martin, R. W. B. Johnson, and M. J. O'Neal, *Spectrochim. Acta* **12**, 12 (1958).

[2] R. N. Jones, A. F. McKay, and R. G. Sinclair, *J. Am. Chem. Soc.* **74**, 2575 (1952).

[3] R. A. Meikeljohn, R. J. Meyer, S. M. Aronouic, H. A. Schnette, and V. W. Meloch, *Anal. Chem.* **29**, 329 (1953).

[4] Solution spectra of straight-chain fatty acids and their methyl esters: R. N. Jones, *Can. J. Chem.* **40**, 301, 321 (1962).

[5] N. Sheppard and D. M. Simpson, *Quart. Revs.* **7**, 19 (1953).

[6] T. Shimanouchi, M. Tsubai, T. Takenishi, N. Iwata, *Spectrochim. Acta* **16**, 1328 (1960).

PROBLEM 55 / What is the following compound that contains sodium?

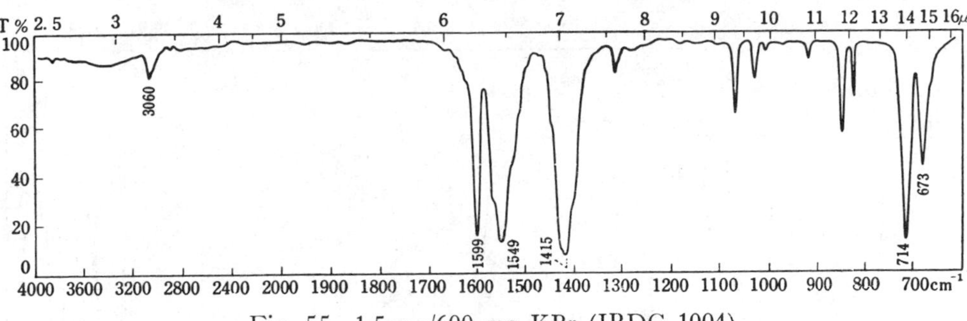

Fig. 55 1.5 mg/600 mg KBr (IRDC 1004)

ANSWER

Sodium benzoate.

-COONa

Presence of mono-substituted phenyl and carboxylate groups are evident, and this leads to sodium benzoate.

Mono-substituted phenyl: 3060, 1599, 714, and 673 cm^{-1}.
Carboxylate ion: 1549 and 1415 cm^{-1} (1610~1550 and 1400 cm^{-1})

PROBLEM 56 / Identify the functional groups present in the gibberellin derivative whose molecular formula is $C_{20}H_{22}O_6$.

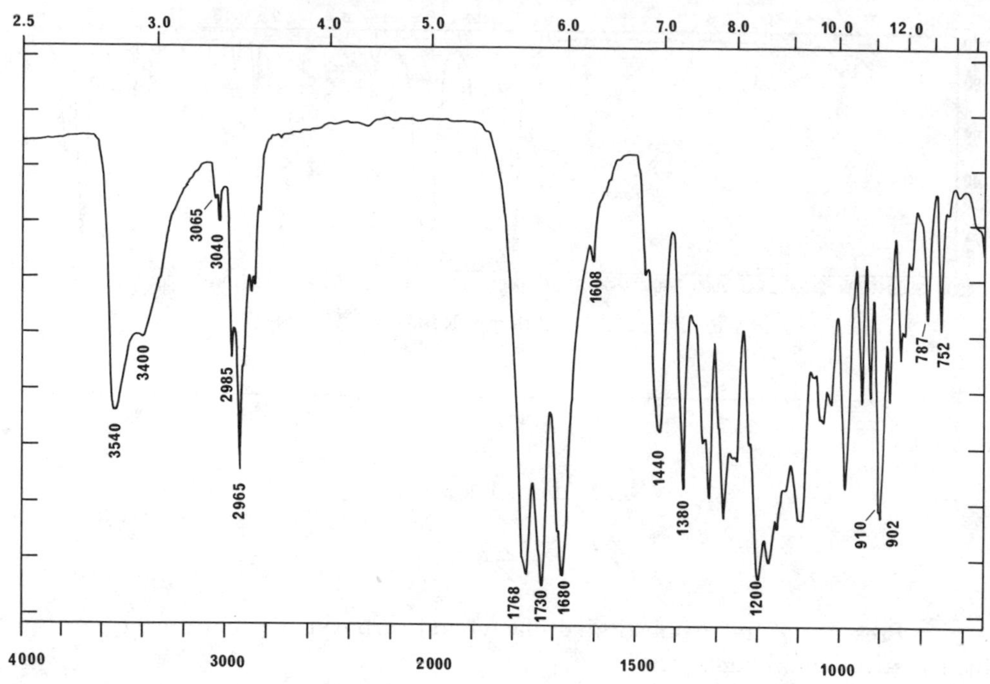

Fig. 56 3 mg/200 mg KBr

ANSWER

O

O

O

OH

COOMe

3540, 3400 cm^{-1}: ν OH and contaminant H_2O.
3065 cm^{-1}: ν_{as} CH_2 of terminal methylene.
3040 cm^{-1}: ν CH aromatic.
1768 cm^{-1}: ν C=O of γ-lactone.
1730 cm^{-1}: ν C=O of methyl ester.
1680 cm^{-1}: ν C=O of α, β-unsaturated ketone.
1608 cm^{-1}: ν C=C conjugated to C=O.
1440 cm^{-1}: δ_{as} $CH_3{-}O{-}\overset{\overset{O}{\|}}{C}{-}$
1380 cm^{-1}: δ_s CH_3 of COOMe and 4-Me groups.
1200 cm^{-1}: ν C–O of ester and alcohol groups.
910, 902 cm^{-1}: δ CH_2 of terminal methylene.

PROBLEM 57 / Below are the spectra for *o*-aminobenzoic acid and *m*-aminobenzoic acid. The spectra appear quite different. Explain.

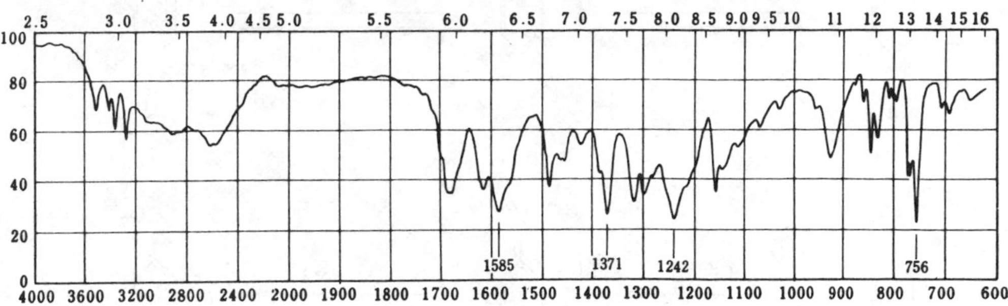

Fig. 57-a *o*-aminobenzoic acid, 1.3 mg/600 mg KBr (IRDC 3490)

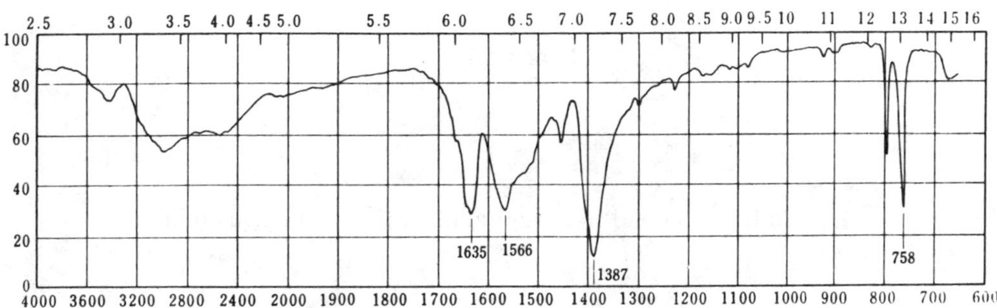

Fig. 57-b *m*-aminobenzoic acid, 0.8 mg/600 mg KBr (IRDC 7452)

ANSWER

The IR spectra show that the ortho isomer 1 exists in the neutral form 1a, whereas the meta isomer 2 exists in the zwitterionic form 2a (in the solid state, KBr disk).

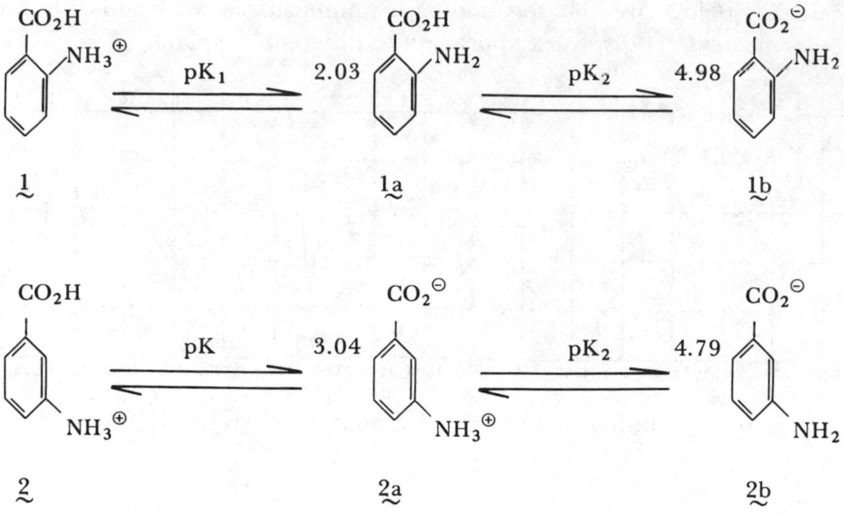

$\mathbf{1}$ (anthranilic acid cation, CO_2H, $NH_3^{\oplus}$) $\underset{}{\overset{pK_1\ \ 2.03}{\rightleftharpoons}}$ $\mathbf{1a}$ (CO_2H, NH_2) $\overset{pK_2\ \ 4.98}{\rightleftharpoons}$ $\mathbf{1b}$ ($CO_2^{\ominus}$, NH_2)

$\mathbf{2}$ (CO_2H, $NH_3^{\oplus}$) $\overset{pK\ \ 3.04}{\rightleftharpoons}$ $\mathbf{2a}$ ($CO_2^{\ominus}$, $NH_3^{\oplus}$) $\overset{pK_2\ \ 4.79}{\rightleftharpoons}$ $\mathbf{2b}$ ($CO_2^{\ominus}$, NH_2)

Scheme 1

In solution this can be explained in a semiquantitative manner as follows.

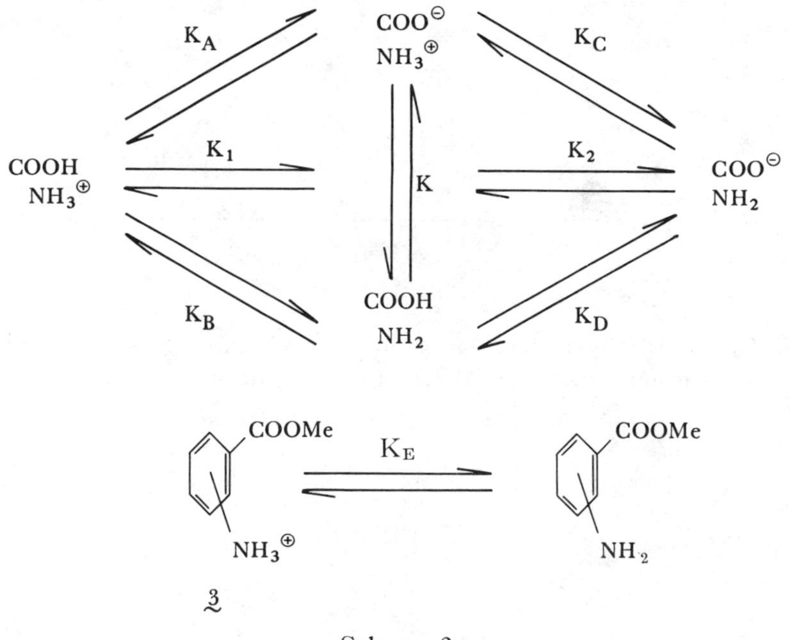

Scheme 2

It can be seen that:

$K_1 = K_A + K_B$ and $K = K_A/K_B$, where K_1 is the first dissociation constant and K is the ratio between zwitterionic and neutral forms. It is safe to assume that the equilibrium constant K_B, which cannot be measured experimentally by pK studies, is approximately equal to that of the corresponding methyl ester 3. Hence,

$$K = K_A/K_B = K_A/K_E = K_1/K_E - 1 \quad \ldots\ldots (1)$$

The pK_E values for the *o*- and *m*-aminobenzoates are 2.09 and 3.56, respectively. Substitution of these values and the K_1 values into equation (1) gives the following results:

ortho: $K = 0.15$
meta: $K = 2.38$

That is, in solution, the neutral form is favored for the *o*-isomer, whereas the zwitterionic form is favored for the *m*-isomer, a tendency which is in agreement with the solid IR results.

Fig. 57-a

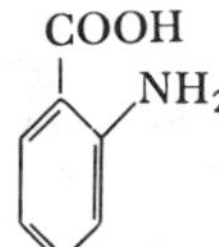

3600∼3200 cm^{-1}: ν NH_2; ordinarily two sharp bands, ν_{as} and ν_s, would be visible. The reason for the appearance of four bands is not clear. Possibly it is due to the presence of a small amount of water or to polymorphic forms.
3200∼2200 cm^{-1}: ν OH of carboxylic acid.
1680 cm^{-1}: ν C=O of aromatic acid.
1615 cm^{-1}: In plane δ NH.
1585 cm^{-1}: Phenyl C=C stretch, intensified by conjugation with C=O and $-NH_2$.
1371, 1242 cm^{-1}: Characteristic combination bands of the $-CO_2H$ group. The higher frequency band is here shifted below the usual position (1420 cm^{-1}).
940 cm^{-1}: Dimeric carboxyl.
756 cm^{-1}: *o*-di-substituted phenyl (770–735 cm^{-1}).

Fig. 57b

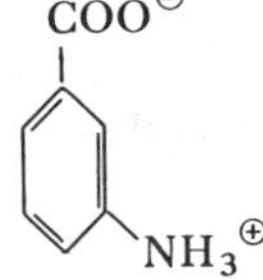

3200∼2200 cm^{-1}: $-NH_3^+$; ν_{as}, ν_s, and combination bands.
1635 cm^{-1}: δ_{as} $-NH_3^+$; distinguishes ammonium salt from carboxylic acid which also absorbs 3200∼2200 cm^{-1}.
1566, 1387 cm^{-1}: ν_{as} and ν_s of CO_2^-.
800, 758 cm^{-1}: The two bands identify the compound as an *m*-substituted phenyl, although they are shifted from their characteristic position (770∼730, ∼700 cm^{-1}) due to the presence of electron-withdrawing groups. Both *o*- and *p*-substituted phenyls usually show only one strong band in this region.

PROBLEM 58 / Deduce the structure of $C_5H_8O_2$.

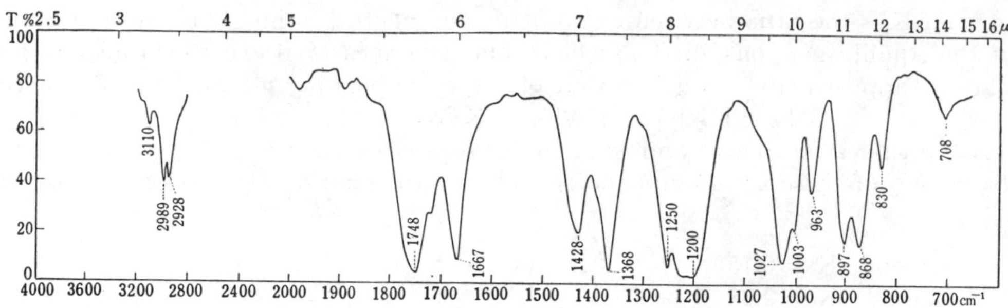

Fig. 58 Liquid film

ANSWER

Isopropenyl acetate

$$\begin{matrix} CH_3 \\ CH_2 \end{matrix}\!\!>\!C\text{-}O\text{-}CO\text{-}CH_3$$

3110 cm^{-1}: =CH_2.
2989 and 2928 cm^{-1}: –CH_3.
1748 cm^{-1}: ν C=O (vinyl esters generally at 1760 cm^{-1}).

The carbonyl stretching frequency is shifted to higher wave-numbers in ordinary esters (1735 cm^{-1}) as compared to aliphatic ketones because the $-I$ effect of the oxygen atom is larger than its $+M$ effect, thus producing an overall electron-attracting effect. In vinyl esters the $+M$ effect is internally compensated due to overlap with π-electrons of both C=C and C=O bonds, and the net effect is operation of the oxygenic $-I$ effect with consequent increase in the double bond character of the carbonyl.

$$-\overset{\frown}{O}\leftarrow \overset{\frown}{C=O} \qquad\qquad \overset{\frown}{C=C}-\overset{\frown\frown}{O}\leftarrow\overset{\frown}{C=O}$$

ordinary esters — vinyl esters

1667 cm^{-1}: ν C=C, conjugation with oxygen increases intensity.
1428 and 1368 cm^{-1}: CH_3 and CH_2. Lower band is stronger as with other acetates.
1250 and 1200 cm^{-1}: Antisym. stretching of =C–O–C=.

The usual " acetate band " at 1240 cm^{-1} is shifted lower by 20 cm^{-1} in vinyl acetates[1]. This is attributed to decreased double bond character in the –O–C portion of the –O–C=O group, and is in accord with explanation given above for enhancement of double bond character of carbonyl group.

1027 cm^{-1}: Sym. stretching of =C–O–C=.
868 cm^{-1}: Out-of-plane δ CH of =CH_2 (890 cm^{-1}).

---

[1] R. N. Jones and C. Sandorfy, "Chemical Applications of Spectroscopy," p. 482 (1956).

PROBLEM 59 / Deduce the structure of $C_{12}H_{14}O_4$.

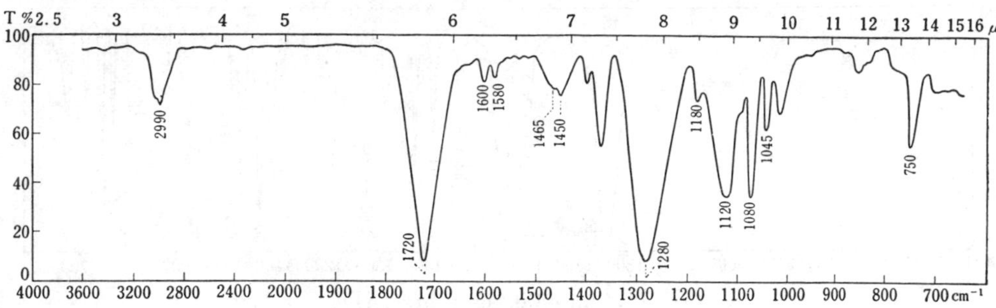

Fig. 59. Liquid film

ANSWER

Diethyl phthalate

$COOC_2H_5$

$COOC_2H_5$

1720 cm^{-1}: C=O (1720 cm^{-1}).
1600, 1580, 1465, 1180, 1080, 1045, and 750 cm^{-1}: *o*-Substituted phenyl.
1280 and 1120 cm^{-1}: Ester (two bands between 1300~1050 cm^{-1}). Note that the 1280 cm^{-1} =C–O–C asymmetric stretching band is stronger than 1720 cm^{-1}.

PROBLEM 60 / Deduce the structure of $C_9H_7NO_4$, m.p. 130~131°.

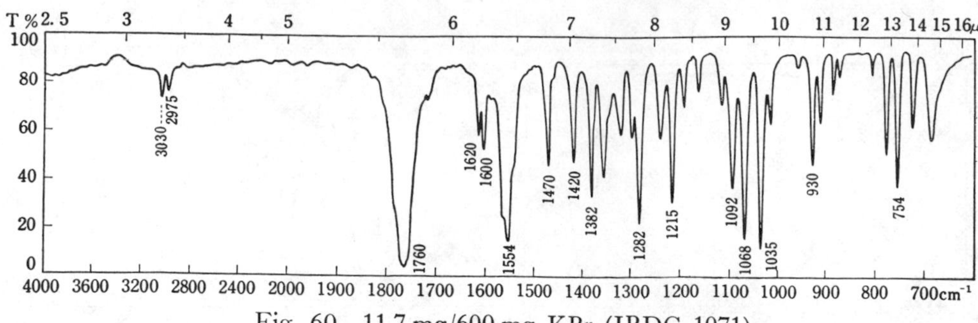

Fig. 60 11.7 mg/600 mg KBr (IRDC 1071)

ANSWER

3–(Nitromethyl) phthalide

CH_2NO_2

O

O

Degree of unsaturation is eight. An *o*-substituted phenyl is indicated by the 754 cm^{-1} band, and a nitro group by the 1554 and 1382 cm^{-1} bands. The two remaining unsaturations are accounted for by a phthalide ring and this leads to the structure shown. The perturbed methylene peak at 1420 cm^{-1} supports the nitromethyl side-chain.

3030, 1620, 1600, 1470, and 754 cm^{-1}: *o*-Substituted phenyl; also bands at 1282, 1092 (or 1068), and 1035 cm^{-1}.

1760 cm^{-1}: Phthalide ν C=O (1760 cm^{-1} in polar solvents and solid phase, 1780 cm^{-1} in nonpolar solvents).

1420 cm^{-1}: Methylene.

1215 and 1068 (or 1092) cm^{-1}: Asymmetric and symmetric stretching of =C–O–C. When this moiety is contained in strained rings, as in the present five-membered ring, the former band is shifted lower from its usual position of ca. 1240 cm^{-1}.

1554 and 1382 cm^{-1}: Asymmetric and symmetric stretching of nitro group. The primary and secondary nitro compounds (RCH_2NO_2 and $RR'CHNO_2$) absorb at 1600~1530 cm^{-1} and 1380–1360 cm^{-1}, while tertiary nitro compounds absorb in the lower frequency range of 1545~1530 cm^{-1} and 1360~1340 cm^{-1}.[1] When strong electron-attracting groups are attached, the two bands become separated, e.g., $CClF_2$–CF_2–NO_2 at 1618 and 1274 cm^{-1}[2]. In conjugated nitro-olefins the asymmetric stretching is shifted to the lower range of 1550~1500 cm^{-1} (ν_s at 1360~1330 cm^{-1}).[1]

930 cm^{-1}: C–N stretching (?). Usually this is between 870~850 cm^{-1}, mostly around 870 cm^{-1}.

---

[1] J. F. Brown, Jr., *J. Am. Chem. Soc.* **77**, 6341 (1955).
[2] R. N. Haszeldine, *J. Chem. Soc.* 2525 (1953).

PROBLEM 61 / Correlate figures with compounds. Figs. a, c, d, and f lack bands between 1600∼1500 cm^{-1}.

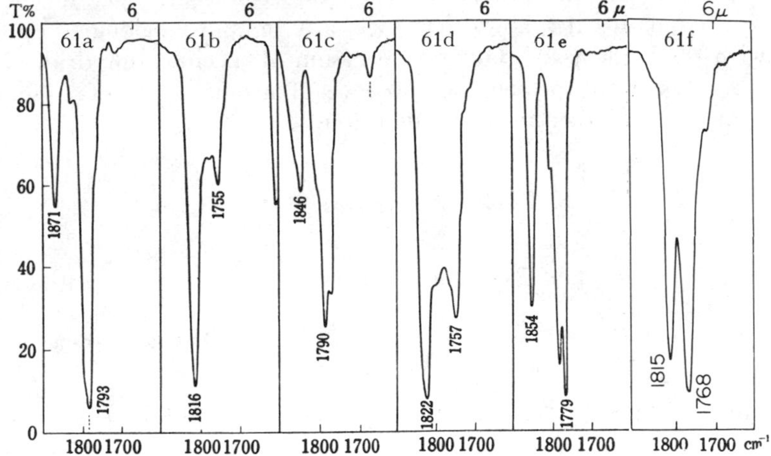

Fig. 54 ν C–O of various anhydrides in $CHCl_3$ (0.1 mm cell)

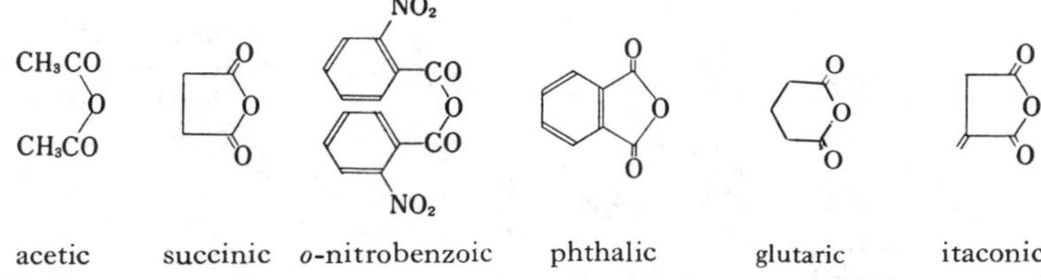

ANSWER

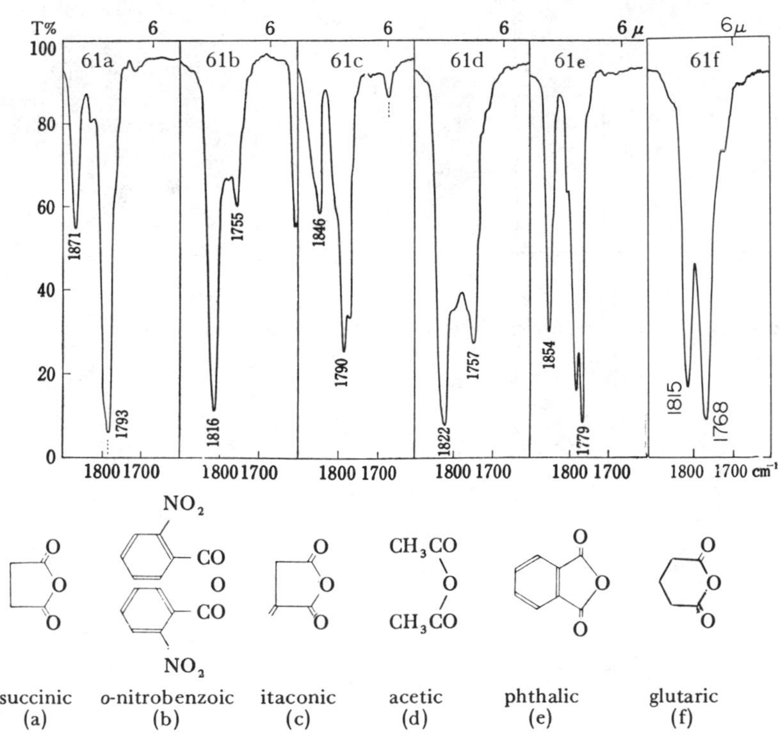

Of the two anhydride bands, the higher frequency one is the stronger in acyclic compounds, whereas the lower frequency one is the stronger in cyclic compounds (Table 8). From this it follows that Figs. a, c, e, and f belong to cyclic compounds; Fig. a with the highest ν C=O is the spectrum of succinic anhydride. The rest is self-explanatory when the presence or absence of aromatic skeletal vibrations between 1600∼1500 cm^{-1} is taken into consideration.

Table Absorptions of Anhydrides

Anhydride	νC=O	Mean νC=O	$\Delta\nu$C=O	-CO-O-CO-*	Remarks
a) succinic	1871 1793	1832	77	1214 1057	At high wave-numbers.
b) *o*-nitro-benzoic	1816	1782	61	1225	Mean νC=O is in same region with (c); effects of *o*-nitro and phenyl groups seem to counterbalance. Both νC=O bands shifted +50 cm^{-1} in comparison to benzoic anhydride.
c) itaconic	1846 1790	1818	56	1235	Mean νC=O similar to (e). νC=C at 1620 cm^{-1}.
d) acetic	1822 1757	1790	65	1125 1000	
e) phthalic	1854 1779	1817	75	1257	No other strong band between 1600–900 cm^{-1} except at 1257 and 907 cm^{-1} (KBr). cf. Table 10-6f for splitting of 1779 cm^{-1} band.
f) glutaric	1815 1768	1791	47	—	Frequency of C=O is similar to that of open-chain anhydrides. Intensities of two carbonyl bands are almost equal.

* Values of KBr disk spectra, because 1250 cm^{-1} region is distorted in $CHCl_3$ solutions by solvent absorption.

PROBLEM 62 / Deduce the structure of $C_8H_{14}O_3$.

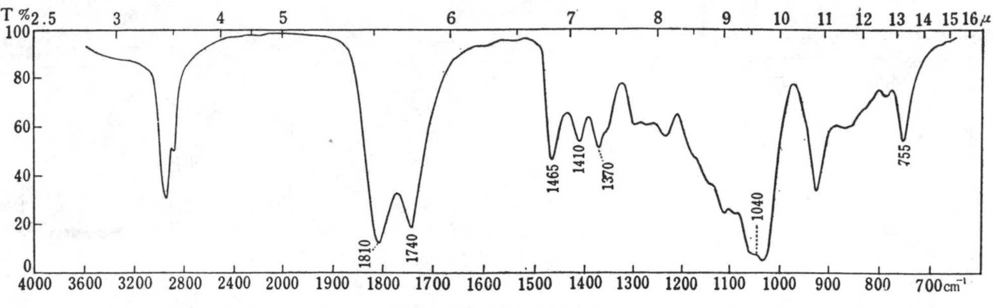

Fig. 62 Liquid film

ANSWER

Butyric anhydride

$$\begin{matrix} CH_3CH_2CH_2CO \\ CH_3CH_2CH_2CO \end{matrix} \Big\rangle O$$

1810 and 1740 cm^{-1}: ν C=O. The compound is acyclic and the higher band is stronger (Table 8).

1465 and ~1370 cm^{-1}: CH_2, CH_3.

1410 cm^{-1}: CH_2 adjacent to C=O.

~1040 cm^{-1}: Stretching of –CO–O–CO– unit. The various stretching vibrations connected with the unit are coupled extensively in anhydrides, and appear as one or two broad strong bands between 1200~1000 cm^{-1}.

755 cm^{-1}: CH_2 rocking of propyl group (Table 1–2).

PROBLEM 63 / What are the groups that can be guessed from the spectrum of a compound $C_8H_{12}O_4$?

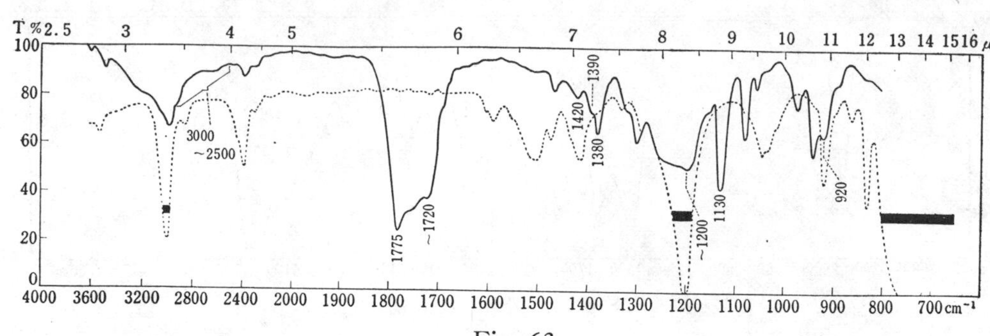

Fig. 63

———: Sample in $CHCl_3$, 0.5 mm cell.

·········: $CHCl_3$, 0.5 mm cell. The region indicated by broad black lines cannot be used because of solvent absorption.

ANSWER

3, 5, 5-Trimethylbutyrolactone-4-carboxylic acid[1]

COOH CH$_3$ / CH$_3$ / CH$_3$ / O / O

3000∼2500, 1720 (sh), 1420, 1200 cm^{-1} (distorted because of solvent absorption): Dimeric COOH. The 920 cm^{-1} band is in the right region for a dimeric carboxyl group; however, it should be assigned to other vibrations since it is too strong for a carboxyl, and still persists in the carboxylate spectrum (cf. Problem 57).

1775 cm^{-1}: γ-lactone ν C=O (1770 cm^{-1}).

1390 and 1380 cm^{-1}: Isopropyl doublet; the separation between the two bands is smaller than usual due to attachment to an electronegative group (see p. 16'). Characteristic frequencies of the isopropyl at 1215 and 1195 cm^{-1} are hidden by solvent absorption.

≃1200 and 1130 cm^{-1}: Antisymmetric and symmetric C–O–C stretching of the lactone (two bands between 1300∼1050 cm^{-1}).

at 1215 and 1195 cm^{-1} are hidden by solvent absorption.

≃1200 and 1130 cm^{-1}: Antisymmetric and symmetric C–O–C stretching of the lactone (two bands between 1300∼1050 cm^{-1}).

---

[1] C. Katsuta, unpublished.

PROBLEM 64 / The following chart was obtained when a drop of diethylamine was added to the previous $CHCl_3$ solution of 3, 5, 5-trimethylbutyrolactone-4-carboxylic acid (Problem 56). Explain changes of bands.

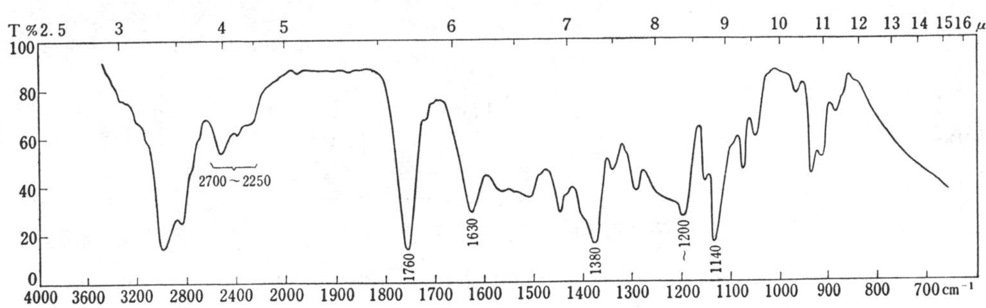

Fig. 64 ($CHCl_3$, 0.5 mm cell)

ANSWER

Diethylammonium salt of 3, 5, 5-trimethylbutyrolactone-4-carboxylic acid[1].

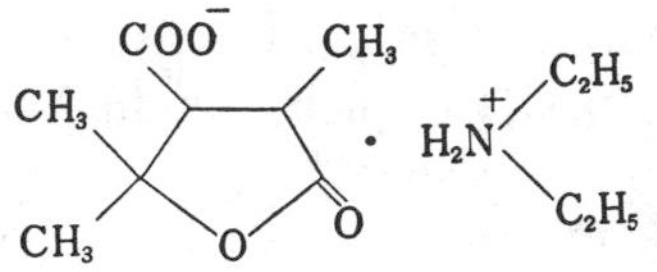

The 1720 cm^{-1} carboxyl peak is replaced by the ammonium band at 2700 ~ 2250 cm^{-1} and the carboxylate bands at 1630 and 1380 cm^{-1}. This seems to provide a simple method for identification of carboxyl groups. However, preliminary studies with models[2] have shown that, in general, this conversion of carboxyl groups to carboxylates by the addition of amines occurs in chloroform but not in carbon tetrachloride; this is probably due to the difference in the nonaqueous acid-base equilibria. The $\delta\, NH_2^+$ of secondary amine salts appear as broad absorptions in the 1550 cm^{-1} region, whereas tertiary amine salts have only very weak absorptions in this region. Thus, addition of triethylamine is to be preferred over diethylamine in order to avoid the appearance of confusing bands.

When an amine is added to a lactone, the latter will be gradually hydrolysed so that the lactone νC=O becomes weaker and will eventually be replaced by bands of the carboxylate. This provides a method for the identification of lactones. For example, mevalonic acid lactone (1) shows a δ-lactone peak at 1730 cm^{-1} in $CHCl_3$; however, when it is dissolved in morpholine and measured for 48 hours, the 1730 cm^{-1} peak becomes weaker and instead a carboxylate peak appears[3] at 1640 cm^{-1}.

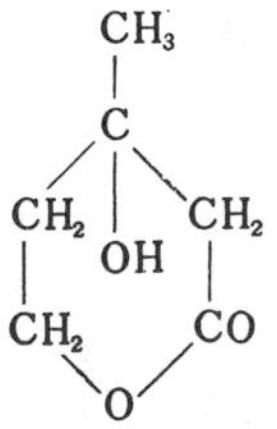

(1) Mevalonic acid lactone

---

[1] C. Katsuta, unpublished.

[2] A. Terahara, unpublished.

[3] D. E. Wolf, C. H. Hoffman, P. E. Aldrich, H. R. Skeggs, L. D. Wright, and K. Folkers, *J. Am. Chem. Soc.* **79**, 1486 (1957).

PROBLEM 65 / Deduce the structure of $C_9H_5NO_4$, m.p. 183~184° (dec.).

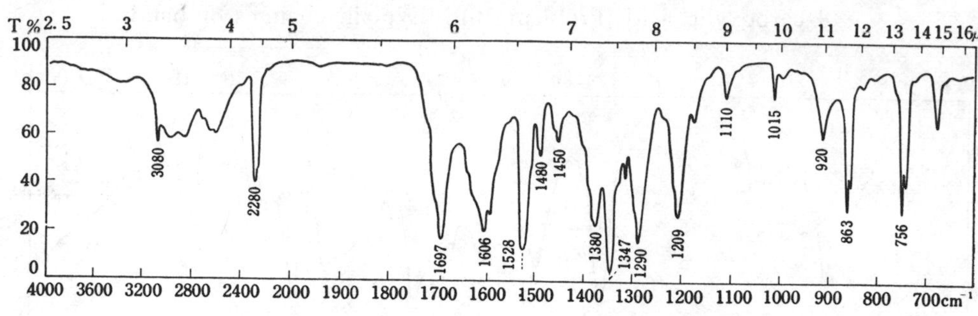

Fig. 65 1.5 mg/600 mg KBr (IRDC 1109)

ANSWER

p-Nitrophenylpropiolic acid

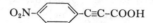

$$O_2N-C_6H_4-C\equiv C-COOH$$

–COOH: Characteristic diffuse absorption between 3000~2500 cm^{-1}, 1697 (ν C=O), 1380 and 1290 (coupling of OH in-plane bending and ν C–O in dimer), 920 cm^{-1} (OH out-of-plane bending in dimer).

–NO_2: 1528, 1347, and 863 cm^{-1}.

p-Substituted phenyl: 3080, 1606, 1480, 1450, and 756 cm^{-1}.

The other possibility is HOOC–C_6H_4–C≡C–NO_2.

PROBLEM 66/ Correlate main bands with vibrations.

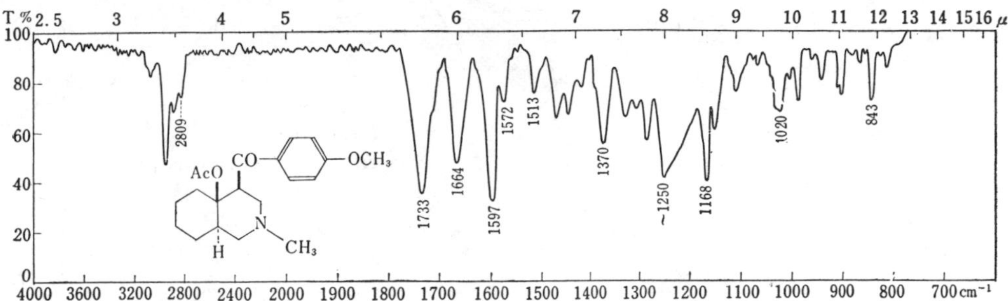

Fig. 66 DL–N–methyl–4–anisoyl–10–acetoxy–decahydroisoquinoline (4,10–*cis*)
(20 mg/ml $CHCl_3$ 0.2 mm)

ANSWER

DL-N-methyl-4-anisoyl-10-acetoxy-decahydroisoquinoline (4, 10-*cis*)[1].

2809 cm^{-1}: NCH_3; small band on higher side is OCH_3.

1733 cm^{-1}: Acetyl.

1664 cm^{-1}: Aromatic ketone. Shifted from standard 1690 cm^{-1} position by $+M$ effect of *p*-methoxyl group.

1597, 1572, 1513, and 843 cm^{-1}: *p*-Substituted phenyl.

1370 cm^{-1}: Symmetric stretching of acetate methyls usually make 1380 cm^{-1} band stronger than 1465 cm^{-1} band (Table 1~11).

ca. 1250 cm^{-1}: Distorted shape because of solvent absorption. Acetates, aromatic ketones, aromatic ethers, etc. (i.e., groups with =C–O– unit), absorb strongly in this region.

1168 cm^{-1}: In-plane bending of aromatic C–H.

1020 cm^{-1}: Aromatic ether.

The ketone band of the deacetylated product is at 1645 cm^{-1} with a shoulder at 1656 cm^{-1} (in $CHCl_3$). The two absorptions can be assigned as shown in the following. The shape (sharp) and position (3460 cm^{-1}) of the ν OH of (2) corresponds to dimeric association, which occurs when steric hindrance is fairly large.

---

[1] T. Omoto, *J. Pharm. Soc. Japan* **80**, 137 (1960).

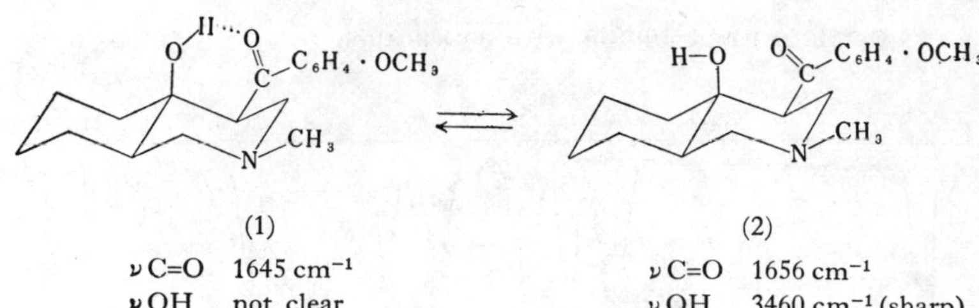

(1)

ν C=O 1645 cm^{-1}
ν OH not clear

(2)

ν C=O 1656 cm^{-1}
ν OH 3460 cm^{-1} (sharp)

PROBLEM 67 / The hydrochloride of an epimer of compound (1) has the following IR. When this hydrochloride is converted into the free form the ν C=O is shifted from 1639 to 1672 cm^{-1}. What is the structure of this epimer?

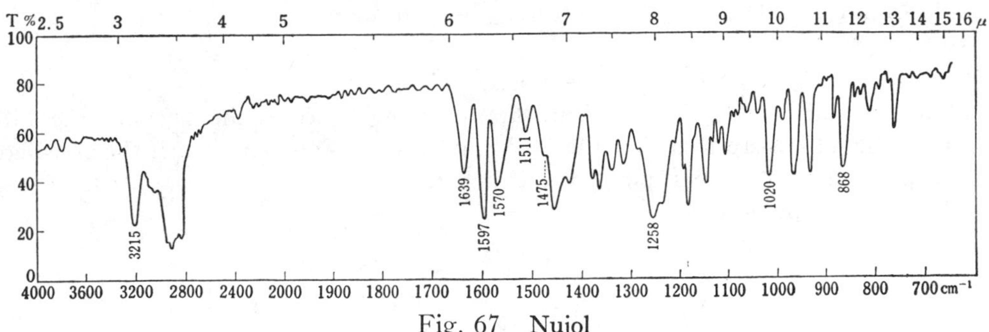

Fig. 67 Nujol

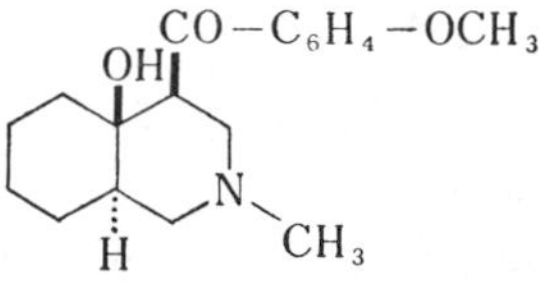

(1)

Bonded ν C=O at 1645 cm^{-1}

Free ν C=O at 1656 cm^{-1} (data in $CHCl_3$)

(cf. Structures 1 and 2 in (59))

ANSWER

DL-N-Methyl-4-anisoyl-10-hydroxydecahydroisoquinoline (4, 10-*trans*)–HCl[1].

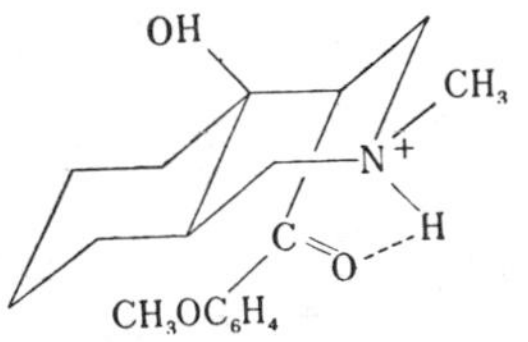

or

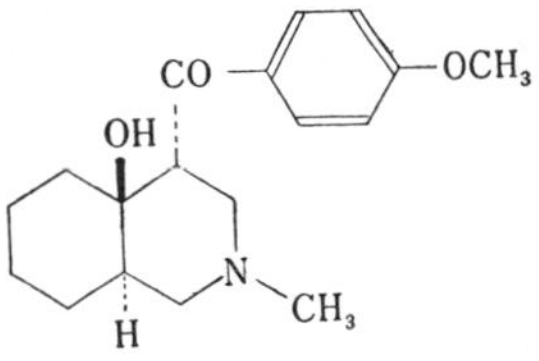

The ν C=O of the hydrochloride at 1639 cm^{-1} is lower than the hydrogen-bonded ν C=O of (1) at 1645 cm^{-1} (cf. structure 1 of preceding problem), whereas that of the free form at 1672 cm^{-1} is higher than the ν C=O of the non-hydrogen-bonded (1) (cf. structure 2 of preceding problem, 1656 cm^{-1}). This implies that a strong hydrogen-bond is involved in the 1639 cm^{-1} band; the only structure that satisfies this requirement is the one shown where 1, 3-diaxial hydrogen-bonds are formed. In the free form, the hydrogen-bond is of course absent and the ν C=O band is shifted higher.

---

[1] T. Omoto, *J. Pharm. Soc. Japan* 80, 137 (1960).

3215 cm^{-1}: Associated ν OH, sharp.
$\simeq$3000 cm^{-1}: Shoulder on higher side of CH band is aromatic ν CH.

Absence of the strong " *tert*-ammonium band " is attributed to the strong intramolecular hydrogen-bond. The hydrochloride of compound 1 (Problem 58), which cannot form such a bond absorbs broadly at $\simeq$2330 cm^{-1}.

1639 cm^{-1}: ν C=O.
1597, 1570, 1511, and 868 cm^{-1}: *p*-Substituted phenyl.
$\simeq$1460 and $\simeq$1380 cm^{-1}: Note that in the preceding figure of the acetate, the 1380 cm^{-1} band was stronger than the 1465 cm^{-1} band. This is no longer the case here.
1258 cm^{-1}: Aromatic ether and aromatic ketone.
1020 cm^{-1}: Aromatic ether.

PROBLEM 68 / What is the structure of $C_7H_7NO_2$, having pK′a values of $\simeq$5.0 and 10.5?

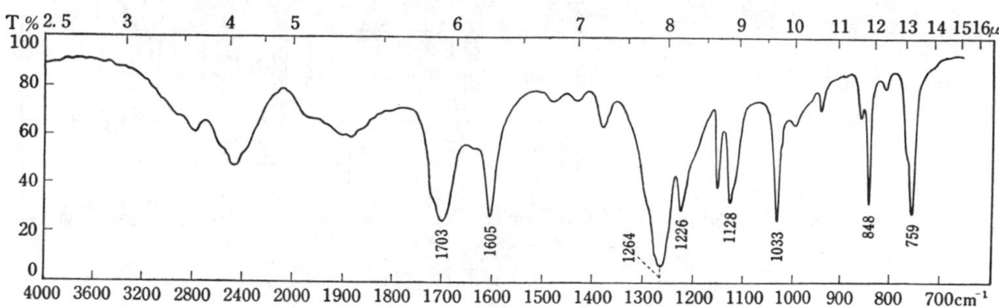

Fig. 68 1.1 mg/600 mg KBr (IRDC 417)

ANSWER

2-Methylpyridine-5-carboxylic acid

COOH
H_3C N

An aromatic ring is suggested by bands at 1605, 848, and 759 cm^{-1}. The pK′a values indicate the presence of two dissociating groups, thus a basic nitrogen function and presumably a carboxyl (to account for the two oxygens). A methyl group is shown by the 1380 cm^{-1} peak. A methylpyridine carboxylic acid is derived but the substitution type is unclear.

ca. 2450, 1900, 1703, 1264 cm^{-1}: Pyridine carboxylic acids differ greatly from ordinary dimeric carboxyl groups.[1] Thus the characteristic peaks are: 2450 (broad), 1900 (broad), 1710, and 1300 cm^{-1}, while the 1420 and 920 cm^{-1} peaks are lacking. The hydrogen-bonded structure (1) has been proposed to account for the anomaly.[1] Actually it could be a resonance hybrid between (1) and (2).

—C(=O)O—H···N (1) —C(=O)O⁻···H—N⁺ (2)

1605 cm^{-1}: Aromatic ring.

848 and 759 cm^{-1}: Out-of-plane CH bending. The pattern is complicated by the presence of electron-withdrawing groups (see Table 10b).

---

[1] S. Yoshida and M. Asai, *Pharm. Bull.* **7**, 162 (1959).
[2] H. Shindo and N. Ikekawa, *ibid.* **4**, 192 (1956).
[3] H. Shindo, *ibid.* **5**, 472 (1957).

PROBLEM 69 / Correlate the IR peaks with structure of 2-cyano-6-methoxyquinoline.

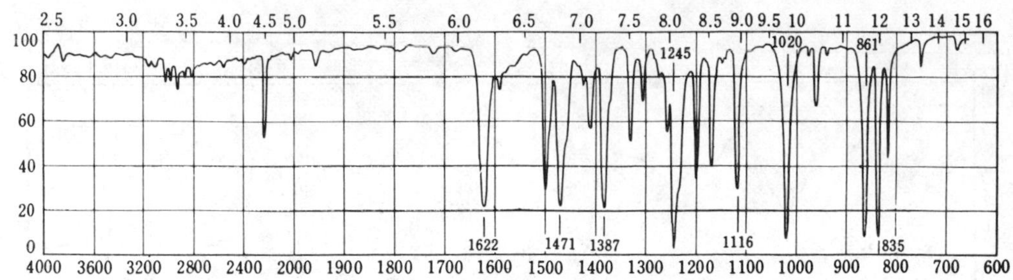

Fig. 69 1.8 mg/600 mg KBr (IRDC 424)

ANSWER

2-cyano-6-methyoxyquinoline

CH_3O N CN

2250 cm^{-1}: ν C≡N.
1622 cm^{-1}: Quinoline ring vibration.
1471 cm^{-1}: δ_{as} CH_3.
1387 cm^{-1}: δ_s CH_3.
1245 cm^{-1}: ν_{as} C–O–$C_{arom.}$
1116 cm^{-1}: Characteristic of aromatic –OMe.
1020 cm^{-1}: ν_s C–O–$C_{arom.}$
861, 835 cm^{-1}: δ CH of two adjacent hydrogens at 3,4 and 7,8 positions (usual position 860~810 cm^{-1}).

PROBLEM 70 / What is the structure of $C_3H_7O_2N$?

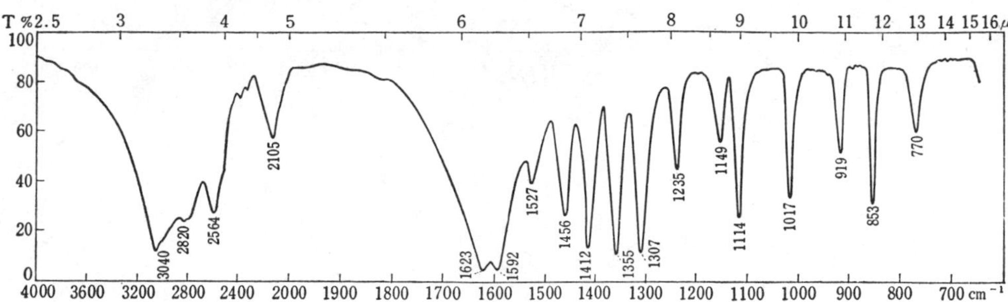

Fig. 70 KBr disk

ANSWER

L-Alanine

$$CH_3\text{-}\underset{\displaystyle |\atop \displaystyle NH_3^+}{CH}\text{-}COO^-$$

Free amino acids exist in the form of dipolar ions, and the main absorptions are those arising from amine hydrochlorides and carboxylate ions. The racemic and optically active isomers show identical spectra in solution but usually differ in the solid state. The magnitude of difference depends on the amino acids; for example, alanine differs only in the finger-print region, whereas serine differs in regions higher than 1500 cm^{-1}. The small differences are attributed to differences in crystal forms, but the large differences are attributed to dimer formation between the D- and L-isomers, rotational isomerism, etc.[1] The IR spectra of amino acids are summarized in references [2] and [3]. The following assignments are quoted from the former reference.

3040∼2820, 2564, 2105 cm^{-1}: $-NH_3^+$
1623 cm^{-1}: NH_3^+ asym. bending.
1592 cm^{-1}: COO^- asym. stretching.
1527 cm^{-1}: NH_3^+ sym. stretching.
1456 cm^{-1}: CH_3 asym. bending.
1412 cm^{-1}: COO^- sym. stretching.
1355 cm^{-1}: CH_3 sym. bending.
1307 cm^{-1}: CH bending.
1235 cm^{-1}: NH_3^+ rocking.
1149 cm^{-1}: CCN asym. stretching
1114 cm^{-1}: NH_3^+ rocking.
1017 cm^{-1}: CH_3 rocking.
919 and 853 cm^{-1}: $C–CH_3$ stretching and CCN sym. stretching.
770 cm^{-1}: COO^- scissoring.

---

[1] H. Brockmann and H. Musso, *Ber.* **89**, 241 (1956).
[2] M. Tsuboi and T. Takenishi, " IR Spectra," Vol. 7, p. 41 (1959).
[3] R. J. Kögl, R. A. McCallum, J. P. Greentein, M. Winitz, and S. M. Birnbaum, *Ann. N. Y. Acad. Sci.* **69**, 94 (1957).

PROBLEM 71 / Deduce the structure of $C_4H_9NO_2S \cdot HCl$.

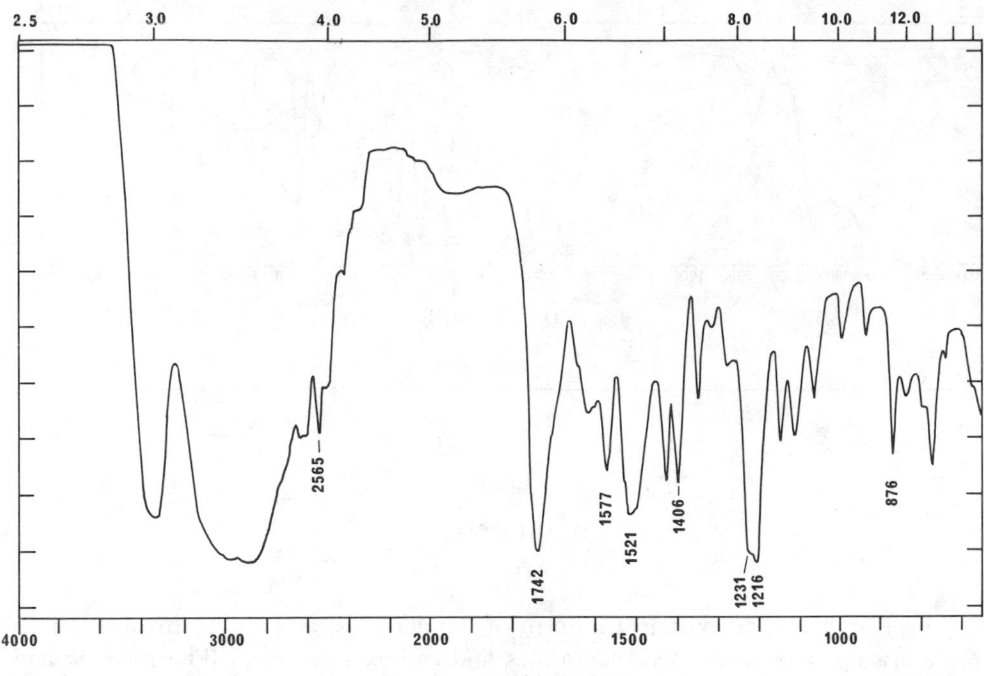

Fig. 71 1.5% in KBr

ANSWER

L-Cysteine hydrochloride

$$\text{H–S–CH}_2\ \underset{}{\overset{\text{NH}_3^+\ \text{Cl}^-}{\overset{|}{\text{CH}}}}-\text{COOH}$$

Main absorptions of amino acid hydrochlorides are due to amine hydrochloride and carboxylic acid groups. The differences between the racemic and optically active forms are smaller than in the case of free amino acids.

$\simeq$2900 cm^{-1}: NH_3^+ and COOH.
2565 cm^{-1}: S–H stretch.
1950 cm^{-1}: NH_3^+.
1742 cm^{-1}: ν C=O of COOH. This value is unusually high.
1577, 1521 cm^{-1}: δ_{as} and δ_s, respectively, of NH_3^+ (1600$\sim$1575 and 1500).
1231, 1216 cm^{-1}: COOH.

PROBLEM 72/ Deduce the structure of $C_4H_9O_2N \cdot HCl$ (optically inactive, m.p. 141~142°).

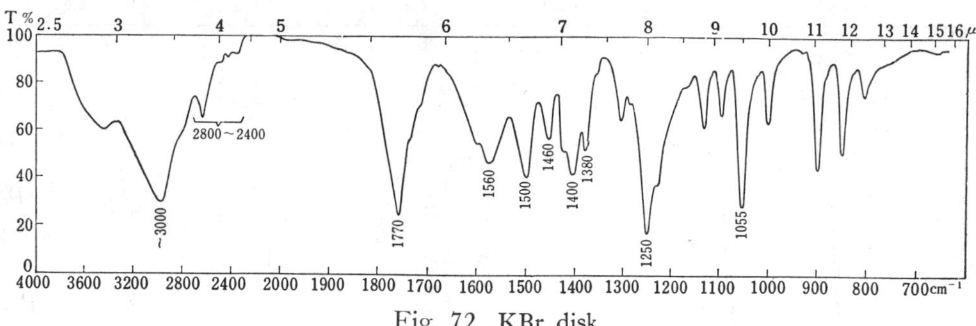

Fig. 72 KBr disk

ANSWER

Glycine ethyl ester monohydrochloride

$CH_3CH_2OOC \cdot CH_2 \cdot NH_3^+Cl^-$
(a)

Degree of unsaturation is one. An ester group is suggested by bands at 1770, 1250, and 1055 cm^{-1}. However, the unusually high ν C=O position remains to be explained by some reason other than ring strain (degree of unsaturation excludes cyclic carbonyl compounds). The ≃3000, 2800~2400, 1560, and 1500 cm^{-1} bands are typical for primary amine salts. The conspicuous peak at 1400 cm^{-1} in the present case can only be assigned to a perturbed methylene that is adjacent to the ester group, the ammonium nitrogen, or both of them. These considerations lead to either structure (a) or (b).

$CH_3OOC \cdot CH_2 \cdot CH_2 \cdot NH_3^+Cl^-$
(b)

In order to account for the high wave-number of the ester band, the former structure (a) having the ester and N groups in β-relation appears to be the more plausible. The 1460 and 1380 bands are CH_2 and CH_3 vibrations of the ethyl group.

PROBLEM 73 / Identify the structure of the *t*-butyl-containing compound with molecular formula $C_6H_{13}NO$.

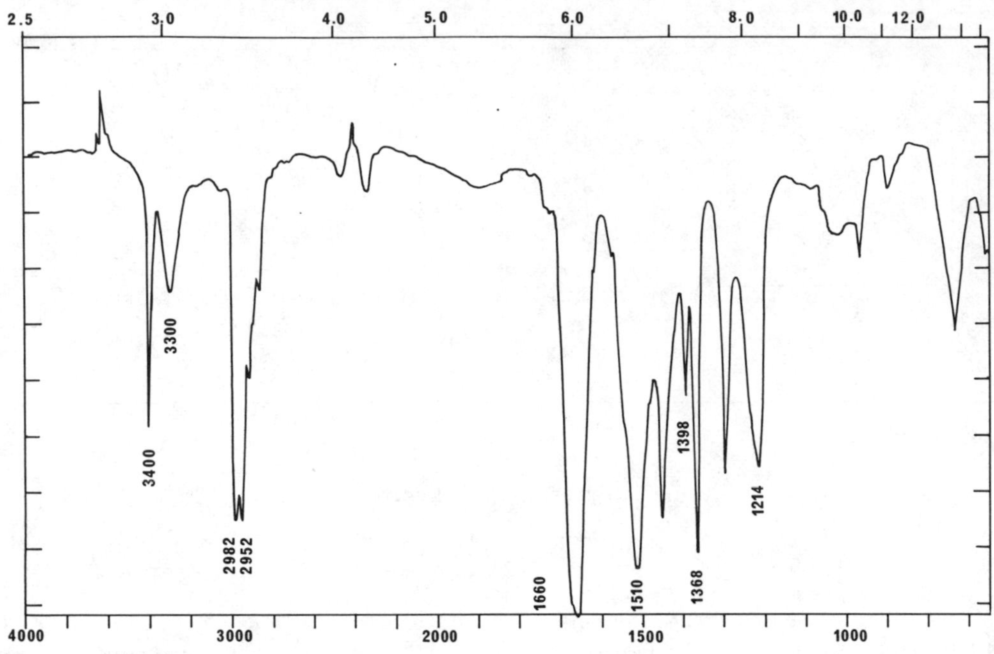

Fig. 73-a $CHCl_3$ solution

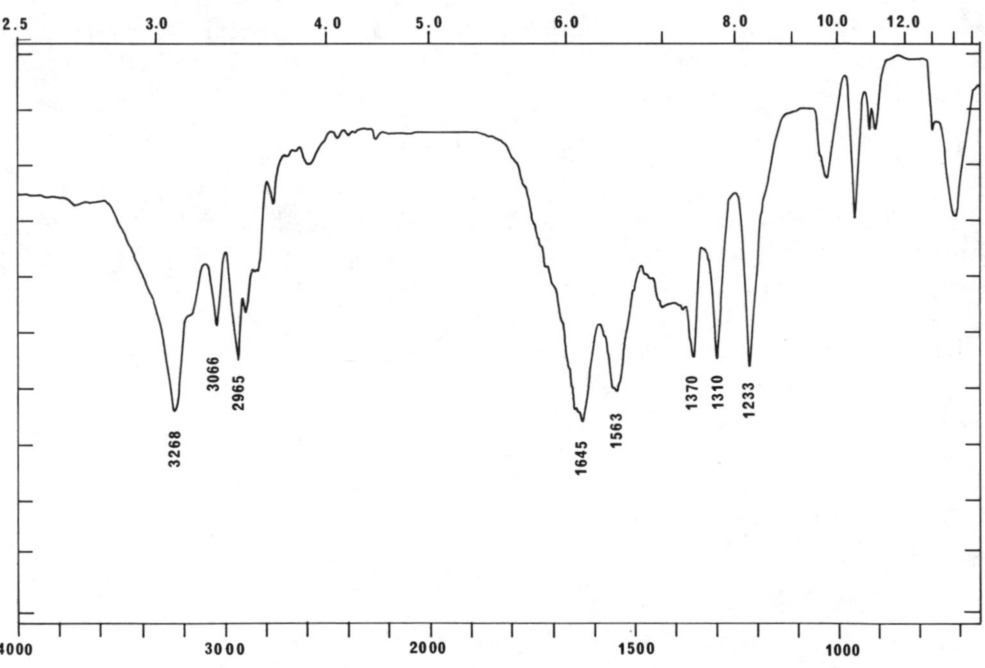

Fig. 73-b 3 mg/200 mg KBr

N-*t*-butyl acetamide

$$\begin{array}{ccccccc} & & CH_3 & & O & & \\ & & | & & \| & & \\ CH_3 & - & C & -N- & C & - & CH_3 \\ & & | & | & & & \\ & & CH_3 & H & & & \end{array}$$

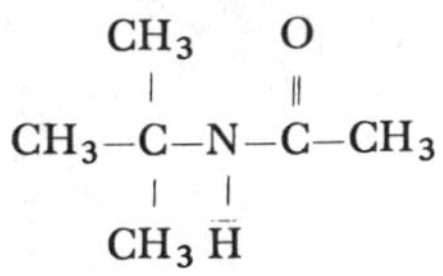

$CHCl_3$ solution:

3400 cm^{-1}: ν_{NH} unassociated.
3300 cm^{-1}: ν_{NH} associated.
1660 cm^{-1}: amide I band.
1510 cm^{-1}: amide II band.
1398, 1368 cm^{-1}: δ_{as} CH_3 of *t*-butyl group, split as usual when more than one CH_3 is attached to a single carbon.
1214 cm^{-1}: Skeletal *t*-butyl (1210 cm^{-1}) is probably overlaid by the stronger amide III band.

KBr spectrum: The shifts for the Amide I, II, and III bands are as expected on going from solution to solid state.

3268 cm^{-1}: ν_{NH} associated.
3066 cm^{-1}: overtone of N—H bend, characteristic of secondary amides.
1645 cm^{-1}: amide I band.
1563 cm^{-1}: amide II band.
1310 cm^{-1}: amide III band.

PROBLEM 74 / Deduce the structure of C_2H_4ClNO, m.p. 118~119°.

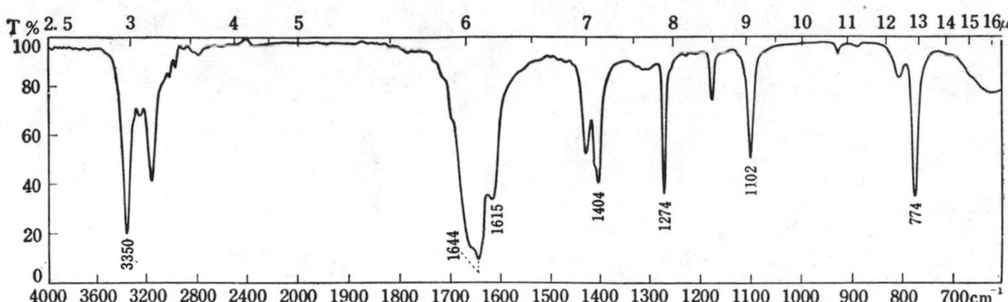

Fig. 74 1.0 mg/600 mg KBr (IRDC 976)

ANSWER

Chloracetamide

$Cl\text{-}CH_2\text{-}CONH_2$

3350 cm^{-1}: NH_2 stretching.

1644 and 1615 cm^{-1}: "Amide I" and "Amide II" bands (1650 and 1640 cm^{-1}). The two bands overlap in solid phase spectra. The ν C=O of an α-chlorocarbonyl moiety is shifted to higher frequencies when the C=O and C–Cl bonds are uniplanar, because of the interplay of dipole-dipole interactions, which tend to suppress bond polarizations (1). In the two stable conformers of monochloracetone, the *cis* (2) and the *gauche* (3), the mentioned electric repulsion (1) operates in the *cis*-form, while steric repulsion (Cl and CH_3) operates in the *gauche*-form. It is generally accepted that electric repulsion is the dominating factor in gases, and that steric repulsion is the dominating factor in fluid and solid states.[1]

(1)	(2) *cis*	(3) *gauche*
$\overset{\delta+}{C}=\overset{\delta-}{O}$; $\overset{\delta+}{C}-\overset{\delta-}{Cl}$	Cl, O, C, C, CH_3, H, H	H, O, C, C, CH_3, H, Cl
	gas ——	1743 cm^{-1}
	liq. 1743 cm^{-1}	1725 cm^{-1}

Thus, the two bands shown by chloracetone (liquid state) have been assigned by Bellamy as shown in the figures.[2] The α-bromocyclohexanones have been studied extensively by Corey and co-workers.[3] The present chloracetamide presumably exists in the *gauche*-form (3) due to stabilization by the hydrogen-bond Cl···H–N, although measurements under different conditions would have to be carried out to corroborate the presumption.

774 cm^{-1}: ν C–Cl.

---

[1] S. Mizushima, T. Shimanouchi, T. Miyazawa, L. Ichishima, K. Kuratani, I. Nakagawa, and N. Shindo, *J. Chem. Phys.* **21**, 815 (1953).

[2] L. J. Bellamy and R. L. Williams, *J. Chem. Soc.* 4294 (1957).

[3] E. J. Corey, T. H. Topie, and W. A. Wozwiak, *J. Am. Chem. Soc.* **77**, 5415 (1955).

PROBLEM 75 / Deduce the structure of C_8H_9ON.

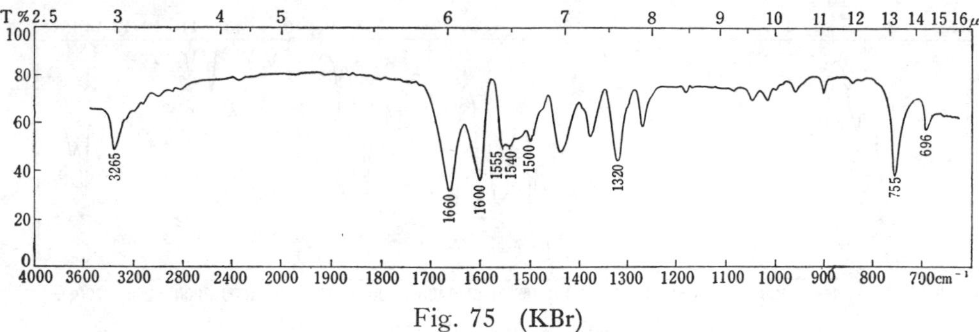

Fig. 75 (KBr)

ANSWER

Acetanilide

$$\begin{matrix} C_6H_5 \\ H \end{matrix} \!\!> N{-}C \begin{matrix} {}^{\nearrow\!O} \\ {}_{\searrow CH_3} \end{matrix}$$

3265 cm^{-1}: νNH of *trans*-form shown. The other secondary amide νNH at 3070 cm^{-1} and the ν CH bands are not clear.

1660, 1540, 1320 cm^{-1}: Amide I, II, and III bands of secondary amides (1665, 1550, and 1300 cm^{-1}). The Amide I band, which is mostly due to ν C=O, is at slightly higher frequencies than in saturated amides; in contrast to the behavior of other carbonyl containing groups, the carbonyl stretching frequency of amides is shifted higher when an unsaturated group is introduced in conjugation with the nitrogen atom or the carbonyl group (Table 8–9 a and 9 b).

1600, 1555, 1500, 755, 696 cm^{-1}: Mono-substituted phenyl.

PROBLEM 76 / Deduce the structure of C_9H_7NO, m.p. 85∼86°.

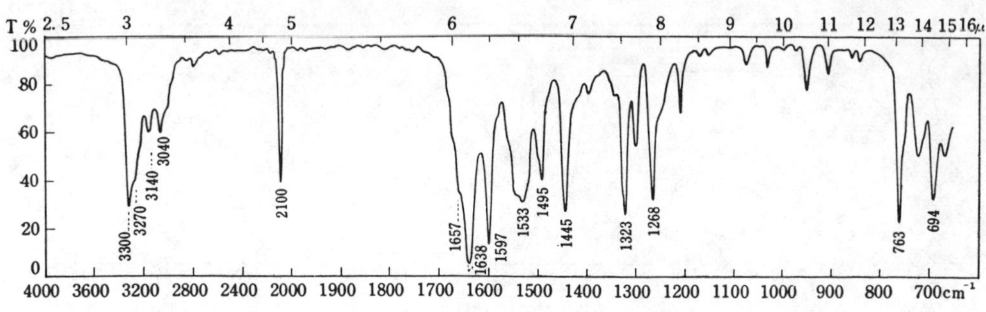

Fig. 76 1.3 mg/600 mg KBr (IRDC 513)

ANSWER

Propiolanilide

The presence of mono-substituted phenyl, terminal acetylene, and secondary amide groups can be inferred.

3300 cm^{-1}: ≡CH stretching (3300 cm^{-1}, sharp).
3270 cm^{-1}: ν NH of associated secondary amide (*trans*-type, 3300 cm^{-1}).
3040 cm^{-1}: Aromatic ν CH.
2100 cm^{-1}: C≡CH stretching (terminal C≡CH, 2140∼2100 cm^{-1}).
1597, 1495, 1445, 763, and 694 cm^{-1}: Mono-substituted phenyl.
1638, 1533, and 1323 cm^{-1}: Amide I, II, and III bands of *sec*-amides (1650, 1550, and 1300 cm^{-1}).

PROBLEM 77 / Deduce the structure of $C_{14}H_{13}ON$. Benzoic acid is obtained upon hydrolysis.

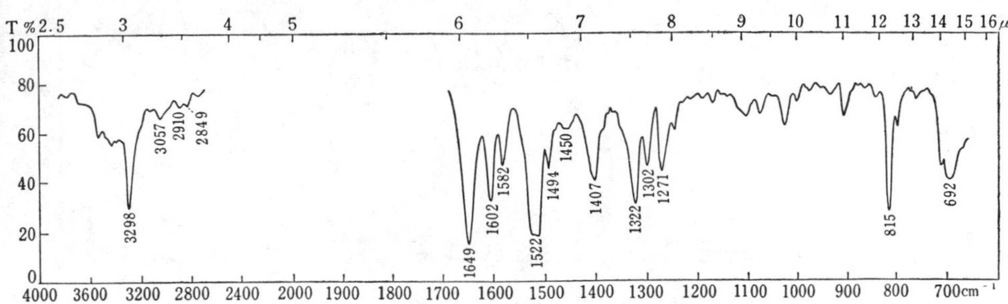

Fig. 77 KBr disk

ANSWER

N-Benzoyl-*p*-toluidine

CH_3-⟨benzene ring⟩-NH-CO-⟨benzene ring⟩

The three strong bands at 1649, 1522, and 1322 cm^{-1} immediately suggest a *sec*-amide; the deduction could be ascertained by observing the shift of the three bands in solution.

3298 cm^{-1}: ν NH. Small shoulders on high frequency side are water bands.
3057 cm^{-1}: Phenyl.
2910 and 2849 cm^{-1}: Methyl.
1649, 1522, and 1322 cm^{-1}: "Amide-I, II, and III" bands of *sec*-amides (1655, 1550, and 1300 cm^{-1} in associated form). They are shifted to 1688, 1514, and 1317 cm^{-1}, respectively, in CCl_4 (1680, 1530, and 1260 cm^{-1} in free form); the shifts are thus in the expected directions for the amide bands. The Amide-I band, which is mainly ν C=O, of the unit C=C–NH–CO–C=C is usually lower by $\simeq$15 cm^{-1} than that of the saturated amide; in the present example it is higher by 8 cm^{-1}. In any case, the difference between the saturated and unsaturated amides (including C=C–NH–CO and –NH–CO–C=C), −15 to +15 cm^{-1}, is rather small. Thus, because of the extensive mesomerism present in the amide group itself, the additional effect of conjugated double bonds on amide vibrations is not so pronounced as with simple ketones (−40 cm^{-1}).

$$\underset{\text{N—C}}{\overset{\text{O}}{\|}} \longleftrightarrow \underset{\text{N}^+\text{=C}}{\overset{\text{O}^-}{|}}$$

1602, 1582, and 1494 cm^{-1}: Phenyl.
1450 cm^{-1}: Methyl.
815 cm^{-1}: *p*-Substituted phenyl.
692 cm^{-1}: Mono-substituted phenyl. The other band usually between 770~730 cm^{-1} is not apparent.

PROBLEM 78 / Correlate bands with bond vibrations.

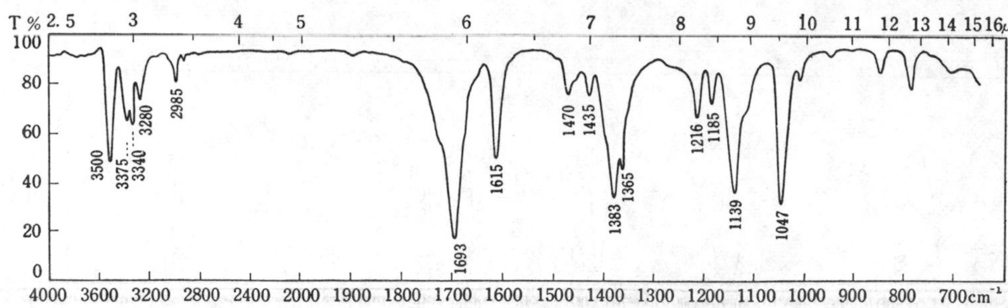

Fig. 78 1,1-Dimethyl-2-propynyl carbamate (0.8 mg/600 mg KBr)

$$(H_3C)_2C(C{\equiv}CH)(O\text{–}CO\text{–}NH_2)$$

ANSWER

Vibrations of the –O–CO–N– group can be roughly regarded as a summation of ester and amide groups, although undoubtedly coupling occurs to a great extent.

3500, 3375, and 3340 cm^{-1}: ν NH.
1693 cm^{-1}: ν C=O.
1615 cm^{-1}: δ NH_2.
1470 and 1435 cm^{-1}: Methyl.
1383 and 1365 cm^{-1}: *gem*-Dimethyl doublet on ν C–N (?).
1139 and 1047 cm^{-1}: C–O–C= asymmetric and symmetric stretching.

PROBLEM 79 / Which of the two pyrones would be more basic as deduced from the IR spectra?

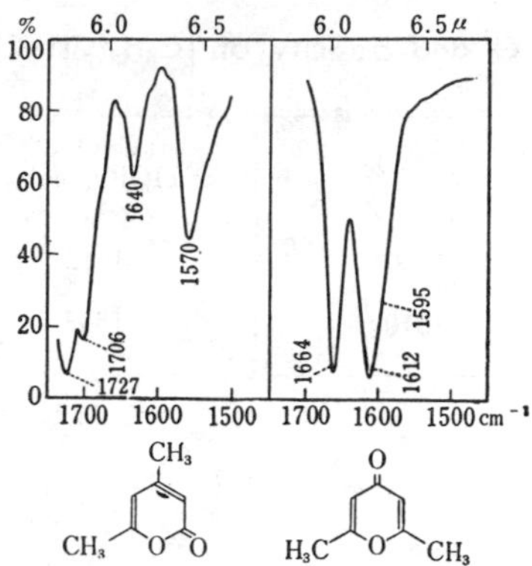

Fig. 79 4, 6–Dimethyl–α–pyrone and 2, 6–dimethyl–γ–pyrone ($CHCl_3$ solution)

ANSWER

α-Pyrone and γ-pyrone

(1) ⟷ (2) (3) ⟷ (4)

The two pyrones can be expressed as hybrids of (1) and (2), and (3) and (4), respectively. γ-Pyrones are more basic than α-pyrones. Namely, the contribution of canonical structure (4) to the hybrid (3) and (4) is greater than that of (2) to the hybrid (1) and (2), and thus the proton-accepting power of the γ-pyrone is stronger. This is reflected in the difference between the ν C=O bands of the two compounds. The stronger the basicity of the carbonyl group, usually the lower the ν C=O band. The effect of substituents on the ν C=O position is smaller in the case of γ-pyrones, and this can also be attributed to the relatively large contribution of (4), which has an aromatic structure (6 π-electrons); namely, if contribution from this aromatic structure (4) is large, the bond order of the exocyclic C=O bond would not be influenced to a great extent.

62 a

1727 and 1706 cm^{-1}: ν C=O, split (1740∼1720 cm^{-1}, frequently split)[1], [2].

1640 and 1570 cm^{-1}: ν C=C (1650∼1620 and 1570∼1540 cm^{-1})[1].

62 b

1664 cm^{-1}: ν C=O (1660∼1650 cm^{-1})[1], [2].

1612 and 1595 cm^{-1}: ν C=C (1650∼1600 and 1560∼1590 cm^{-1})[1].

---

[1] K. Yamada, " IR Spectra," Vol. 7, p. 106 (1959).

[2] A. Fujino and M. Yamaguchi, *ibid.* Vol. 3, p. 123 (1958).

Data in the following table also demonstrate the parallel relation between basicity and νC=O band position.

Table νC=O and basicity of $[C_6H_5(CH{=}CH)_n]_2C{=}O$[3]

Compound	νC=O	νC=O of $HClO_4$ salt	Stability of $HClO_4$ salt
n=1, Dibenzalacetone	1660	1612	Hydrolysed in AcOH
n=2, Dicinnamylacetone	1610	1553	Stable in AcOH

[3] H. Brockman and B. Franck, *Natrwiss.* **42**, 70 (1955).

PROBLEM 80 / Assign the bands at 1680 and 1625 cm^{-1} to the two carbonyl groups of 2-aminoanthraquinone. In contrast, only a single band at 1673 cm^{-1} is seen in the νC=O region of 2-hydroxyanthraquinone (in nujol). How could this difference be rationalized?

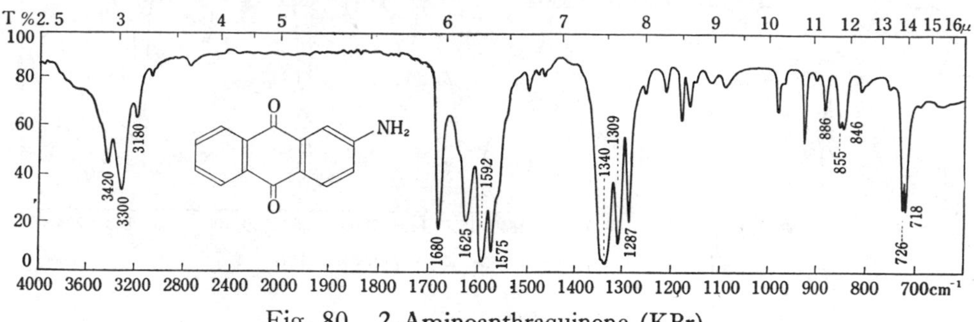

Fig. 80 2–Aminoanthraquinone (KBr)

ANSWER

Aminoanthraquinones.

(2) ⟷ (2) (3)

3330 cm^{-1}: Together with the shoulders at 3420 and 3180 cm^{-1}, ν NH_2.

1680 cm^{-1}: C=O at C–9.

1625 cm^{-1}: C=O at C–10. Aminoquinones can be regarded as vinylogous amides, and the ν C=O is shifted lower because of contribution from limiting structure (2). The ν C=O of 1-amino-anthraquinone is also split into 1670 and 1610 cm^{-1} (KBr), and the lower location of the latter as compared with the 1625 cm^{-1} band of the 2-amino isomer is due to formation of the hydrogen-bond (3). However, since the NH_2 frequency is shown clearly at the rather high position of 3420 and 3300 cm^{-1}, the extent of type (3) conjugate chelation is not appreciable. In 2-hydroxyanthraquinone the C=O is at 1673 cm^{-1} and the OH at 3320 cm^{-1}, whereas in 1-hydroxyanthraquinone the C=O appears as a doublet at 1673 and 1636 cm^{-1} and the OH as a broad band centered around 2700 cm^{-1} (nujol)[1]. The appearance of only a single C=O band in the 2-hydroxy isomer indicates the negligible contribution of the canonical structure corresponding to (2), and this is in accord with the smaller contribution of (4) in ester groups as compared with amide groups.

1592 and 1575 cm^{-1}: Aromatic ring.

1340 cm^{-1}: C–N stretching.

886 cm^{-1}: Isolated H.

855 and 846 cm^{-1}: Adjacent 2H.

726 and 718 cm^{-1}: Adjacent 4H (doublet).

$-O-\overset{O}{\overset{\|}{C}}-$ ⟷ $-O^{+}=\overset{O^{-}}{\overset{|}{C}}-$

(4)

[1] M. St. Flett, *J. Chem. Soc.* 1441 (1948).

PROBLEM 81 / Deduce the structure of C_5H_5NO, m.p. 66~67°.

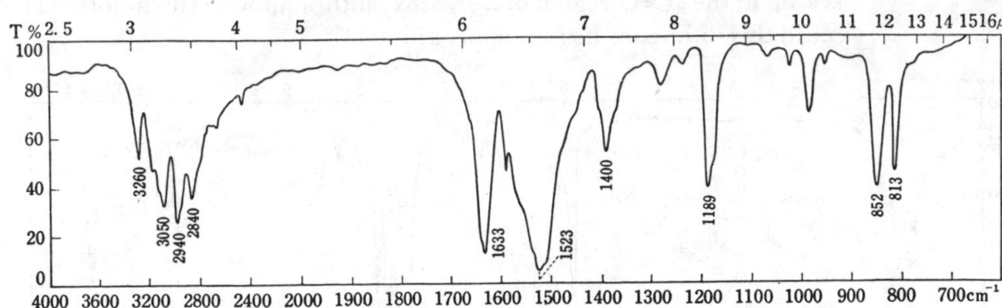

Fig. 81 0.8 mg/600 mg KBr (IRDC 310)

ANSWER

γ-Pyridone.

H–N=O

3260~2840 cm⁻¹: N–H stretching. Reason for this complexity is not clear. The ν NH bands of α-pyridones and γ-quinolones have been studied in detail;[1] the complex broad bands between 3200~2400 cm⁻¹ have been attributed to associated NH groups, and the sharp band at 3400 cm⁻¹ (in dilute CCl_4 solutions of concentrations less than 0.025 mole) to monomeric NH groups.

1633 cm⁻¹: ν C=O. The solution spectra of α-pyridones and α-quinolones show bands between 1690~1650 cm⁻¹, and those of γ-pyridones and γ-quinolones between 1650~1630 cm⁻¹; in the solid phase the bands shift to slightly lower wavenumbers.[2] The cause for the lower location of the γ-isomers has been atributed to the fact that since the distance between N^+ and CO^- is greater in (4) as compared to (2), the $-I$ effect of N^+ is transmitted to a smaller extent. The lower location of the γ-isomer ν C=O bands can be correlated with their stronger basicities. Thus, pKa of γ-pyridonium is 3.27, and that of α-pyridonium is 0.75.[3]

1523 cm⁻¹: ν C=C coupled with ν C–N.

1400 cm⁻¹: ν C=N coupled with ν C=C.

852 cm⁻¹: Out-of-plane bending of CH. Corresponds to 860~810 cm⁻¹ band of adjacent two hydrogens on phenyl ring.

(1) ⟷ (2) (3) ⟷ (4)

---

[1] H. Shindo, *Chem. and Pharm. Bull.* **7**, 407 (1959).
[2] S. F. Mason, *J. Chem. Soc.* 4874 (1957).
[3] A. Albert and J. N. Phillips, *J. Chem. Soc.* 1924 (1956).

PROBLEM 82/ Deduce the structure of $C_9H_6O_4$.

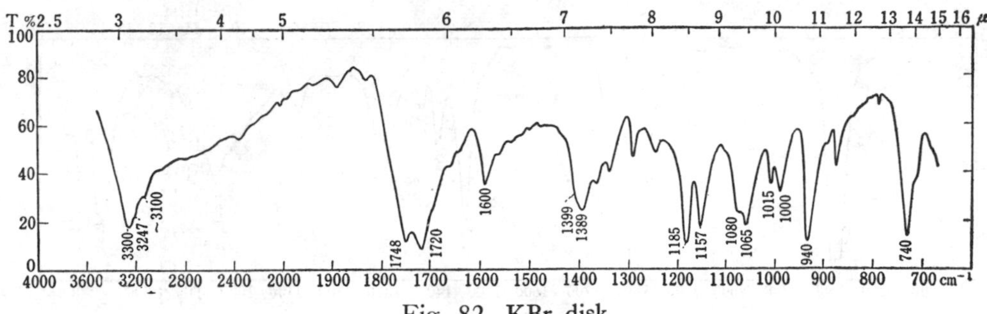

Fig. 82 KBr disk

ANSWER

Triketohydrindene hydrate

O
OH
OH
O

Assignments are based on reference [1].

3300 and 3247 cm^{-1}: Associated ν OH (3400~3200 cm^{-1}). The position is in the region of normal hydrogen-bonds, and therefcre the O–H bending bands are also expected to appear in the usual range. Namely, if the ν X–H band is displaced to lower frequencies in X–H type links because of strong hydrogen-bonds, the X–H bending vibrations will require higher energy and the δX–H bands will be shifted higher.

$$\cdots X\text{–}H\cdots X\text{–}H\cdots X\text{–}H$$

1399 and 1389 cm^{-1}: In-plane bending of associated O–H; they are in the normal range. Lithium pyruvate hydrate, $CH_3C(OH)_2COOLi$, forms extremely strong hydrogen-bonds, and the νOH is observed at the low wave-number of 3000 cm^{-1}; the corresponding δOH is at the high wave-number of 1551 cm^{-1} (nujol)[1]. The two O–H stretching bands are replaced by bands at 2439 and 2140 cm^{-1} upon deuteration (frequency ratios of the two O–H and O–D bands are 1.35 and 1.36). The two O–H bending bands are also shifted to 941 cm^{-1} (ratio, 1.49)* and 930 cm^{-1} (ratio, 1.50)*.

1200~1000 cm^{-1}: The six peaks are thought to arise from the coupling between the C–O and C–C links. The view is supported by the fact that they are not affected much upon deuteration (hence, not related directly to O–H links).

1748 and 1720 cm^{-1}: ν C=O. The compound should be regarded as a vinylogous α-diketone rather than an aromatic ketone.

3100, 1600, and 740 cm^{-1}: *o*-substituted phenyl.

---

[1] D. M. W. Anderson, L. J. Bellamy and R. L. William *Spectrochim. Acta* **12**, 233 (1958).

* The ratios, 1.49 and 1.50, deviate considerably form the expected value of 1.37 (p. 63). Furthermore since a band originally exists at 940 cm^{-1}, it is doubtful whether the 941 cm^{-1} band is actually related to the OD link. However, the disappearance of the two bands near 1400 cm^{-1} indicate that these are attributable to OH links.

PROBLEM 83 / Deduce the structure of $C_{18}H_{35}OCl$.

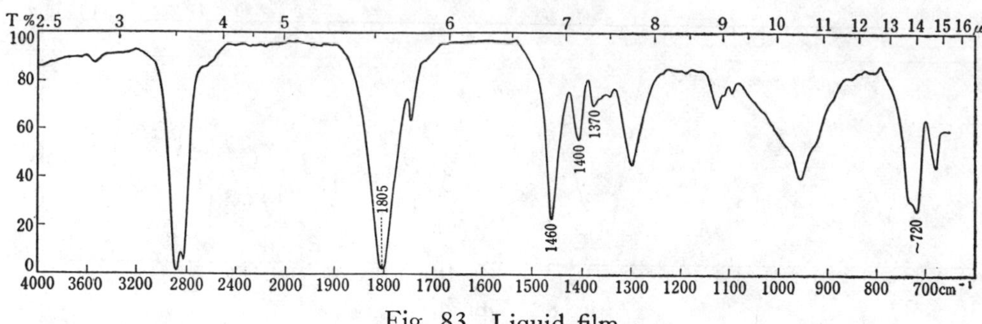

Fig. 83 Liquid film

ANSWER

Stearoyl chloride

$CH_3(CH_2)_{16}COCl$

1805 cm^{-1}: ν C=O, wave-number increased by $-I$ effect of chloride (1800 cm^{-1}). The shoulder at 1740 cm^{-1} is present in the spectra of other acid chlorides.

1460 and 1370 cm^{-1}: CH_2 and CH_3.

1400 cm^{-1}: CH_2 adjacent to CO.

$\simeq$720 cm^{-1}: CH_2 chain.

The band progressions between 1350~1180 cm^{-1} are not apparent due to the fluid state of the sample (compare with palmitic acid, Fig. 54).

PROBLEM 84/ Deduce the structure of the hydrate, $C_6H_7NO \cdot H_2O$.

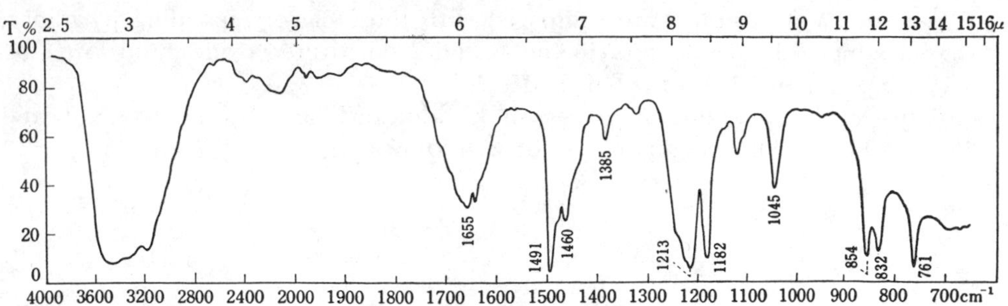

Fig. 84 Liquid film (IRDC 418)

ANSWER

4-Methylpyridine 1-oxide monohydrate.

H_3C–⟨ ⟩N→O·H_2O

Hydrates have a band (usually sharp) at 3600∼3100 cm^{-1} (OH stretching) and a weaker band at 1640∼1615 cm^{-1} (H–O–H bending). Water absorptions in the present figure are broader than ordinary hydrates. Other solvents of crystallization also show their characteristic bands; the spectra of the sample will be affected by solvents of crystallization, especially in solid phase spectra due to the greater extent of interaction. Aromatic and methyl peaks are present, and the strong absorption at 1213 cm^{-1} suggests an N-oxide group. Thus methylpyridine 1-oxide is derived.

ca. 3500 and 1655 cm^{-1}: H_2O.
1491 cm^{-1}: Aromatic.
1460 and 1385 cm^{-1}: Methyl.
1213 cm^{-1}: N–O stretching. The frequency is sensitive to the electronic effect of substituents and varies from 1300 to 1200 cm^{-1} according to its electron-attracting or electron-donating properties. The dehydrated γ-picoline N-oxide absorbs at 1270∼1260 cm^{-1} in solvents, and the low frequency of its hydrate (1213 cm^{-1}) is due to lowering of the N-O bond order by formation of the hydrogen-bond: N–O···H_2O.
1182 and 1045 cm^{-1}: Pyridine N-oxides give absorptions between 1190∼1150 cm^{-1} (strong) and near 1020 cm^{-1} (meduim)[1] that are presumably the in-plane CH bendings. The former appears as a shoulder on the very strong N–O stretching band, and helps to characterize the group.

832 cm^{-1}: Two adjacent hydrogens. The CH out-of-plane bending is similar to phenyl derivatives,[1] and substituted pyridines (see Table 10 b).

When the band positions are compared with those of corresponding pyridines, they are higher by 4∼30 cm^{-1} in the 4- and 2-substituted oxides, and lower by 4∼36 cm^{-1} in the 3-substituted oxides.[1, 2]

854 and 761 cm^{-1}: Pyridine N-oxides show additional bands at 880∼840 and 780∼720 cm^{-1}. The assignment is not well known.

[1] H. Shindo, *Pharm. Bull.* **6**, 117 (1958); **7**, 407, 791 (1959).
[2] D. B. Cunliffe-Jones, *Spectrochim. Acta* **21**, 747 (1965).

PROBLEM 85 / Deduce the structure of the nitrate, $C_3H_9N_3O \cdot HNO_3$, m.p. 108°

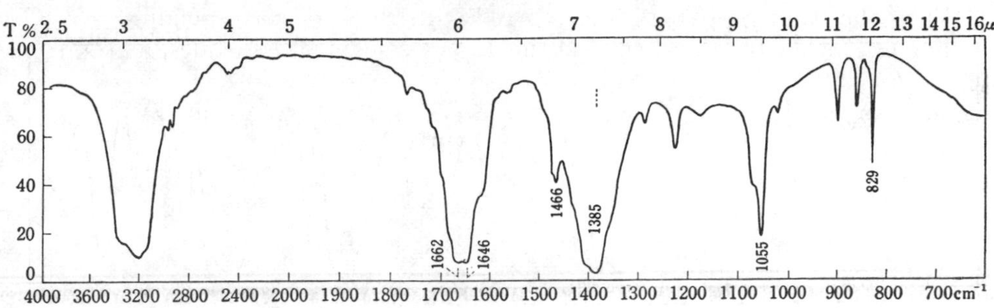

Fig. 85 1.0 mg/600 mg KBr (IRDC 1319)

ANSWER

2-Guanidinoethanol nitrate

$$(H_2N)(HN{=})C\text{-}NH\text{-}CH_2\text{-}CH_2\text{-}OH \cdot HNO_3$$

The high nitrogen content suggests a guanidine group, the presence of which is supported by the 1662 and 1646 cm^{-1} bands; furthermore, the position of these guanidinium bands corresponds to mono-substituted compounds. Although the OH stretching is overlaid by strong NH absorptions, the low position of the band at 1055 cm^{-1} can be correlated more easily with a primary alcohol group than with an ether group.

3300~3200 cm^{-1}: ν NH of guanidinium. (3300 cm^{-1}, broad).
1662 and 1646 cm^{-1}: Mono-substituted " Guanidinium I " and " Guanidinium II " band (1660 and 1630 cm^{-1}).
1466 cm^{-1}: δ CH_2.
1385 cm^{-1}: Asymmetric stretching of NO_3^- (1410~1340 cm^{-1} for organic salts,
1055 cm^{-1}: ν OH of primary hydroxyl.
829 cm^{-1}: O–N–O bending of NO_3^- (860~800 cm^{-1} for organic salts, Table 9~15).

PROBLEM 86 / Deduce the structure of $C_7H_8O_3$.

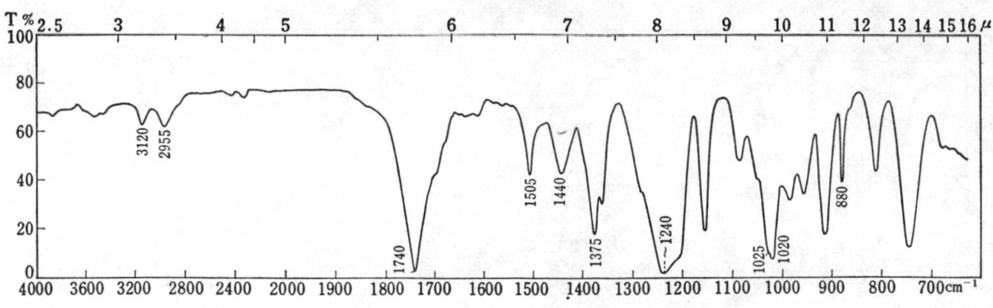

Fig. 86 Liquid film

ANSWER

Furfuryl acetate

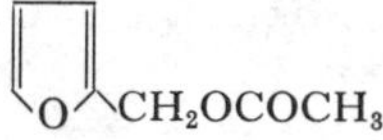

3120 cm^{-1}: ν CH of furan (higher than 3000 cm^{-1}).
2955 cm^{-1}: ν CH of acetoxyl.
1740 cm^{-1}: ν C=O of acetoxyl.
1505 cm^{-1}: ν C=O of furan ($\simeq$1500 cm^{-1}). The $\simeq$1565 cm^{-1} is weak in the present case.
1440 and 1375 cm^{-1}: Side-chain methylene and methyl. Intensity of the 1380 cm^{-1} peak is greatly enhanced in acetates, and usually becomes stronger than the 1460 cm^{-1} peak; this assists in characterization of acetates (Table 1~11).
1240 cm^{-1}: Asymmetric stretching of acetate =C–O–C; so-called " acetate band."
1025 and 1020 cm^{-1}: Furan 1030~1015 cm^{-1} band and symmetric stretching of acetate =C–O–C.
880 cm^{-1}: Most characteristic peak of furans,[1] sharp.
745 cm^{-1}: CH out-of-plane bending of furans (800~740 cm^{-1}).

---

[1] M. Yamaguchi, *Bunseki-kagaku* (*Analytical Chem.*) **7**, 210 (1958).

PROBLEM 87 / Deduce the structure of $C_{11}H_{11}NO_2$, m.p. 134°.

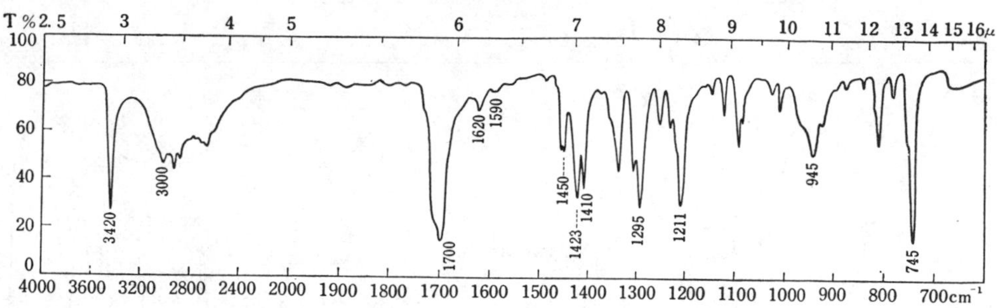

Fig. 87 1.5 mg/600 mg KBr (IRDC 1333)

ANSWER

3-(3-Indolyl) propionic acid

CH_2–CH_2–COOH

N
H

The presence of a carboxyl group is obvious by the broad absorption between 3000~2500 cm^{-1}, the band at 1700 cm^{-1}, and the characteristic medium band at 945 cm^{-1}. Four adjacent aromatic hydrogens are present, and the nitrogen exists as a pyrrole or indole NH. The degree of unsaturation requires it to be an indole. The IR spectrum thus suggests the shown structure or the 2-indolyl isomer.

3420 cm^{-1}: ν N–H. Lowered by hydrogen-bonding from 3490 cm^{-1}, but is still sharp.
3000 cm^{-1}: Indole ν C–H and carboxylic ν O–H.
3000~2500 cm^{-1}: Combination tones of dimeric –COOH.
1620, 1590, and 1450 cm^{-1}: Indole ring.
1423 and 1211 cm^{-1}: Coupled O–H in-plane bending and C–O stretching of –COOH dimer.
1295 cm^{-1}: C–N stretching.
945 cm^{-1}: O–H out-of-plane bending of acid dimer.
745 cm^{-1}: Four adjacent hydrogens on aromatic ring.

The pyrrole or indole rings lack characteristic frequencies to distinguish them from carbocyclic aromatics, except that the NH appears as a sharp band around 3400 cm^{-1} (in solution).

PROBLEM 88/ Identify the functional groups present in the hygroscopic organic salt with molecular formula $C_{21}H_{33}N_3SO_4$.

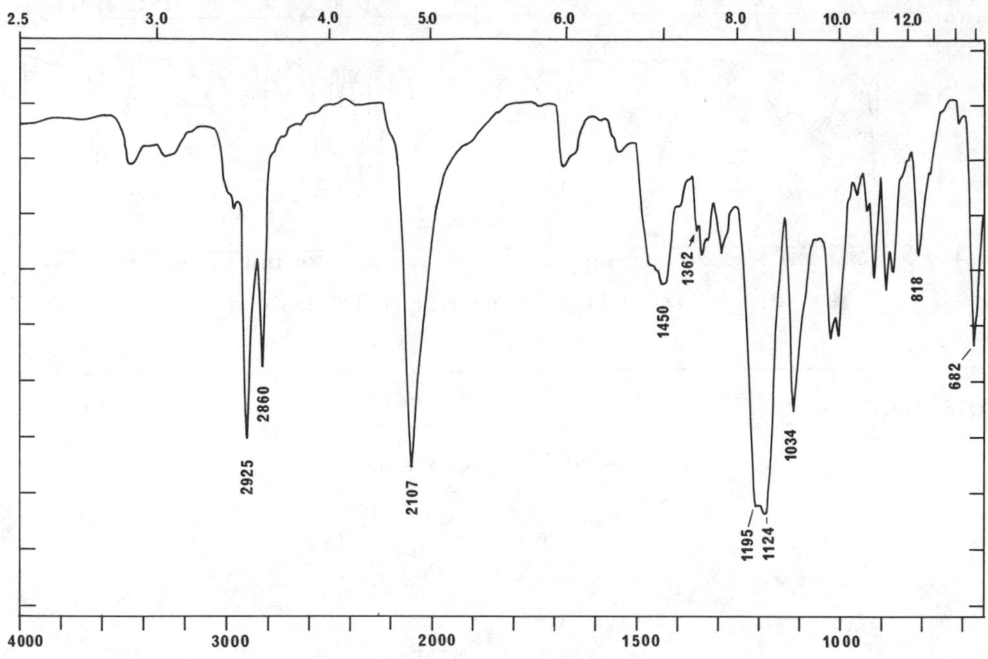

Fig. 88 3 mg/200 mg KBr

ANSWER

1-cyclohexyl-3-(2-morpholinoethyl)-carbodiimide metho-*p*-toluenesulfonate.

$$\text{C}_6\text{H}_{11}\text{–N=C=N–CH}_2\text{CH}_2\text{–}\overset{\oplus}{\text{N}}(\text{CH}_3)\text{(CH}_2\text{CH}_2)_2\text{O}$$

3550, 3300 cm^{-1}: These absorptions are due to water absorbed by the hygroscopic salt; their intensities increase with the passage of time.
3000~2800 cm^{-1}: Aliphatic C–H stretch.
2107 cm^{-1}: ν N=C=N.
1450 cm^{-1}: CH_2–N^+ scissoring.
1362 cm^{-1}: CH_3–N^+, δ_s CH_3 lowered by presence of electron-attracting group.
1195 cm^{-1}: ν_{as} SO_3^- (~1200 cm^{-1}).
1124 cm^{-1}: ν C–O.
1034 cm^{-1}: ν_s SO_3^- (~1050 cm^{-1}).

Identification of the cation as a quaternary ammonium salt, which has no characteristic bands, is suggested by the absence of strong broad ν $\overset{+}{N}$–H bands between 3000 and 2250 cm^{-1}.

The spectra of sulfonates in aqueous solution (KRS-5, AgCl cells) are often much simplified and, as the ν_{as} and ν_s SO_3^- bands are still intense and very little shifted from values obtained in solid or non-polar solution spectra, such aqueous solution spectra may be useful for identification.[1]

[1] B. J. Lindberg, *Acta Chem. Scand.* **21** 2215 (1967).

PROBLEM 89/ The spectrum is that of a steroid derivative containing three oxygen atoms and one sulfur. What is the nature of the derivative?

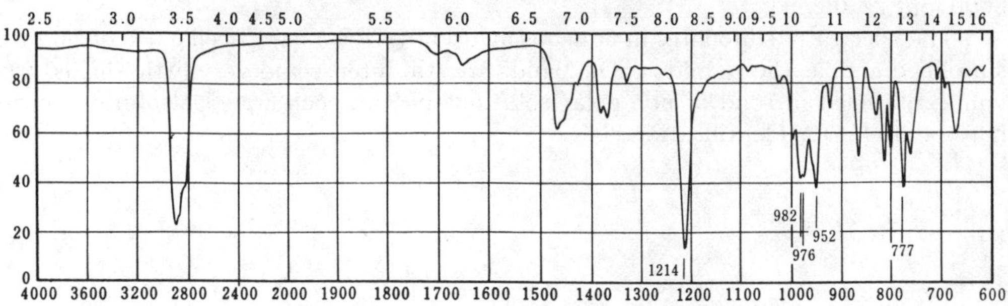

Fig. 89 0.9 mg/300 mg Kr (IRDC 6544)

ANSWER

Cholest-5-ene-3β,4β-diolsulfite

1214 cm^{-1}: ν S=O.

The strong bands between 1000 and 950 cm^{-1} are due to ν S—O.

PROBLEM 90 / What groups can be detected in the spectrum of $C_{17}H_{26}O_2S$?

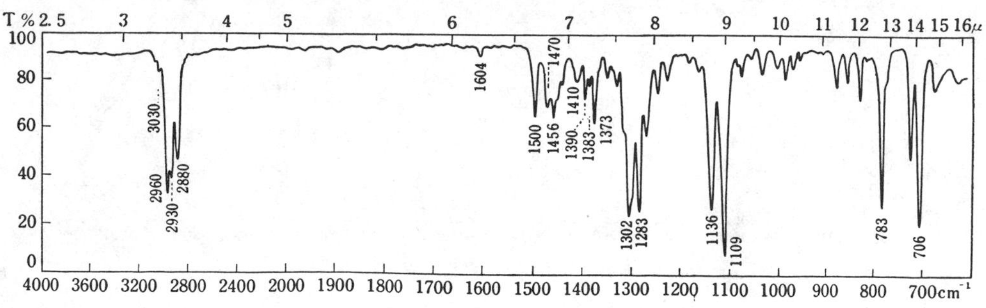

Fig. 90 2.6 mg/600 g KBr (IRDC 1157)

ANSWER

2-(Benzylsulfonyl)-*p*-menthane.

SO_2-CH_2-

The two strong band groups near 1290 and 1120 cm⁻¹ are characteristic for the sulfone group; a mono-substituted phenyl, a *gem*-dimethyl, and an active methylene can also be detected.

3030, 1604, 1500, 783, and 706 cm^{-1}: Mono-substituted phenyl.
1410 cm^{-1}: Active methylene.
≃1380 cm^{-1} multiplet: *gem*-Dimethyl.
1302 and 1283 cm^{-1}: Asymmetric SO_2 stretching (1350∼1310 cm^{-1}, lowered by 10∼20 cm^{-1} and frequently split in solid spectra).
1136 and 1109 cm^{-1}: Symmetric SO_2 stretching (1160∼1120 cm^{-1}, lowered by 10∼20 cm^{-1} and frequently split in solid spectra).

PROBLEM 91 / Deduce the structure of $C_6H_8N_2O_2S$.

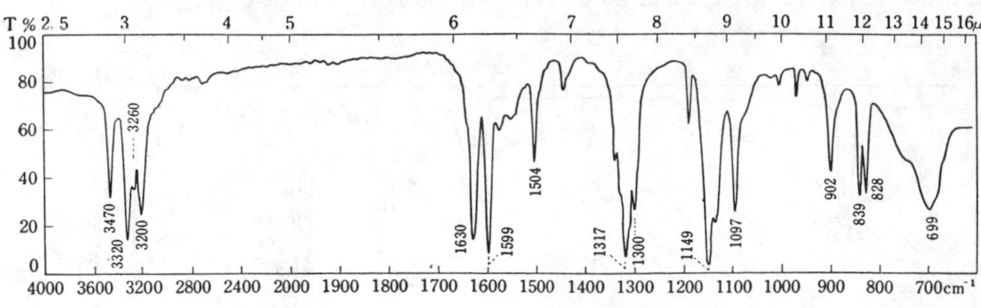

Fig. 91 1.5 mg/600 mg KBr (IRDC 1165)

ANSWER

Sulfanilamide.

$$H_2N-\bigcirc-SO_2-NH_2$$

Primary amine (complex absorption in NH region), phenyl (probably *para*-substituted); and sulfonamide (1317 and 1149 cm^{-1}) groups are present. See reference [1] for the IR of sulfonamides.

3470 and 3320 cm^{-1}: ν NH of amino group.

3200 and 3260 cm^{-1}: ν NH of sulfonamide group. The *mono*-substituted RSO_2–NHR′ show two bands at 3390 (free NH) and 3270 cm^{-1} (bonded NH) in CCl_4, both of which are sharp; the latter still persist in 0.001 M solutions. These observations have been attributed to strong dimeric hydrogen-bonds[1] shown in (1).

$$\begin{array}{c} -SO\cdots{}_2HN- \\ \;\;| \qquad\quad | \\ -HN\cdots O_2S- \end{array}$$

(1)

1630 cm^{-1}: δ NH_2.

1599 and 1504 cm^{-1}: Phenyl ring.

1317 cm^{-1}: Asymmetric stretching of SO_2 (1370~1330 cm^{-1}) overlapping ν C–N (1360~1250 cm^{-1}).

1149 cm^{-1}: Symmetric stretching of SO_2 (1180~1160 cm^{-1}). The fact that the two SO_2 stretching frequencies are higher rather than lower in comparison to sulfones (1350~1310 and 1160~1120 cm^{-1}) shows the negligible contribution of canonical structure (2)[1].

---

[1] J. N. Baxter, J. Cymerman-Craig, and J.B. Willis, *J. Chem. Soc.* 609 (1955).

[2] D. Hadži, *J. Chem. Soc.* 847 (1957).

$$-\overset{\overset{O^-}{\uparrow}}{\underset{\underset{O^-}{|}}{S^+}}=N^+\Big\langle$$

(2)

1097 cm^{-1}: Aromatic CH in-plane bending.
902 cm^{-1}: S–N stretching[2].
839 or 828 cm^{-1}: Adjacent 2H.
699 cm^{-1}: NH_2 out-of-plane bending (900∼650 cm^{-1}).

PROBLEM 92 / Deduce the structure of $C_{14}H_{14}OS$, m.p. 133~4°.

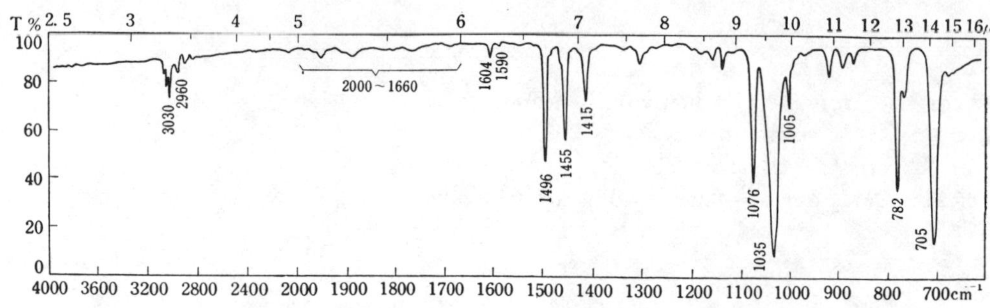

Fig. 92 2.0 mg/600 mg KBr (IRDC 1146)

ANSWER

Dibenzyl sulfoxide

C_6H_5—CH_2
S=O
C_6H_5—CH_2

Degree of unsaturation is nine. A *mono*-substituted phenyl is present, and the rather simple IR curve suggests that the molecule is symmetric. The sulfur is contained as a sulfoxide group in view of the strong band at 1035 cm^{-1}; the derived structure is supported by the active methylene peak at 1415 cm^{-1}.

3030, 1604, 1590, 1496, 1455, 1076 (in-plane), 782, and 705 cm^{-1}: Aromatic.
2960 and 1415 cm^{-1}: Methylene.
1035 cm^{-1}: S=O stretching of sulfoxide (1060~1040 cm^{-1}).

Compare the spectrum with the more complex spectrum of the asymmetric 2-mercaptodiphenylmethane (Fig. 96).

PROBLEM 93 / Deduce the structure of $C_7H_7ClO_2S$, m.p. 92°.

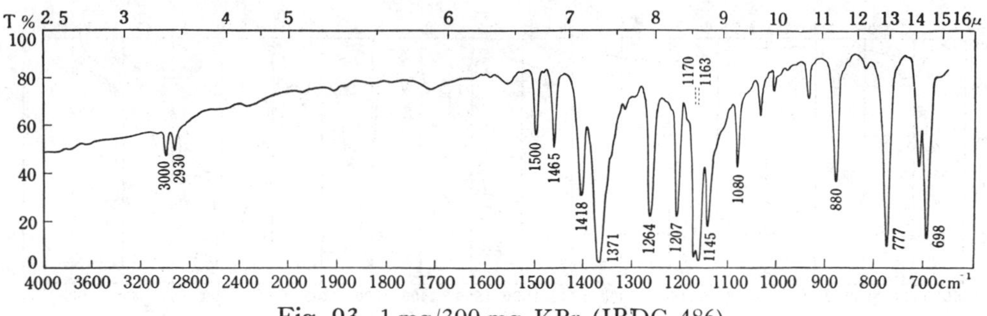

Fig. 93 1 mg/300 mg KBr (IRDC 486)

ANSWER

Toluene-α-sulfonyl chloride

(phenyl)-CH_2-SO_2Cl

Mono-substituted phenyl and sulfochloride groups can be characterized.

3000, 1500, 1465, 777, and 698 cm^{-1}: Mono-substituted phenyl.
1418 cm^{-1}: Perturbed methylene.
1371 cm^{-1}: Asymmetric SO_2 stretching (1370∼1365 cm^{-1}).
1170 cm^{-1}: Symmetric SO_2 stretching, split (1190∼1170 cm^{-1}).
1163, 1145, 1080 cm^{-1}: Probably in-plane C–H bending of phenyl.
1264 cm^{-1}: Methylene wagging or twisting?

PROBLEM 94 / Deduce the structure of the derivative of allyl alcohol with formula $C_9H_{15}PO_4$.

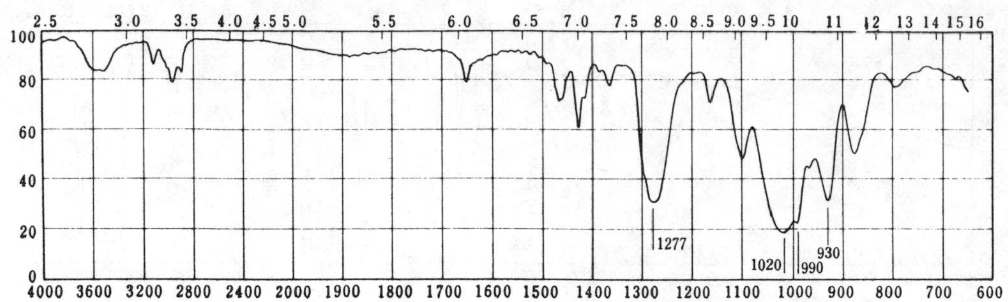

Fig. 94 Liquid film (IRDC 4306)

ANSWER

Allyl phosphate

$$(CH_2{=}CH{-}CH_2{-}O{-})_3\,P{=}O$$

3500 cm^{-1}: Contaminants; probably due either to water or hydrolysis of the phosphate to the alcohol.

3200~2800 cm^{-1}: Absorption on high and low frequency sides of 3000 cm^{-1} is due to unsaturated and saturated ν CH, respectively.

1650 cm^{-1}: ν C=C.

1440 cm^{-1}: CH_2 scissoring; lowered from normal position (1470 cm^{-1}) by adjacent —C=C and —O—P=O.

1277 cm^{-1}: ν P=O of phosphate.

1020 cm^{-1}: ν P—O—C_{alkyl}.

990, 930 cm^{-1}: δ CH of

H H
C=C
H

PROBLEM 95 / What functional groups can be discerned in the compound with molecular formula $C_8H_{17}PO_5$?

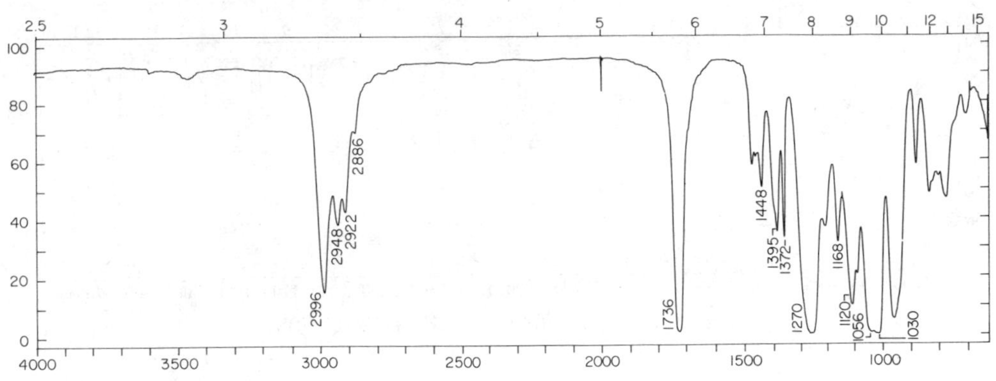

Fig. 95 Liquid film

ANSWER

Ethyl diethylphosphonoacetate

$$(CH_3CH_2\text{–}O\text{–})_2\overset{\displaystyle O}{\overset{\|}{P}}CH_2CO_2CH_2CH_3$$

1736 cm^{-1}: ν C=O of ester function.
1448, 1395 cm^{-1}: bending of active methylene (see Table 1∼14).
1372 cm^{-1}: δ_s CH_3.
1270 cm^{-1}: ν P=O of phosphonate ester (1265∼1230 cm^{-1} in solid state); also ν C–O carboxylic acid ester.
1056, 1030 cm^{-1}: ν P–O–C; the two bands often seen are usually ascribed to rotational isomerism.

PROBLEM 96 / Deduce the structure of $C_{13}H_{12}S$, m.p. 55°.

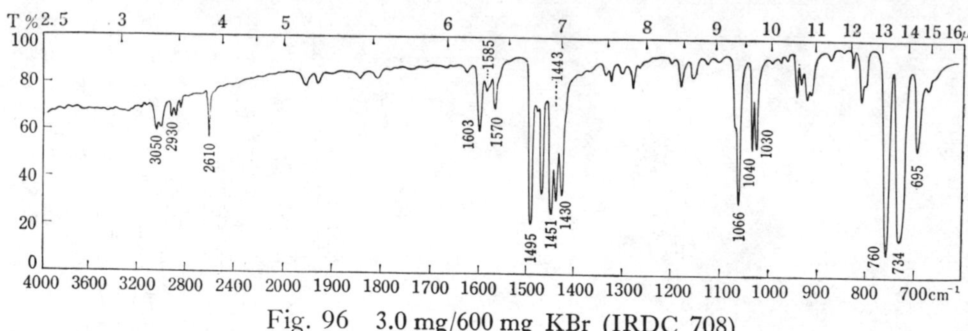

Fig. 96 3.0 mg/600 mg KBr (IRDC 708)

ANSWER

2-Mercaptodiphenylmethane

SH

$-CH_2-$

The sharp band at 2610 cm^{-1} shows the sulfur to be contained in the form of a thiol group. The eight unsaturations are accounted for by mono- and *ortho*-substituted phenyls. No methyl group is present (1380 cm^{-1} band lacking).

3050, 1603, 1585, 1570, 1451, 1443, 1066, 1040, 1030, 760 (adjacent 4H) 734, and 695 cm^{-1} (mono-substituted phenyl): Aromatic. Many bands appear because of the asymmetric structure of the two phenyl groups; compare with spectrum of dibenzyl sulfoxide (Fig. 81).

2610 cm^{-1}: S–H stretching. (2600～2550, weak but sharp).

1430 cm^{-1}: Perturbed methylene.

PROBLEM 97 / Deduce the structure of $C_7H_4ClF_3$, b.p. 135~136° (745 mm Hg).

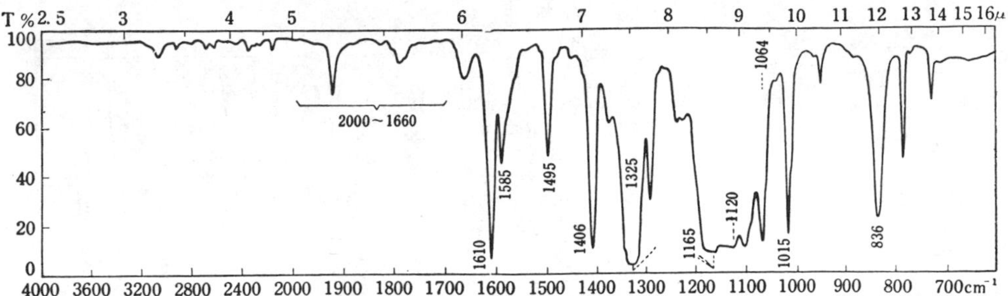

Fig. 97 Liquid film (IRDC 1381)

ANSWER

p-Chlorobenzotrifluoride

Cl-⟨benzene ring⟩-CF_3

The C–F bands absorb intensely between 1400~1000 cm^{-1} (usually 1100~1000 cm^{-1}), but in view of the considerable coupling between C–C and C–F bonds no general correlations can be formulated. The aromatic CF_3 groups, however, can be recognized fairly easily by three strong peaks at 1320 (sym. bending), 1180 and 1140 cm^{-1} (asym. bending, split into doublet).[1]

2000~1660, 1610, 1585, 1495, 1406, and 836 cm^{-1}: *para*-Substituted phenyl.
1325, 1165, and 1120 cm^{-1}: $-CF_3$ stretchings.

Fluorines linked *alpha* or *beta* to carbonyl groups raise the ν C=O considerably because of the strong *–I* effect: R_FCOR*, +50 cm^{-1} (CF_3COCH_3 at 1780 cm^{-1}); R_FCOR_F, +80 cm^{-1} (CF_3COCF_3 at 1802 cm^{-1}); R_FCOF at 1885 cm^{-1} (CF_3COF at 1901 cm^{-1}).

---

[1] R. R. Randle and D. W. Whiffen, *J. Chem. Soc.* 1311 (1955).
[*] R_F denotes perfluoroalkyl groups. Review of organic fluorine IR: J. K. Brown, K. J. Morgan, "The Vibration Spectra of Organic Fluorine Compounds" in *Advances in Fluorine Chemistry*, v. 4, M. Stacey, J. C. Tatlow, A. G. Sharpe, eds., Butterworths, Washington, D.C., 1965, pp. 253-313.

PROBLEM 98 / Deduce the structure of $C_{13}H_{11}N$, m.p. 56°.

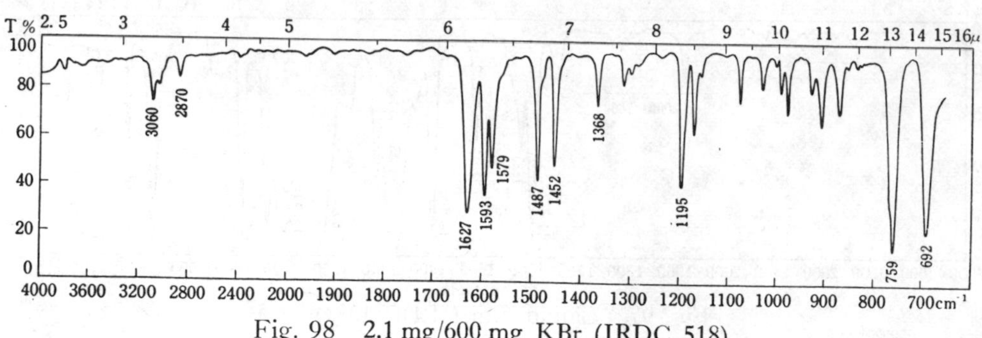

Fig. 98 2.1 mg/600 mg KBr (IRDC 518)

ANSWER

Benzylideneaniline

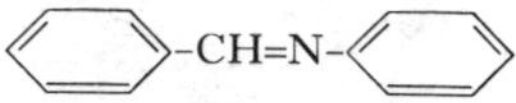

No NH stretching vibrations are present. The given structure is derived in conjunction with the molecular formula and presence of mono-substituted phenyl and C=N groups.

3060 and 3040 cm^{-1}: CH stretching.
1593, 1579, 1487, 1452, 1195, 759, and 692 cm^{-1}: Phenyl.
1627 cm^{-1}: C=N stretching (1690∼1640 cm^{-1}; shifted lower by conjugation).
1368 cm^{-1}: Probably C–N stretching (1360∼1250 cm^{-1}).

PROBLEM 99 / Identify the structure of the hygroscopic compound with formula $C_4H_7NO_2$.

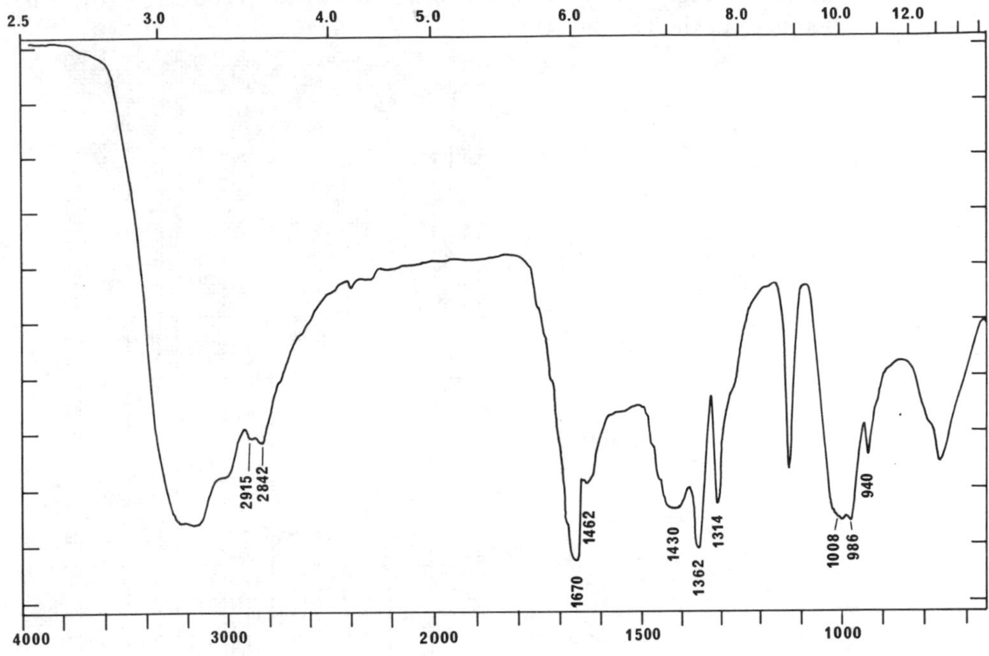

Fig. 99 3 mg/200 mg KBr

ANSWER

Biacetyl monoxime

$$\begin{array}{cc} \text{H} & \!\!-\text{O} \\ \vdots & | \\ \text{O} & \text{N} \\ \| & \| \\ CH_3C & \!\!-\!\!-C\!-\!CH_3 \end{array}$$

3240～2600 cm^{-1}: ν OH, intramolecular hydrogen bonding; obscured by absorbed H_2O.

1670 cm^{-1}: ν C=O, frequency lowered by H-bonding and conjugation with C=N.

1642 cm^{-1}: ν C=N, shift to lower than usual frequency (1680～1650) and increase in intensity due to conjugation with C=O.

1362 cm^{-1}: δ_s CH_3, lowered by position adjacent to C=O.

PROBLEM 100 / A sample of cyclohexane was flash heated under reduced pressure at 800°C and trapped on a CsI plate cooled to 20°K. The infrared spectrum was measured at 10°K (Fig. 100-a) and then remeasured at 10°K (Fig. 100-b) after warming to 74°K. Account for the differences in the two spectra.

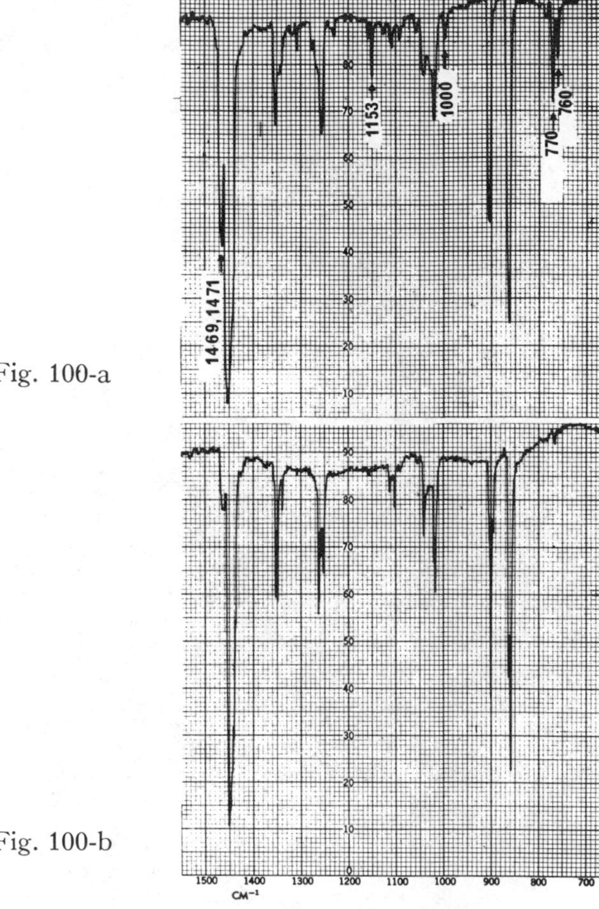

Fig. 100-a

Fig. 100-b

ANSWER

The additional bands which appear in Fig. 100-a at 760, 770, 772, 869, 1000, 1153, 1469, and 1471 cm^{-1} are ascribed to the presence of the twist-boat form of cyclohexane, which reverts to the more stable chair conformation at temperatures over 70°K.

Theoretical calculations have predicted an appreciable percentage of the higher energy twist-boat conformer should be present at 800°C, and this prediction is supported by the appearance of methylene scissoring bands at 1469 and 1471 cm^{-1}. These values agree with those observed for cyclohexane derivatives constrained in a nonchair (presumably twist-boat) form by bulky substituents.[1]

The percentage of the twist-boat form may be calculated from the relative areas of the bands at 1469 and 1471 cm^{-1} and those at 1454 and 1457 cm^{-1}. The twist boat is present to the extent of 25% at 800°C. The free energy difference for the two forms ($\Delta G_{800°}$) is 1.3 kcal/mole, as calculated from the formula $\Delta G = -RT \ln K$.

[1] M. Squillacote, R. S. Sheridan, O. L. Chapman, Y. A. L. Anet, *J. Amer. Chem. Soc.* **97**, 3244 (1975).

APPENDIX 1 / NMR Data

Table 1 Approximate Chemical Shift of Methyl, Methylene, and Methine Protons, in δ or τ values, Me_4Si internal reference

Most of the data have been adapted from: N. S. Bhacca, L. F. Johnson and J. N. Shoolery, "NMR Spectra Catalog," Varian Associates, Palo Alto, California, 1962.
The peaks usually appear within ±0.2 ppm of the values quoted, unless inductive or anisotropic effects of nearby groups are operating.

Methyl Protons		Methylene Protons		Methine Protons	
Proton	δ values (τ)	Proton	δ values (τ)	Proton	δ values (τ)
CH_3-C	0.9 (9.1)	-C-CH_2-C	1.3 (8.7)	C-CH-C	1.5 (8.5)
		" (cyclic)	1.5 (8.5)	" (bridgehead)	2.2 (7.8)
CH_3-C-C=C	1.1 (8.9)	-C-CH_2-C-C=C	1.7 (8.3)		
CH_3-C-O	1.4 (8.6)	-C-CH_2-C-O	1.9 (8.1)	-C-CH-C-O	2.0 (8.0)
CH_3-C=C	1.6 (8.4)	-C-CH_2-C=C	2.3 (7.7)		
CH_3-Ar	2.3 (7.7)	-C-CH_2-Ar	2.7 (7.3)	-CH-Ar	3.0 (7.0)
CH_3-CO-R	2.2 (7.8)	-C-CH_2-CO-R	2.4 (7.6)	-C-CH-CO-R	2.7 (7.3)
CH_3-CO-Ar	2.6 (7.4)				
CH_3-CO-O-R	2.0 (8.0)	-C-CH_2-CO-O-R	2.2 (7.8)		
CH_3-CO-O-Ar	2.4 (7.6)				
CH_3-CO-N-R	2.0 (8.0)				
CH_3-O-R	3.3 (6.7)	-C-CH_2-O-R	3.4 (6.6)	-C-CH-O-R	3.7 (6.3)
		-C-CH_2-O-H	3.6 (6.4)	-C-CH-O-H	3.9 (6.1)
CH_3-OAr	3.8 (6.2)	-C-CH_2-OAr	4.3 (5.7)		
CH_3-O-CO-R	3.7 (6.3)	-C-CH_2-O-CO-R	4.1 (5.9)	-C-CH-O-CO-R	4.8 (5.2)
CH_3-N	2.3 (7.7)	-C-CH_2-N	2.5 (7.5)	-C-CH-N	2.8 (7.2)
CH_3-N^+	3.3 (6.7)				
CH_3-S	2.1 (7.9)	-C-CH_2-S	2.4 (7.6)		
CH_3-C-NO_2	1.6 (8.4)	-C-CH_2-NO_2	4.4 (5.6)	-C-CH-NO_2	4.7 (5.3)
		-C-CH_2-C-NO_2	2.1 (7.9)		
CH_3-C=C-CO	2.0 (8.0)	-C-CH_2-C=C-CO	2.4 (7.6)		
-C=C(CH_3)-CO	1.8 (8.2)	-C=C(CH_2)-CO	2.4 (7.6)		
		O CH_2 O	5.9 (4.1)		
		CH_2	0.3 (9.7)	CH-	0.7 (9.3)
		O CH_2	2.6 (7.4)	O CH-	3.1 (6.9)

Table 2 Chemical Shift of Miscellaneous Protons

	δ values	(τ)		δ values	(τ)
(cyclic) C, C >C=CH_2	4.6*	(5.4)	benzene –H	7.246	(2.754)
–C=CH_2	5.3*	(4.7)			
–C=CH–	5.1*	(4.9)	CH_3 (toluene)	2.337	(7.663)
–C=CH–(cyclic)	5.3*	(4.7)			
R–C≡C–H	3.1*	(6.9)	toluene ring –H	7.095	(2.905)
Ar–H	7.0~9.0**	(3.0~1.0)			
–C=CH–CO	5.9	(4.1)	dihydropyran H	4.65	(5.35)
–CH=C–CO	6.8	(3.2)	dihydropyran H (next to O)	6.37	(3.36)
R–CHO	9.9	(0.1)			
Ar–CHO	9.9	(0.1)	dihydropyran OCH_2	3.97	(6.03)
H–CO–O–	8.0	(2.0)			
H–CO–N	8.0	(2.0)	cyclohexane H	1.44	(8.56)
			$CHCl_3$	7.25	(2.75)
			H_2O	≃5.0	(≃5.0)

* Olefinic protons are subject to shifts larger than ±0.2.
** Similar range for heterocyclic aromatics.

APPENDIX II / Wave-number Wavelength Conversion Table.

Wavelength(μ)	Wave-number(cm^{-1}) 0	1	2	3	4	5	6	7	8	9
2.0	5000	4975	4950	4926	4902	4878	4854	4831	4808	4785
2.1	4762	4739	4717	4695	4673	4651	4630	4608	4587	4566
2.2	4545	4525	4505	4484	4464	4444	4425	4405	4386	4367
2.3	4348	4329	4310	4292	4274	4255	4237	4219	4202	4184
2.4	4167	4149	4132	4115	4098	4082	4065	4049	4032	4016
2.5	4000	3984	3968	3953	3937	3922	3906	3891	3876	3861
2.6	3846	3831	3817	3802	3788	3774	3759	3745	3731	3717
2.7	3704	3690	3676	3663	3650	3636	3623	3610	3597	3584
2.8	3571	3559	3546	3534	3521	3509	3497	3484	3472	3460
2.9	3448	3436	3425	3413	3401	3390	3378	3367	3356	3344
3.0	3333	3322	3311	3300	3289	3279	3268	3257	3247	3236
3.1	3226	3215	3205	3195	3185	3175	3165	3155	3145	3135
3.2	3125	3115	3106	3096	3086	3077	3067	3058	3049	3040
3.3	3030	3021	3012	3003	2994	2985	2976	2967	2959	2950
3.4	2941	2933	2924	2915	2907	2899	2890	2882	2874	2865
3.5	2857	2849	2841	2833	2825	2817	2809	2801	2793	2786
3.6	2778	2770	2762	2755	2747	2740	2732	2725	2717	2710
3.7	2703	2695	2688	2681	2674	2667	2660	2653	2646	2639
3.8	2632	2625	2618	2611	2604	2597	2591	2584	2577	2571
3.9	2564	2558	2551	2545	2538	2532	2525	2519	2513	2506
4.0	2500	2494	2488	2481	2475	2469	2463	2457	2451	2445
4.1	2439	2433	2427	2421	2415	2410	2404	2398	2392	2387
4.2	2381	2375	2370	2364	2358	2353	2347	2342	2336	2331
4.3	2326	2320	2315	2309	2304	2299	2294	2288	2283	2278
4.4	2273	2268	2262	2257	2252	2247	2242	2237	2232	2227
4.5	2222	2217	2212	2208	2203	2198	2193	2188	2183	2179
4.6	2174	2169	2165	2160	2155	2151	2146	2141	2137	2132
4.7	2128	2123	2119	2114	2110	2105	2101	2096	2092	2088
4.8	2083	2079	2075	2070	2066	2062	2058	2053	2049	2045
4.9	2041	2037	2033	2028	2024	2020	2016	2012	2008	2004
5.0	2000	1996	1992	1988	1984	1980	1976	1972	1969	1965
5.1	1961	1957	1953	1949	1946	1942	1938	1934	1931	1927
5.2	1923	1919	1916	1912	1908	1905	1901	1898	1894	1890
5.3	1887	1883	1880	1876	1873	1869	1866	1862	1859	1855
5.4	1852	1848	1845	1842	1838	1835	1832	1828	1825	1821
5.5	1818	1815	1812	1808	1805	1802	1799	1795	1792	1789
5.6	1786	1783	1779	1776	1773	1770	1767	1764	1761	1757
5.7	1754	1751	1748	1745	1742	1739	1736	1733	1730	1727
5.8	1724	1721	1718	1715	1712	1709	1706	1704	1701	1698
5.9	1695	1692	1689	1686	1684	1681	1678	1675	1672	1669
	0	1	2	3	4	5	6	7	8	9

(by the courtesy of Drs. M. Kishita and Y. Kuroda, Nagoya University)

Wavelength (μ)	Wave-number (cm⁻¹)									
	0	1	2	3	4	5	6	7	8	9
6.0	1667	1664	1661	1658	1656	1653	1650	1647	1645	1642
6.1	1639	1637	1634	1631	1629	1626	1623	1621	1618	1616
6.2	1613	1610	1608	1605	1603	1600	1597	1595	1592	1590
6.3	1587	1585	1582	1580	1577	1575	1572	1570	1567	1565
6.4	1563	1560	1558	1555	1553	1550	1548	1546	1543	1541
6.5	1538	1536	1534	1531	1529	1527	1524	1522	1520	1517
6.6	1515	1513	1511	1508	1506	1504	1502	1499	1497	1495
6.7	1493	1490	1488	1486	1484	1481	1479	1477	1475	1473
6.8	1471	1468	1466	1464	1462	1460	1458	1456	1453	1451
6.9	1449	1447	1445	1443	1441	1439	1437	1435	1433	1431
7.0	1429	1427	1425	1422	1420	1418	1416	1414	1412	1410
7.1	1408	1406	1404	1403	1401	1399	1397	1395	1393	1391
7.2	1389	1387	1385	1383	1381	1379	1377	1376	1374	1372
7.3	1370	1368	1366	1364	1362	1361	1359	1357	1355	1353
7.4	1351	1350	1348	1346	1344	1342	1340	1339	1337	1335
7.5	1333	1332	1330	1328	1326	1325	1323	1321	1319	1318
7.6	1316	1314	1312	1311	1309	1307	1305	1304	1302	1300
7.7	1299	1297	1295	1294	1292	1290	1289	1287	1285	1284
7.8	1282	1280	1279	1277	1276	1274	1272	1271	1269	1267
7.9	1266	1264	1263	1261	1259	1258	1256	1255	1253	1252
8.0	1250	1248	1247	1245	1244	1242	1241	1239	1238	1236
8.1	1235	1233	1232	1230	1229	1227	1225	1224	1222	1221
8.2	1220	1218	1217	1215	1214	1212	1211	1209	1208	1206
8.3	1205	1203	1202	1200	1199	1198	1196	1195	1193	1192
8.4	1190	1189	1188	1186	1185	1183	1182	1181	1179	1178
8.5	1176	1175	1174	1172	1171	1170	1168	1167	1166	1164
8.6	1163	1161	1160	1159	1157	1156	1155	1153	1152	1151
8.7	1149	1148	1147	1145	1144	1143	1142	1140	1139	1138
8.8	1136	1135	1134	1133	1131	1130	1129	1127	1126	1125
8.9	1124	1122	1121	1120	1119	1117	1116	1115	1114	1112
9.0	1111	1110	1109	1107	1106	1105	1104	1103	1101	1100
9.1	1099	1098	1096	1095	1094	1093	1092	1091	1089	1088
9.2	1087	1086	1085	1083	1082	1031	1080	1079	1078	1076
9.3	1075	1074	1073	1072	1071	1070	1068	1067	1066	1065
9.4	1064	1063	1062	1060	1059	1058	1057	1056	1055	1054
9.5	1053	1052	1050	1049	1048	1047	1046	1045	1044	1043
9.6	1042	1041	1040	1038	1037	1036	1035	1034	1033	1032
9.7	1031	1030	1029	1028	1027	1026	1025	1024	1022	1021
9.8	1020	1019	1018	1017	1016	1015	1014	1013	1012	1011
9.9	1010	1009	1008	1007	1006	1005	1004	1003	1002	1001
	0	1	2	3	4	5	6	7	8	9

Wavelength(μ)	Wave-number(cm^{-1})									
	0	1	2	3	4	5	6	7	8	9
10.0	1000.0	999.0	998.0	997.0	996.0	995.0	994.0	993.0	992.1	991.1
10.1	990.1	989.1	988.1	987.2	986.2	985.2	984.3	983.3	982.3	981.4
10.2	980.4	979.4	978.5	977.5	976.6	975.6	974.7	973.7	972.8	971.8
10.3	970.9	969.9	969.0	968.1	967.1	966.2	965.3	964.3	963.4	962.5
10.4	961.5	960.6	959.7	958.8	957.9	956.9	956.0	955.1	954.2	953.3
10.5	952.4	951.5	950.6	949.7	948.8	947.9	947.0	946.1	945.2	944.3
10.6	943.4	942.5	941.6	940.7	939.8	939.0	938.1	937.2	936.3	935.5
10.7	934.6	933.7	932.8	932.0	931.1	930.2	929.4	928.5	927.6	926.8
10.8	925.9	925.1	924.2	923.4	922.5	921.7	920.8	920.0	919.1	918.3
10.9	917.4	916.6	915.8	914.9	914.1	913.2	912.4	911.6	910.7	909.9
11.0	909.1	908.3	907.4	906.6	905.8	905.0	904.2	903.3	902.5	901.7
11.1	900.9	900.1	899.3	898.5	897.7	896.9	896.1	895.3	894.5	893.7
11.2	892.9	892.1	891.3	890.5	889.7	888.9	888.1	887.3	886.5	885.7
11.3	885.0	884.2	883.4	882.6	881.8	881.1	880.3	879.5	878.7	878.0
11.4	877.2	876.4	875.7	874.9	874.1	873.4	872.6	871.8	871.1	870.3
11.5	869.6	868.8	868.1	867.3	866.6	865.8	865.1	864.3	863.6	862.8
11.6	862.1	861.3	860.6	859.8	859.1	858.4	857.6	856.9	856.2	855.4
11.7	854.7	854.0	853.2	852.5	851.8	851.1	850.3	849.6	848.9	848.2
11.8	847.5	846.7	846.0	845.3	844.6	843.9	843.2	842.5	841.8	841.0
11.9	840.3	839.6	838.9	838.2	837.5	836.8	836.1	835.4	834.7	834.0
12.0	833.3	832.6	831.9	831.3	830.6	829.9	829.2	828.5	827.8	827.1
12.1	826.4	825.8	825.1	824.4	823.7	823.0	822.4	821.7	821.0	820.3
12.2	819.7	819.0	818.3	817.7	817.0	816.3	815.7	815.0	814.3	813.7
12.3	813.0	812.3	811.7	811.0	810.4	809.7	809.1	808.4	807.8	807.1
12.4	806.5	805.8	805.2	804.5	803.9	803.2	802.6	801.9	801.3	800.6
12.5	800.0	799.4	798.7	798.1	797.4	796.8	796.2	795.5	794.9	794.3
12.6	793.7	793.0	792.4	791.8	791.1	790.5	789.9	789.3	788.6	788.0
12.7	787.4	786.8	786.2	785.5	784.9	784.3	783.7	783.1	782.5	781.9
12.8	781.3	780.6	780.0	779.4	778.8	778.2	777.6	777.0	776.4	775.8
12.9	775.2	774.6	774.0	773.4	772.8	772.2	771.6	771.0	770.4	769.8
13.0	769.2	768.6	768.0	767.5	766.9	766.3	765.7	765.1	764.5	763.9
13.1	763.4	762.8	762.2	761.6	761.0	760.5	759.9	759.3	758.7	758.2
13.2	757.6	757.0	756.4	755.9	755.3	754.7	754.1	753.6	753.0	752.4
13.3	751.9	751.3	750.8	750.2	749.6	749.1	748.5	747.9	747.4	746.8
13.4	746.3	745.7	745.2	744.6	744.0	743.5	742.9	742.4	741.8	741.3
13.5	740.7	740.2	739.6	739.1	738.6	738.0	737.5	736.9	736.4	735.8
13.6	735.3	734.8	734.2	733.7	733.1	732.6	732.1	731.5	731.0	730.5
13.7	729.9	729.4	728.9	728.3	727.8	727.3	726.7	726.2	725.7	725.2
13.8	724.6	724.1	723.6	723.1	722.5	722.0	721.5	721.0	720.5	719.9
13.9	719.4	718.9	718.4	717.9	717.4	716.8	716.3	715.8	715.3	714.8
14.0	714.3	713.8	713.3	712.8	712.3	711.7	711.2	710.7	710.2	709.7
14.1	709.2	708.7	708.2	707.7	707.2	706.7	706.2	705.7	705.2	704.7
14.2	704.2	703.7	703.2	702.7	702.2	701.8	701.3	700.8	700.3	699.8
14.3	699.3	698.8	698.3	697.8	697.4	696.9	696.4	695.9	695.4	694.9
14.4	694.4	694.0	693.5	693.0	692.5	692.0	691.6	691.1	690.6	690.1
14.5	689.7	689.2	688.7	688.2	687.8	687.3	686.8	686.3	685.9	685.4
14.6	684.9	684.5	684.0	683.5	683.1	682.6	682.1	681.7	681.2	680.7
14.7	680.3	679.8	679.3	678.9	678.4	678.0	677.5	677.0	676.6	676.1
14.8	675.7	675.2	674.8	674.3	673.9	673.4	672.9	672.5	672.0	671.6
14.9	671.1	670.7	670.2	669.8	669.3	668.9	668.4	668.0	667.6	667.1
	0	1	2	3	4	5	6	7	8	9

Wavelength(μ)	Wave-number(cm^{-1}) 0	1	2	3	4	5	6	7	8	9
15.0	666.7	666.2	665.8	665.3	664.9	664.5	664.0	663.6	663.1	662.7
15.1	662.3	661.8	661.4	660.9	660.5	660.1	659.6	659.2	658.8	658.3
15.2	*657.9*	*657.5*	*657.0*	*656.6*	*656.2*	*655.7*	*655.3*	*654.9*	*654.5*	*654.0*
15.3	653.6	653.2	652.7	652.3	651.9	651.5	651.0	650.6	650.2	649.8
15.4	649.4	648.9	648.5	648.1	647.7	647.2	646.8	646.4	646.0	645.6
15.5	645.2	644.7	644.3	643.9	643.5	643.1	642.7	642.3	641.8	641.4
15.6	641.0	640.6	640.2	639.8	639.4	639.0	638.6	638.2	637.8	637.3
15.7	*636.9*	*636.5*	*636.1*	*635.7*	*635.3*	*634.9*	*634.5*	*634.1*	*633.7*	*633.3*
15.8	632.9	632.5	632.1	631.7	631.3	630.9	630.5	630.1	629.7	629.3
15.9	628.9	628.5	628.1	627.7	627.4	627.0	626.6	626.2	625.8	625.4
16.0	625.0	624.6	624.2	623.8	623.4	623.1	622.7	622.3	621.9	621.5
16.1	621.1	620.7	620.3	620.0	619.6	619.2	618.8	618.4	618.0	617.7
16.2	*617.3*	*616.9*	*616.5*	*616.1*	*615.8*	*615.4*	*615.0*	*614.6*	*614.3*	*613.9*
16.3	613.5	613.1	612.7	612.4	612.0	611.6	611.2	610.9	610.5	610.1
16.4	609.8	609.4	609.0	608.6	608.3	607.9	607.5	607.2	606.8	606.4
16.5	606.1	605.7	605.3	605.0	604.6	604.2	603.9	603.5	603.1	602.8
16.6	602.4	602.0	601.7	601.3	601.0	600.6	600.2	599.9	599.5	599.2
16.7	*598.8*	*598.4*	*598.1*	*597.7*	*597.4*	*597.0*	*596.7*	*596.3*	*595.9*	*595.6*
16.8	595.2	594.9	594.5	594.2	593.8	593.5	593.1	592.8	592.4	592.1
16.9	591.7	591.4	591.0	590.7	590.3	590.0	589.6	589.3	588.9	588.6
17.0	588.2	587.9	587.5	587.2	586.9	586.5	586.2	585.8	585.5	585.1
17.1	584.8	584.5	584.1	583.8	583.4	583.1	582.8	582.4	582.1	581.7
17.2	*581.4*	*581.1*	*580.7*	*580.4*	*580.0*	*579.7*	*579.4*	*579.0*	*578.7*	*578.4*
17.3	578.0	577.7	577.4	577.0	576.7	576.4	576.0	575.7	575.4	575.0
17.4	574.7	574.4	574.1	573.7	573.4	573.1	572.7	572.4	572.1	571.8
17.5	571.4	571.1	570.8	570.5	570.1	569.8	569.5	569.2	568.8	568.5
17.6	568.2	567.9	567.5	567.2	566.9	566.6	566.3	565.9	565.6	565.3
17.7	*565.0*	*564.7*	*564.3*	*564.0*	*563.7*	*563.4*	*563.1*	*562.7*	*562.4*	*562.1*
17.8	561.8	561.5	561.2	560.9	560.5	560.2	559.9	559.6	559.3	559.0
17.9	558.7	558.3	558.0	557.7	557.4	557.1	556.8	556.5	556.2	555.9
18.0	555.6	555.2	554.9	554.6	554.3	554.0	553.7	553.4	553.1	552.8
18.1	552.5	552.2	551.9	551.6	551.3	551.0	550.7	550.4	550.1	549.8
18.2	*549.5*	*549.1*	*548.8*	*548.5*	*548.2*	*547.9*	*547.6*	*547.3*	*547.0*	*546.7*
18.3	546.4	546.1	545.9	545.6	545.3	545.0	544.7	544.4	544.1	543.8
18.4	543.5	543.2	542.9	542.6	542.3	542.0	541.7	541.4	541.1	540.8
18.5	540.5	540.2	540.0	539.7	539.4	539.1	538.8	538.5	538.2	537.9
18.6	537.6	537.3	537.1	536.8	536.5	536.2	535.9	535.6	535.3	535.0
18.7	*534.8*	*534.5*	*534.2*	*533.9*	*533.6*	*533.3*	*533.0*	*532.8*	*532.5*	*532.2*
18.8	531.9	531.6	531.3	531.1	530.8	530.5	530.2	529.9	529.7	529.4
18.9	529.1	528.8	528.5	528.3	528.0	527.7	527.4	527.1	526.9	526.6
19.0	526.3	526.0	525.8	525.5	525.2	524.9	524.7	524.4	524.1	523.8
19.1	523.6	523.3	523.0	522.7	522.5	522.2	521.9	521.6	521.4	521.1
19.2	*520.8*	*520.6*	*520.3*	*520.0*	*519.8*	*519.5*	*519.2*	*518.9*	*518.7*	*518.4*
19.3	518.1	517.9	517.6	517.3	517.1	516.8	516.5	516.3	516.0	515.7
19.4	515.5	515.2	514.9	514.7	514.4	514.1	513.9	513.6	513.3	513.1
19.5	512.8	512.6	512.3	512.0	511.8	511.5	511.2	511.0	510.7	510.5
19.6	510.2	509.9	509.7	509.4	509.2	508.9	508.6	508.4	508.1	507.9
19.7	*507.6*	*507.4*	*507.1*	*506.8*	*506.6*	*506.3*	*506.1*	*505.8*	*505.6*	*505.3*
19.8	505.1	504.8	504.5	504.3	504.0	503.8	503.5	503.3	503.0	502.8
19.9	502.5	502.3	502.0	501.8	501.5	501.3	501.0	500.8	500.5	500.3
	0	1	2	3	4	5	6	7	8	9

Wavelength (μ)	Wave-number (cm^{-1}) 0	1	2	3	4	5	6	7	8	9
20.0	500.0	499.8	499.5	499.3	499.0	498.8	498.5	498.3	498.0	497.8
20.1	497.5	497.3	497.0	496.8	496.5	496.3	496.0	495.8	495.5	495.3
20.2	495.0	494.8	494.6	494.3	494.1	493.8	493.6	493.3	493.1	492.9
20.3	492.6	492.4	492.1	491.9	491.6	491.4	491.2	490.9	490.7	490.4
20.4	490.2	490.0	489.7	489.5	489.2	489.0	488.8	488.5	488.3	488.0
20.5	487.8	487.6	487.3	487.1	486.9	486.6	486.4	486.1	485.9	485.7
20.6	485.4	485.2	485.0	484.7	484.5	484.3	484.0	483.8	483.6	483.3
20.7	483.1	482.9	482.6	482.4	482.2	481.9	481.7	481.5	481.2	481.0
20.8	480.8	480.5	480.3	480.1	479.8	479.6	479.4	479.2	478.9	478.7
20.9	478.5	478.2	478.0	477.8	477.6	477.3	477.1	476.9	476.6	476.4
21.0	476.2	476.0	475.7	475.5	475.3	475.1	474.8	474.6	474.4	474.2
21.1	473.9	473.7	473.5	473.3	473.0	472.8	472.6	472.4	472.1	471.9
21.2	471.7	471.5	471.3	471.0	470.8	470.6	470.4	470.1	469.9	469.7
21.3	469.5	469.3	469.0	468.8	468.6	468.4	468.2	467.9	467.7	467.5
21.4	467.3	467.1	466.9	466.6	466.4	466.2	466.0	465.8	465.5	465.3
21.5	465.1	464.9	464.7	464.5	464.3	464.0	463.8	463.6	463.4	463.2
21.6	463.0	462.7	462.5	462.3	462.1	461.9	461.7	461.5	461.3	461.0
21.7	460.8	460.6	460.4	460.2	460.0	459.8	459.6	459.3	459.1	458.9
21.8	458.7	458.5	458.3	458.1	457.9	457.7	457.5	457.2	457.0	456.8
21.9	456.6	456.4	456.2	456.0	455.8	455.6	455.4	455.2	455.0	454.8
22.0	454.5	454.3	454.1	453.9	453.7	453.5	453.3	453.1	452.9	452.7
22.1	452.5	452.3	452.1	451.9	451.7	451.5	451.3	451.1	450.9	450.7
22.2	450.5	450.2	450.0	449.8	449.6	449.4	449.2	449.0	448.8	448.6
22.3	448.4	448.2	448.0	447.8	447.6	447.4	447.2	447.0	446.8	446.6
22.4	446.4	446.2	446.0	445.8	445.6	445.4	445.2	445.0	444.8	444.6
22.5	444.4	444.2	444.0	443.9	443.7	443.5	443.3	443.1	442.9	442.7
22.6	442.5	442.3	442.1	441.9	441.7	441.5	441.3	441.1	440.9	440.7
22.7	440.5	440.3	440.1	439.9	439.8	439.6	439.4	439.2	439.0	438.8
22.8	438.6	438.4	438.2	438.0	437.8	437.6	437.4	437.3	437.1	436.9
22.9	436.7	436.5	436.3	436.1	435.9	435.7	435.5	435.4	435.2	435.0
23.0	434.8	434.6	434.4	434.2	434.0	433.8	433.7	433.5	433.3	433.1
23.1	432.9	432.7	432.5	432.3	432.2	432.0	431.8	431.6	431.4	431.2
23.2	431.0	430.8	430.7	430.5	430.3	430.1	429.9	429.7	429.6	429.4
23.3	429.2	429.0	428.8	428.6	428.4	428.3	428.1	427.9	427.7	427.5
23.4	427.4	427.2	427.0	426.8	426.6	426.4	426.3	426.1	425.9	425.7
23.5	425.5	425.4	425.2	425.0	424.8	424.6	424.4	424.3	424.1	423.9
23.6	423.7	423.5	423.4	423.2	423.0	422.8	422.7	422.5	422.3	422.1
23.7	421.9	421.8	421.6	421.4	421.2	421.1	420.9	420.7	420.5	420.3
23.8	420.2	420.0	419.8	419.6	419.5	419.3	419.1	418.9	418.8	418.6
23.9	418.4	418.2	418.1	417.9	417.7	417.5	417.4	417.2	417.0	416.8
24.0	416.7	416.5	416.3	416.1	416.0	415.8	415.6	415.5	415.3	415.1
24.1	414.9	414.8	414.6	414.4	414.3	414.1	413.9	413.7	413.6	413.4
24.2	413.2	413.1	412.9	412.7	412.5	412.4	412.2	412.0	411.9	411.7
24.3	411.5	411.4	411.2	411.0	410.8	410.7	410.5	410.3	410.2	410.0
24.4	409.8	409.7	409.5	409.3	409.2	409.0	408.8	408.7	408.5	408.3
24.5	408.2	408.0	407.8	407.7	407.5	407.3	407.2	407.0	406.8	406.7
24.6	406.5	406.3	406.2	406.0	405.8	405.7	405.5	405.4	405.2	405.0
24.7	404.9	404.7	404.5	404.4	404.2	404.0	403.9	403.7	403.6	403.4
24.8	403.2	403.1	402.9	402.7	402.6	402.4	402.3	402.1	401.9	401.8
24.9	401.6	401.4	401.3	401.1	401.0	400.8	400.6	400.5	400.3	400.2
	0	1	2	3	4	5	6	7	8	9

GENERAL INDEX

A
Apparent molecular absorption
  coefficient 10, 11
Axial substituent 28
B
Bending vibration 1–4, 12
"Bohlman Band" 35
C
Calibration 5, 6, 8
Carbon magnetic resonance 15
Cell thickness 5, 11
Characteristic absorption 4
Chelation 66
Combination tone 12–13
Concentration
  effect of 11, 26–27, 58, 68
Coupling 12–13
D
Deformation vibration 12
Depolarization ratio 76
Deuteration 59, 63
Dimedone 66–72
  Enol form 66, 68–70
  Keto form 66–69
Dipole moment 11, 32, 74, 75
Doubling of band 13

E
Elastic collision 73
Electronegativity 59
  Electronegative group 18
  Electronegative substituent 16
Equatorial substituent 28
Equilibrium 67
Esters 61
F
Far–infrared 1
Fermi resonance 13
Force constant 63
Fundamental vibration
  (frequency) 12, 13
  Variations in 59
G
Gem-dimethyl group 16
Guanidines 37
H
Hammett value 61
Hooke's Law 59
Hydrogen bonding 4, 8, 10,
  ....26, 29, 38, 58, 67
I
Inductive effect 60
Inelastic collision 74
Infrared absorption band 3
  Assignment of 11–13, 63
  Intensity of ... 10–11, 32, 64, 74
Isobestic point 68
L
Light scattering 73–77

M
Mesomeric effect 60, 62
Methyl deformation frequency ... 59
Moisture, effect of 27, 58
Molecular flow resonance
  Raman spectra 88–90
Multiple bond frequency ... 3, 60–62
N
Near-infrared 1
Nitriles 24, 61, 64
NMR spectroscopy 28
O
Oscillating dipole 73
Overtone 12–13, 18, 42

P
Polarizability 11, 75, 76
Polymorphism 3, 57, 58
Potassium bromide spectra 58
Prism materials 5
Proton magnetic resonance 15

R
Raman spectrum 11, 73-77
  Intensity of 75
Rayleigh Scattering 73, 74
Resonance laser Raman
  spectroscopy 77
Ring strain 62
Rotational isomerism 57

S
Skeletal vibration 10
Slit width, effect of 11
Solids 5, 6, 58
Solutions 10, 58
  Solvation 8, 67, 69
  Solvent 4, 5, 7, 8, 10, 11, 58
Steric inhibition of resonance 62
Stokes band 74
  Anti-stokes band 74
Stretching vibration (frequency)
  .... 1-4, 12, 44, 59, 60, 61
  Antisymmetric 12
  Asymmetric 12
Sugars 29-30

T
Tautomerism 36, 58
Temperature, effect of ... 11, 26, 58
Tetrasubstituted double bond 18
Thiol esters 60
Trans-annular effect 62

V
Vinylogous system 71
Vinyl ethers 13

INDEX OF COMPOUNDS

Infrared Spectra

A

Acetanilide . . . . (75) (237)
Acetophenone. . . . . (44) (195)
L-Alanine . . . . (70) (231)
Allyl alcohol . . . . (22) (175)
Allyl acetate . . . . (23) (176)
Allyl cyanide. . . . . (24) (177)
Allyl phosphate. . . . . (94) (260)
Aminoanthraquinone . . . . . . (80) (243)
Aminobenzoic acid (*o*–, *m*). . (57) (213)
p-Anisaldehyde . . . . (46) (197)
Anisole . . . . (32) (189)

B

N-Benzoyl-*p*-toluidine . . . . . (77) (239)
Benzylideneaniline . . . . (98) (264)
2-(Benzylsulfonyl)-*p*-menthane
. . . . (90) (255)
Biacetyl monoxime . . . . (99) (265)
p-Bromotoluene . . . . (8) (160)
N-*tert*-Butylacetamide. . . . . . (73) (234)
n-Butyl vinyl ether . . . . (30) (187)
Butyric anhydride . . . . (62) (221)

C

Camphor . . . . (40) (198)
Capillene . . . . (17) (170)
Carbon disulfide . . . . Fig. 1.7 (7)
Carbon tetrachloride . . . . . Fig. 1.6 (7)
15-*cis*: 15′-*cis*-β-Carotene. . . . (6) (157)
Chloracetamide . . . . (74) (236)
Chloroform . . . . Fig. 1.8 (7)
p-Chlorobenzotrifluoride. . . . (97) (263)
Cholest-5-ene-3β, 4β-
diolsulfite . . . . (89) (254)
Cholesteryl benzoate. . . . . . . (49) (200)
2-Cyano-6-methoxy-
quinoline. . . . . (69) (230)
Cyclohexane-1,2-diol. . . . . . . (26) (180)
Cyclohexane . . . . (100) (266)
Cyclohexanol . . . . (27) (182)
L-Cysteine hydrochloride . . . (71) (232)

D

Dibenzyl sulfoxide . . . . (92) (258)
N,N-dicyclohexylcarbo-
diimide . . . . (12) (164)
Diethyl phthalate . . . . (59) (217)
N,N-Diethyl-*p*-phenyl-
enediamine monohydro-
chloride. . . . . (37) (195)
N,N-Dimethylaniline . . . . (32) (189)
1,1-Dimethyl-2-propynyl-
carbamate . . . . (78) (240)
Dioxane. . . . . Fig. 1.9 (8)
1,2-Diphenylethane. . . . . (9) (161)

E

1,2-Epoxydodecane. . . . . (34) (192)
Estra-1:3:5(10)-triene-3,17β-
diol . . . . (25) (178)
Ethyl diethylphosphono-
acetate. . . . . (95) (261)
Ethyl *p*-nitrosobenzoate . . . . (50) (201)

F

4-Fluoro-2-nitrophenylazide . (15) (167)
Furfuryl acetate . . . . (86) (250)

G

Gibberellin A_3 derivative. . . . (56) (212)
Glycine ethyl ester mono-
hydrochloride . . . . (72) (233)
2-Guanidinoethanol nitrate . . (85) (249)

H

Hepta-1:*trans*-3:*cis*-5-triene. . (5) (156)
n-Hexylamine . . . . (35) (193)

2-Hydroperoxycumene . . . . . (33) (191)
2-Hydroxy-4-methoxy-
acetophenone . . . . (45) (196)

I

3-(3-Indolyl) propionic acid . (87) (251)
Isopropenyl acetate. . . . . (58) (216)
2-Isopropylphenanthrene
(retene) . . . . (10) (162)

L

d-Limonene. . . . . (3) (153)
(−)-Linalool. . . . . (19) (172)

M

2-Mercaptodiphenyl-
methane . . . . (96) (262)
Mesityl oxide. . . . . (43) (194)
DL-N-methyl-4-anisoyl-10-
acetoxy decahydroisoquino-
line-(4,10-*trans*)-HC1. . . . . . (66), (67)
(225) (227)
Methyl-α-D-(+)-glucoside. . . . (29) (185)
p-Methylbenzonitrile. . . . . . . (13) (165)
3,4-Methylenedioxybenzal-
dehyde . . . . (47) (198)
3-Methylpentane . . . . (2) (152)
2-Methylpyridine-5-carboxylic
acid . . . . (68) (229)
4-Methylpyridine 1-oxide
monohydrate. . . . . (84) (247)

N

α-Naphthol methyl ether. . . . (31) (188)
p-Nitrobenzaldehyde . . . . . . (48) (199)
3-(Nitromethyl) phthalide. . . (60) (218)
p-Nitrophenylpropiolic acid . (65) (224)
Nona-1:4-diyne . . . . (16) (168)
Nujol. . . . . Fig. 1.5 (6)

P

Palmitic acid . . . . (54) (209)
Phenyl isocyanate . . . . (11) (163)
2-Phenyl-3-butyn-2-ol . . . . . . (28) (184)
Phenylisothiocyanate . . . . . . (14) (166)
Polystyrene . . . . Fig. 1.4 (6)
Propionic acid . . . . (52) (206)
Propiolanilide . . . . (76) (238)
2-(2-Pyridyl) ethanol. . . . . . . (21) (174)
α- and γ-Pyrone. . . . . (79) (241)
γ-Pyridone. . . . . (81) (244)

S

Sodium benzoate. . . . . (55) (211)
Stearoyl chloride. . . . . (83) (246)
Sulfanilamide . . . . (91) (256)

T

Tetrahydrofuran . . . . Fig. 1.10 (8)
Thioanisole . . . . (32) (189)
Thujopsene anhydrodihydro-
ketone. . . . . (42) (193)
Toluene-α-sulfonylchloride . . (93) (259)
o-Toluidine . . . . (36) (194)
Triketohydrindene hydrate . . (82) (245)
2,6,6-Trimethyl-9-methylene-
bicyclo-[3:3:1]-nonan-
2-ol . . . . (20) (173)
3,5,5-Trimethylbutyrolactone-
4-carboxyliç acid. . . . . . . (63) (222)
Diethylammonium salt . . . (64) (223)

W

Water. . . . . Fig. 1.3 (6)

X

Xylene (*o*-, *m*-, *p*) . . . . (7) (158)

Y

Yohimbine . . . . (38) (196)
–HC1. . . . . (39) (197)

Raman Spectra

Acetone. . . . . Fig. 5.3 (77)
L-Alanine . . . . Fig. 5.22 (87)
Anisole . . . . Fig. 5.15 (83)
n-Butyl disulfide . . . . Fig. 23 (88)
Carbon disulfide . . . . Fig. 5.4 (77)
Carbon tetrachloride . . . . Fig 5.5 (78)
Chloroform . . . . Fig. 5.6 (78)
Cyclohexane . . . . Fig. 5.11 (81)
Cyclohexanol . . . . Fig. 5.12 (81)
Cyclohexyl amine . . . . . Fig. 5.13 (81)
N,N-Dimethylaniline . . . Fig. 5.17 (84)
2,3-Dimethyl-2-butene . Fig. 5.14 (83)
Dimethyl sulfoxide . . . . . Fig. 5.9 (79)
Dioxane. . . . . Fig. 5.7 (78)
Methyl alcohol . . . . Fig. 5.8 (79)
Thioanisole . . . . Fig. 5.16 (83)
Water. . . . . Fig. 5.10 (79)
Xylene (*o*-, *m*-, *p*) . . . . . Figs. 5.18–20 (85–87)

HOLDEN-DAY
500 Sansome Street
San Francisco, CA 94111

0-8162-6251-9